90 0416137

D1756554

7 Day

University of Plymouth Library

Subject to status this item may be renewed
via your Voyager account

http://voyager.plymouth.ac.uk

Exeter tel: (01392) 475049
Exmouth tel: (01395) 255331
Plymouth tel: (01752) 232323

European Wet Grasslands

LANDSCAPE ECOLOGY SERIES

Ecology and Management of Invasive Riverside Plants
Edited by Louise C. de Waal, Lois E. Child, P. Max Wade and John H. Brock

European Wet Grasslands
Edited by Chris B. Joyce and P. Max Wade

European Wet Grasslands

Biodiversity, Management and Restoration

Edited by

CHRIS B. JOYCE *and* **P. MAX WADE**

International Centre of Landscape Ecology, Department of Geography, Loughborough University, UK

JOHN WILEY & SONS

Chichester · New York · Weinheim · Brisbane · Singapore · Toronto

Other Wiley Editorial Offices

John Wiley & Sons, Inc., 605 Third Avenue,
New York, NY 10158–0012, USA

WILEY-VCH Verlags GmbH, Pappelallee 3,
D-69469 Weinheim, Germany

Jacaranda Wiley Ltd, 33 Park Road, Milton,
Queensland 4064, Australia

John Wiley & Sons (Asia) Pte Ltd, 2 Clementi Loop #02-01,
Jin Xing Distripark, Singapore 129809

John Wiley & Sons (Canada) Ltd, 22 Worcester Road,
Rexdale, Ontario M9W 1L1, Canada

Library of Congress Cataloging-in-Publication Data

European wet grasslands : biodiversity, management, and restoration /
edited by Max Wade and Chris Joyce.
p.   cm. — (Landscape ecology series)
Includes bibliographical references and index.
ISBN 0-471-97619-9 (alk. paper)
1. Grassland ecology—Europe.   2. Grassland conservation—Europe.   3.
Grasslands—Europe—Management.   I. Wade, Max.   II. Joyce, Chris.   III. Series.
QH135.E94   1998
639.9′094′09153—dc21   97-46061
                                                                                 CIP

British Library Cataloguing in Publication Data

A catalogue record for this book is available from the British Library

ISBN 0-471-97619-9

Typeset in 10/12pt Times from the authors' disks by Vision Typesetting, Manchester, UK
Printed and bound in Great Britain by Bookcraft (Bath) Ltd, Midsomer Norton, Somerset

This book is printed on acid-free paper responsibly manufactured from sustainable forestry, for which at
least two trees are planted for each one used for paper production.

Floodplain grasslands of the Morava River, Slovakia
Source: Viera Banásová, Slovak Academy of Sciences.

Contents

CONTENTS

Contributors

Adrian Armstrong
ADAS Land Research Centre, Gleadthorpe, Meden Vale, Mansfield, NG20 9PF, UK

Vera Banásová
Institute of Botany, Slovak Academy of Sciences, Dubravská 14, Sk 84223, Bratislava, Slovakia

Shona Blake
The Scottish Agricultural College, Environmental Sciences Department, Auchincruive, Ayr, KA6 5HW, Scotland, UK

Adriano Boscaini
Museo Tridentino di Scienze Naturali, Via Calepina, 14, I 38100 Trento, Italy

Wim Braakhekke
Wageningen Agricultural University, Department of Terrestrial Ecology and Nature Conservation, Bornsesteeg 69, 6708 PD Wageningen, The Netherlands

Martin Drake
English Nature, Northminster House, Peterborough, PE1 1UA, UK

Garth N. Foster
The Scottish Agricultural College, Environmental Sciences Department, Auchincruive, Ayr KA6 5HW, Scotland, UK

Alessandra Franceschini
Museo Tridentino di Scienze Naturali, Via Calepina, 14, I 38100 Trento, Italy

David J. Glaves
Farming and Rural Conservation Agency, Staplake Mount, Starcross, Exeter, EX6 8PE, UK

David J.G. Gowing
Silsoe College, Cranfield University, Silsoe, MK45 4DT, UK

Philip V. Grice
English Nature, Northminster House, Peterborough, PE1 1UA, UK

Dick van der Hoek
Wageningen Agricultural University, Department of Terrestrial Ecology and Nature Conservation, Bornsesteeg 69, 6708 PD Wageningen, The Netherlands

Ivan Jarolimek
Institute of Botany, Slovak Academy of Sciences, Dubravská 14, Sk 84223, Bratislava, Slovakia

Richard G. Jefferson
English Nature, Northminster House, Peterborough, PE1 1UA, UK

Bill Jenman
Sussex Wildlife Trust, Woods Mill, Henfield, BN5 9SD, UK

Chris B. Joyce
International Centre of Landscape Ecology, Department of Geography, Loughborough University, Loughborough, LE11 3TU, UK

Francis Kirkham
ADAS, Research and Development, Woodthorne, Wergs Road, Wolverhampton, WV6 8TQ, UK

Charlie Kitchin
Royal Society for the Protection of Birds, 21a East Delph, Whittlesey, Peterborough, PE7 1RH, UK

Catherine Levassor
Departmento de Ecología, Facultad de Ciencias, Universidad Autónoma de Madrid, E-28049 Cantoblanco, Spain

Bruno Maiolini
Museo Tridentino di Scienze Naturali, Via Calepina, 14, I 38100 Trento, Italy

Sarah Manchester
Institute of Terrestrial Ecology, Monks Wood, Abbots Ripton, Huntingdon, PE17 2LS, UK

Christer Nilsson
Riparian Ecology Group, Department of Ecological Botany, Umeå University, S-901 87 Umeå, Sweden

Owen Mountford
Institute of Terrestrial Ecology, Monks Wood, Abbots Ripton, Huntingdon, PE17 2LS, UK

Helena Otahelová
Institute of Botany, Slovak Academy of Sciences, Dubravská 14, Sk 84223, Bratislava, Slovakia

Karel Prach
Faculty of Biological Sciences, University of South Bohemia, Branišovská 31, CZ-370 05, České Budějovice; and Institute of Botany, Academy of Sciences, CZ-379 82 Třeboň, Czech Republic

Elle Puurmann
Institute of Ecology, Kevade Str. 2, EE-0001 Tallinn, Estonia

Richard Pywell
Institute of Terrestrial Ecology, Monks Wood, Abbots Ripton, Huntingdon, PE17 2LS, UK

Urve Ratas
Institute of Ecology, Kevade Str. 2, EE-0001 Tallinn, Estonia

José M. Rey Benayas
Area de Ecología, Facultad de Ciencias, Universidad de Alcalá, E-28871 Alcalá, Spain

Steve Rose
ADAS Land Research Centre, Gleadthorpe, Meden Vale, Mansfield, NG20 9PF, UK

Manuel G. Sánchez-Colomer
CEDEX, Paseo Bajo de la Virgen del Puerto, 3. E-28005 Madrid, Spain

Roger Smith
Institute of Grassland and Environmental Research, North Wyke Research Station, Okehampton, EX20 2SB, UK

Tim Sparks
Institute of Terrestrial Ecology, Monks Wood, Abbots Ripton, Huntingdon, PE17 2LS, UK

Gordon Spoor
Silsoe College, Cranfield University, Silsoe, MK45 4DT, UK

Jana Straškrabová
Institute of Applied Ecology, Agricultural University Prague, CZ-281 63 Kostelec n.Č.l., Czech Republic

Jerry Tallowin
Institute of Grassland and Environmental Research, North Wyke Research Station, Okehampton, EX20 2SB, UK

Andres Tōnisson
Estonian Institute of Meteorology and Hydrology, Teaduse Str. 2, EE-3400 Saku, Estonia

Jo Treweek
Institute of Terrestrial Ecology, Monks Wood, Abbots Ripton, Huntingdon, PE17 2LS, UK

Laimi Truus
Institute of Ecology, Kevade Str. 2, EE-0001 Tallinn, Estonia

Iñigo Vázquez-Dodero
Area de Ecología, Facultad de Ciencias, Universidad de Alcalá, E-28871 Alcalá, Spain

P. Max Wade
International Centre of Landscape Ecology, Department of Geography, Loughborough University, Loughborough, LE11 3TU, UK

Dirk M. Wascher
European Centre for Nature Conservation, Warandelaan 2, PO Box 1352, 5004 BJ Tilburg, The Netherlands

Shaojun Xiong
*Riparian Ecology Group, Department of Ecological Botany, Umeå University, S-901 87
Umeå, Sweden*

Mária Zaliberová
*Institute of Botany, Slovak Academy of Sciences, Dubravská 14, Sk 84223, Bratislava,
Slovakia*

Preface

The Landscape Ecology series seeks to focus on a range of topics that present challenges to the scientist and practitioner. It presents a spectrum of knowledge and perspectives in order to encourage an integrated approach to contemporary issues. Landscape ecology concerns the interrelationships between the various components of the landscape, including flora, fauna, soil, water and human impact. This focus needs to encompass both the growing body of knowledge concerned with the processes that underlie the landscape, for example ecological, hydrological and climatological, and the experience gained by those managing the landscape such as the farmer, the conservation officer, the engineer and the administrator.

The maintenance of the European wet grassland landscape, including floodplain (or alluvial) meadows, coastal grazing marshes or pastures, and sea-shore grasslands, through traditional low-intensity farming practices has conserved a biodiverse habitat of international importance. Such grasslands are often of high nature conservation value, because they support rare plant species and vegetation types and important concentrations of wildfowl and wading birds. Indeed, the international importance of the habitat for the conservation of biodiversity has been recognized with its inclusion in the European Union Habitats Directive and as themes within the Convention on Biological Diversity fostered by the United Nations. The UK Government, too, has identified the need to focus on wet grasslands at a European scale through the Department of the Environment's funding of a Darwin Initiative Project on implementing sustainable conservation of European wet grasslands. This was undertaken by the International Centre of Landscape Ecology in the Department of Geography at Loughborough University, UK, in collaboration with the Institute of Ecology, Tallinn, Estonia and the Faculty of Biological Sciences, University of South Bohemia, Czech Republic. Two international workshops were organized as part of the project; these nourished a network of researchers and practitioners across Europe concerned with various aspects of wet grassland management and restoration for the conservation of biodiversity. The meetings not only identified the need for a book focussing on the conservation of European wet grasslands but also provided papers that form the core of this book.

Chris B. Joyce and Max Wade
July 1997

Acknowledgements

The publication of this book was supported by the UK Department of the Environment through its Darwin Initiative Programme. The editors would like to thank Valerie Richardson, Paul José, Richard Jefferson and Ian Reid for their valuable advice. Also, thanks to Erica Milwain and Peter Robinson who provided cartographical expertise and to Liz Traynor and Gill Giles for technical support. Thanks, finally, to the anonymous referees who reviewed the chapters so effectively and to the chapter authors for their co-operation throughout the development of this book.

1 Wet Grasslands: A European Perspective

CHRIS B. JOYCE and P. MAX WADE

International Centre of Landscape Ecology, Loughborough University, UK

INTRODUCTION

Wet grasslands are characterized by an abundance of grasses, periodic but not perpetual flooding with fresh or brackish water, or a high water table, and regular management, usually cutting (mowing) or grazing. They often support a mosaic of associated plant communities (e.g. swamp, mire and saltmarsh) and habitats, including wetland features such as drainage channels or ditches and floodplain pools. They can include *Carex*-dominated habitats but not extensive reedbeds (e.g. of *Phragmites australis*). Most European wet grasslands are located on riverine floodplains and lake margins, and in the coastal zone, often behind sea defences. They are therefore primarily a lowland habitat; hence the generic term that has been applied to them of 'lowland wet grasslands' (e.g. Dargie 1993; Jefferson and Grice 1998). This general label covers a number of terms used to identify and describe different types of wet grasslands in Europe. These include inundation, alluvial or flood meadows, washlands, polders, coastal grazing marshes or pastures, and sea-shore grasslands.

This chapter provides a context for the subsequent sections and chapters within this book by introducing the core themes that unify them. These themes provide the basis for the four parts of the book: past and contemporary status, biodiversity and nature conservation value, vegetation and hydrological management, and restoration through creation and rehabilitation. The chapter concludes by identifying key issues affecting the biodiversity and the conservation management of wet grasslands in Europe in the future.

STATUS

Few wet grasslands in Europe are natural. Exceptions may include plant communities of spring-fed sites where frequent inundation restricts vegetation succession, depressions in natural steppe grasslands where accumulations of snow and rain form mesophytic conditions and the ice-governed meadows of northern Europe (Arnqvist and Dynesius 1987; Dziewulska 1990; Rychnovská 1993). Most wet grasslands have been created

See Glossary, p. 305, for explanation of technical terms. Scientific names of vascular plants follow Tutin, T. G. *et al.* (1964–80) *Flora Europaea* Volumes 1–5. Cambridge University Press. See p. 319.

by human activity, usually by forest clearance or the drainage of bogs and marshes (Ellenberg 1988), and are maintained by human intervention, usually grazing by live-stock or cutting. In the past, management was part of a low-intensity agricultural system that was characterized by low fertilizer input, minimal land drainage, cutting for hay and low stocking densities (Beaufoy 1994 *et al*). These wet grasslands provided an important contribution to the agricultural economy and their agricultural value has been appreci-ated and actively managed for since at least the sixteenth century (Sheail 1971). Tradi-tional management created and maintained beneficial conditions for a wide range of flora and fauna (Bignal and McCracken 1996). However, widespread losses and ecological degradation of European wet grasslands have occurred, especially in the last 50 years (van Dijk 1991). This has been largely due to agricultural intensification, often facilitated by flood defence and land drainage in order to increase grass production or convert permanent grassland to arable crops (Fuller 1987; Wells and Sheail 1988; Haury *et al*. 1990). Increased exploitation includes the use of inorganic fertilizers and other ag-rochemicals, increased cutting frequency (e.g. for silage) and increased stocking density (International Union for Conservation of Nature and Natural Resources 1991, 1993; van Dijk 1991). For example, between 1950 and 1990 the area of wet meadows in Hungary declined from 600 000 to 200 000 ha as a result of drainage and agricultural intensifica-tion to increase production. During this period, fertilizer usage on Hungarian grasslands doubled. Instead of increasing productivity, however, this process led to reductions in grass yields and deterioration in soil fertility, partly through excessive trampling by grazing animals, as well as ecological damage through habitat fragmentation and a decrease in biodiversity (International Union for Conservation of Nature and Natural Resources 1990). European wet grasslands have also been eliminated and degraded through pollution (especially eutrophication), mineral extraction, and industrial and urban development (Vermeer 1986; van der Hoek 1987; International Union for Conser-vation of Nature and Natural Resources 1993).

European wet grasslands have environmental functions and values additional to their agricultural and nature conservation importance. Water resource benefits include flood-water retention and control, especially in floodplains, which can result in enhanced sediment storage, fewer erosion problems and improved groundwater recharge (Dister *et al*. 1990). Floodplain vegetation can also maintain or improve water quality through retention of suspended matter and natural purification of nutrients (Brinson *et al*. 1984; Pinay *et al*. 1990). The dense root system of grasses plays a key role in both filtration and reduction of erosion as it contributes substantially to soil formation (Kvet 1996). The recreational, educational/scientific and cultural value of the European wet grassland landscape is widely acknowledged (Beaufoy *et al*. 1994; Rychnovská *et al*. 1994; Bignal and McCracken 1996).

In recent years, heightened scientific and political concerns over the decrease in extent and deterioration of the European wet grassland resource, coupled with recognition of its international importance for biodiversity conservation, have led to its inclusion within a number of environmental treaties, including those fostered by the United Nations and the European Union (EU) as well as national governments. The strategic and policy framework for conserving wet grasslands in Europe is presented in Part One of the book, with a focus on actions that aim to conserve biodiversity. This section also examines the contemporary status of European wet grasslands through studies in Estonia, Spain and England.

BIODIVERSITY

European wet grasslands are often of high nature conservation value, supporting considerable biodiversity including rare and threatened plant species and vegetation types (Rodwell 1991, 1992; Rychnovská 1993; Straškrabová *et al.* 1996), nationally and internationally important bird populations (Hötker 1991a; BirdLife International European Agriculture Task Force 1996) and a range of invertebrates, some of which are also rare (Drake 1998). Botanical diversity can be high. Prach and Straškrabová (1996) recorded a maximum of 38 species m^{-2}, and a total of almost 80 species from a transect of approximately 150 m length from the floodplain grassland of the Lužnice River in the Czech Republic, and Mediterranean pastures and meadows subject to groundwater seepage can display more than 50 species 2.5 m^{-2} (Puerto *et al.* 1990). Approximately 540 species of vascular plants have been recorded from the Morava River floodplain, Slovakia, of which 12% are nationally rare or endangered (Ružička 1994). A number of globally threatened and declining plant species are associated with wet grasslands in Europe, including *Apium repens, Selinum carvifolia* and *Scorzonera humilis* (Jefferson and Grice 1998). Some of the plant communities of wet grassland have become very restricted in extent, such as the *Alopecurus pratensis–Sanguisorba officinalis* flood meadow community, which is specially protected by the EU through the EU Habitats Directive (Council of the European Communities 1992).

Wet grasslands provide breeding or wintering habitat for a number of bird species, particularly wading birds (waders) and wildfowl. Recent estimates of the populations of waders breeding in the countries of the EU found that more than half of all the waders in the region breed on wet grasslands and that seven of the eight species are declining due to habitat loss (Hötker 1991b). In the UK, over 40 bird species of conservation concern are dependent or partly dependent on wet grasslands, including globally endangered species such as *Crex crex* (corncrake), and other internationally important species, e.g. *Limosa limosa* (black-tailed godwit), *Cygnus columbianus bewickii* (Bewick's swan), *Anser fabalis* (bean goose) and *Anas acuta* (pintail) (Royal Society for the Protection of Birds, English Nature and Institute of Terrestrial Ecology 1997).

Wet grasslands can support a high diversity of invertebrates (Kirby 1992) although relatively few rare species appear to be dependent upon the grass sward itself (Drake 1998). Invertebrate diversity and the number of rare species tend to be associated with environmental heterogeneity in the wet grassland landscape, which is provided by habitats such as damp hollows and temporary pools (Kirby 1992), drainage channels (Higler 1976; Hingley 1979; Verdonschot and Higler 1989) and old trees (Drake 1998). Some internationally rare invertebrates are associated with wet grasslands including the butterfly *Eurodryas aurinia* (marsh fritillary), which has a restricted and declining European distribution and is globally threatened (Thomas and Lewington 1991). Brackish marshes are the stronghold for many coastal Coleoptera (water beetles) (Royal Society for the Protection of Birds, English Nature and Institute of Terrestrial Ecology 1997), and Leibak and Lutsar (1996) highlight the value of coastal grasslands in Estonia as migratory corridors for Lepidoptera (butterflies and moths) and Odonata (dragonflies and damselflies).

The invertebrate fauna of wet grasslands is of considerable value in providing prey items for waders. The abundance of these invertebrates is influenced by management, with a reduction of almost 30% as management increases from the lowest to the highest level of agricultural intensity (Blake and Foster 1998).

Other wildlife of importance that utilize wet grasslands and their associated habitats in Europe include amphibians, fish, and mammals such as *Lutra lutra* (otter), *Arvicola terrestris* (water vole) and *Alces alces* (moose) (Kminiak 1994; Bejček and Šťastný 1996; Leibak and Lutsar 1996).

Part Two of the book explores European biodiversity issues. The factors that influence plant diversity are highlighted in chapters covering floodplain grasslands in Slovakia and sea-shore grasslands in Estonia, while the remaining chapters describe the invertebrate communities of European wet grassland habitats and elucidate the role they have in sustaining bird populations.

MANAGEMENT

The distinctive and ecologically valuable communities of European wet grasslands are characterized by periodic inundation (or a high water table) and regular appropriate vegetation management, which is often based on traditional farming practices. The latter is usually as part of an agricultural system that utilizes the primary production to support domestic herbivores either directly through grazing (pastures) or indirectly by harvesting hay (meadows). Pastures are grasslands that are grazed usually by cattle or sheep, a form of management prevalent on many coastal wet grasslands, such as coastal grazing marshes in the UK (Tickner and Evans 1991). Meadow management comprises cutting (mowing) for a hay crop at least once annually. The regrowth (aftermath) can subsequently be used for grazing, which is common in western Europe (e.g. Baker 1937), or, as in many parts of central Europe, cut again for further hay crops (Prach *et al.* 1996).

Most wet grassland plant communities of conservation importance are associated with low soil-nutrient availability and evidence strongly indicates that increasing nutrient supply alters community composition and reduces plant species diversity, favouring taller productive species, especially some grasses (Traczyk *et al.* 1984; Mountford *et al.* 1993). Traczyk *et al.* (1976) found that the use of inorganic fertilizers on mesophilous, floristically rich Polish meadows caused a decrease in species richness of almost 40% in only two years, and Mountford *et al.* (1993) reported that even modest fertilizer applications of 25 kg nitrogen $ha^{-1} yr^{-1}$ led to changes in a wet meadow plant community in England. Eutrophic floodwater can also induce vegetation change, for example by encouraging the development of *Glyceria maxima* swamp (Burgess *et al.* 1990). Whilst the more homogeneous and luxuriant vegetation that develops in grasslands of higher nutrient status is generally less suitable for most wildlife, some species (e.g. grazing geese) can benefit from the increased production or enhanced cover (Norriss and Wilson 1993).

Many remaining European wet grasslands of nature conservation value are deteriorating through a lack of management, with, for example, agricultural overproduction and policy reform in western Europe leading to the withdrawal of marginal areas from agriculture (Bignal and McCracken 1992). Also, many central and eastern European countries have recently undergone change in their agricultural as well as political systems, resulting in uncertainty over land ownership and the neglect or abandonment of many areas (Baldock 1994; Straškrabová *et al.* 1996). Unmanaged wet grasslands tend to exhibit reduced nature conservation value, becoming dominated by just a few robust competitive plant species, such as *Arrhenatherum elatius, Phalaris arundinacea, Rumex* spp. and *Urtica dioica* (Bakker 1989; Rychnovská *et al.* 1994; Guth and Prach 1996; Joyce 1998), at the expense of plant diversity (Regnéll 1980; Oomes and Mooi 1981). They eventually succeed to shrubland or forest, although succession can be arrested if plant litter or competition from the herbaceous field layer excludes woody species (Prach 1994).

Hydrological management for nature conservation objectives is primarily concerned with the maintenance of an appropriate water regime, particularly achieving wet conditions for certain periods of the year. The specific water regime depends on the target wildlife species, groups or habitats. For example, many waterfowl and waders require shallow flooding to provide feeding opportunities and secure roost sites in winter (Thomas 1982; Self *et al.* 1994). For breeding waders, a high water table during the breeding season (approximately March–June) is probably the single most important hydrological factor (Ward 1994). Plant communities are substantially influenced by the depth and annual variation of the water table. The composition of a plant community may be radically altered by minor shifts in water regime, as some species are adapted to survive in anaerobic soil conditions due to high water tables (Ernst 1990). Newbold and Mountford (1997) document the many wet grassland plant communities and species that require or tolerate moist and even waterlogged soil conditions, and periodic inundation, although some wet grasslands are characterized by drought stress in summer due to reduced water supply and free-draining topsoils. In addition, many plant and animal species have exacting water quality requirements, being influenced by factors such as pH, salinity, and nutrient status (Ellenberg 1988; Rimes 1992; Spieksma *et al.* 1995).

Drainage ditches often form an important element of the wet grassland hydrological infrastructure, facilitating manipulation of field water levels through systems of pumps, pipes, sluices and earth dams or bunds (Self *et al.* 1994). Such structures can promote targeted management by dividing sites into hydrologically isolated compartments.

The interactions between hydrological and vegetation management are highlighted by Part Three of the book. This section reviews and analyses factors of importance for maintaining biodiverse European wet grasslands, and includes a practical focus on management techniques.

RESTORATION

Considerable past reduction in the extent and quality of the wet grassland resource in Europe means that protection and sustainable management of the remaining sites of conservation value are essential. However, expansion of the resource is also necessary in order to facilitate the conservation of species and communities that were formerly more widespread.

The aim of any proposed restoration scheme should be carefully considered. Restoration should not be used as a substitute for the *in situ* protection of existing high-quality wet grassland habitat, but as a tool that supports the conservation of the resource. The objectives of rehabilitation or creation will also largely determine the subsequent hydrological and vegetation management of the restored site. For example, wet grassland sites being established for birds require appropriate hydrological conditions, such as a high water table, and may benefit from grazing by cattle. In contrast, cutting for hay (perhaps with grazing of the regrowth) may be the optimal aftercare management for a diverse flora, particularly when there is a need to reduce the soil nutrient status to promote botanical diversity (Berendse *et al.* 1992; Oomes *et al.* 1996).

Restoration of the wet grassland resource can be achieved through the rehabilitation of ecologically degraded sites to a condition similar to their former state and creation of new sites of wildlife interest. The former has been demonstrated in the Czech Republic by Straškrabová and Prach (1998), who reinstated cutting management to a wet grassland abandoned for approximately 20 years. After five years species diversity and composition were restored to a quality comparable with adjacent wet grasslands that had received uninterrupted management. Manchester *et al.* (1998) tested the effectiveness of restoring arable land to floristically diverse wet grassland by natural regeneration and by introducing seeds. They found that on ex-arable sites natural regeneration would take many years to restore a characteristic vegetation type and that sowing seed mixtures resulted in significantly greater numbers of species in the subsequent sward. This highlights the problems of attempting to restore wet grasslands on areas that have been managed intensively, such as arable land, which are characterized by impoverished seedbanks, altered hydrological regimes and high nutrient availability. For example, Bekker *et al.* (1997) found that the soil seedbanks of a range of wet grasslands in western Europe were negatively affected by intensive agricultural management such as drainage and fertilization.

Hydrological rehabilitation is also an integral part of sustainable ecological restoration, since this indirectly governs nutrient availability. Bakker (1994) illustrates this by referring to the restoration management of a wet grassland flora in The Netherlands, which could not be achieved without the reinstatement of high groundwater levels and calcium-rich seepage.

Thus, in order to maximize the effectiveness of restoration initiatives they should be targeted towards areas that supported wet grasslands in the recent past and can continue to sustain them in the future, such as in river valleys with a sufficient supply of clean water. Restoration schemes that link remaining wet grassland habitats or are adjacent to existing sites of wildlife interest, such as botanically diverse grasslands that can provide a seed source, are more likely to succeed and be of strategic and lasting benefit.

Part Four of the book explores the restoration of wet grasslands in Europe with studies from England, The Netherlands and the Czech Republic. These address some of the most important issues confronting the planning and implementation of wet grassland rehabilitation and creation, including the problems caused by fertilizer residues, water management for restoration and effective vegetation establishment and maintenance within the restoration scheme.

THE FUTURE

The European wet grassland resource and the wildlife associated with it are continuing to decline (BirdLife International European Agriculture Task Force 1996), particularly through agricultural intensification or an absence of management. This degradation can be arrested by adopting complementary strategies that protect priority European sites and conserve the habitat in the wider countryside, ensuring that the remaining resource is managed appropriately in order to prevent further habitat loss and fragmentation.

Nature reserves and similar designations have been widely implemented to protect wet grassland sites and species in Europe (Wascher 1998). Protected sites may be relatively small and, given their primary nature conservation function, difficulties may arise in maintaining the management practices (e.g. cutting and grazing) necessary to sustain their wildlife value. Several international initiatives, such as NATURA 2000 and the Convention on Biological Diversity (and its national Biodiversity Action Plans), have emerged recently that build upon the foundation established by existing strategies, such as the Ramsar Convention (Hill *et al.* 1996; Wascher 1998), to protect key European sites. These not only reinforce protective legislation but also offer opportunities for the sustainable management and restoration of the wet grassland resource.

Bignal and McCracken (1996) assert that the only socially acceptable and sustainable management for European landscapes and biotopes of high nature conservation value, including wet grasslands, involves the continuation or reinstatement of low-intensity or traditional farming. Reforms and reviews of international and national environmental and agricultural policies (e.g. the Common Agricultural Policy) are providing opportunities for supporting such practices. Schemes that integrate agricultural and conservation management by offering financial incentives to farmers to undertake management for conservation benefit have been implemented by many European countries, particularly through the EU Agri-environment Regulation 2078/92 (BirdLife International European Agriculture Task Force 1996). In the case of wet grasslands, such environmentally sensitive management prescriptions include limiting the use of agrochemicals such as fertilizers, establishing cutting dates to benefit meadow plants, and raising water levels to encourage wintering or breeding birds (Glaves 1998). Agri-environment schemes may adopt a whole farm, river catchment or regional perspective and are therefore capable of delivering appropriate management over an extensive area of wet grassland landscape.

Monitoring of the environmental performance of strategic initiatives such as international legislation and agri-environment schemes is important, not only to assess their effectiveness against their objectives, but also to provide a measure of the status of the European wet grassland resource. Comprehensive monitoring of the extent and quality of wet grasslands should include characteristic and rare wildlife species and populations. This may provide valuable information in relation to several emerging or predicted threats to the wet grassland ecosystem, such as the possibilities of distribution changes and local extinctions of species due to global climate change (e.g. sea-level rise on coastal grasslands) and the ecological impact of the overexploitation of water resources (van Diggelen *et al.* 1994; Bennett 1996; Campbell *et al.* 1997).

Governments and environmental organizations have made the maintenance and restoration of the biodiversity of wet grasslands at both national and international levels a focus of their conservation efforts (e.g. Royal Society for the Protection of Birds 1993; Ministry of Agriculture, Czech Republic 1994; The UK Steering Group 1995). The challenge of conserving the resource can only be effectively addressed, however, by an approach that embraces the interrelationships between the various components of the wet grassland landscape, including flora, fauna, water, soil, and human intervention and impact, particularly in the context of rural development. Hence, the future protection, management and restoration of wet grasslands in Europe will depend upon integrating the knowledge and commitment of a broad spectrum of scientists and practitioners, such as policy makers, engineers, land managers and ecologists.

ACKNOWLEDGEMENTS

This chapter is partly based on projects funded by the Darwin Initiative (a commitment by the UK Government to safeguard global biodiversity) and Loughborough University Research Fund. Dr Richard Jefferson and Phil Benstead provided valuable comments on an earlier version of the manuscript.

SUMMARY

The conservation of European wet grasslands is reviewed in the context of past and contemporary status, biodiversity, vegetation and hydrological management, and restoration. Wet grasslands are characterized by an abundance of grasses, periodic flooding or a high water table, and regular cutting or grazing management. Nature conservation value is derived both from the plant and animal communities of the grassland itself and from features associated with the grassland habitat, e.g. drainage channels and old trees. The wet grassland resource in Europe and its biodiversity have decreased due to habitat loss and degradation, particularly as a result of land drainage and agricultural intensification. Many remaining areas are still threatened by intensive management, and others by abandonment through the withdrawal of agriculture. Nevertheless, the existing wet grassland resource supports considerable biodiversity, including rare and globally threatened species. Effective conservation of wet grassland sites includes vegetation management, usually through grazing (pastures) or cutting for hay (meadows), limitation of nutrient availability, and hydrological control, particularly maintenance of high water levels and flooding. Recent international and national legislation has highlighted the biodiversity value of European wet grasslands and offers mechanisms for site protection and sustainable management and restoration of the resource. Restoration, either through the rehabilitation of degraded grasslands or through the creation of new sites, needs to be targeted towards areas able to sustain wet grasslands in order to maximize benefits. In the future, priority site protection needs to be combined with environmentally sensitive management of the habitat in the wider countryside to arrest the ongoing decline and begin to recover the biodiversity value of the European wet grassland resource.

Keywords: Biodiversity, Conservation, Ecological restoration, Grassland management.

REFERENCES

Arnqvist, G. and Dynesius, M. (1987) *Råneälven naturinventering och bedömning av vetenskapliga naturvärden*. Länsstyrelsen i Norrbottens län, Luleå.

Baker, H. (1937) Alluvial meadows: a comparative study of grazed and mown meadows. *Journal of Ecology*, **25**, 408–420.

Bakker, J.P. (1989) *Nature management by grazing and cutting*. Kluwer Academic, Dordrecht.

Bakker, J.P. (1994) Nature management in Dutch grasslands. In: Haggar, R.J. and Peel, S. (eds), *Grassland management and nature conservation*, pp. 115–124. British Grassland Society Occasional Symposium No. 28. British Grassland Society, Reading.

Baldock, D. (1994) Possible policy options and their implications for conservation. In: Haggar, R.J. and Peel, S. (eds), *Grassland management and nature conservation*, pp. 167–176. British Grassland Society Occasional Symposium No. 28. British Grassland Society, Reading.

Beaufoy, G., Baldock, D. and Clark, J. (1994) *The nature of farming: low intensity farming systems in nine European countries*. Institute for European Environmental Policy, London.

Bejček, P. and Šťastný, K. (1996) Vertebrates of the Lužnice River floodplain. In: Prach, K., Jenik, J. and Large, A.R.G. (eds), *Floodplain ecology and management. The Lužnice River in the Třeboň Biosphere Reserve, central Europe*, pp. 113–124. SPB Academic. Amsterdam.

Bekker, R.M., Verweij, G.L., Smith, R.E.N., Reine, R., Bakker, J.P. and Schneider, S. (1997) Soil seed banks in European grasslands: does land use affect regeneration perspectives? *Journal of Applied Ecology*, **34**, 1293–1310.

Bennett, S. (1996) *Impact of water abstraction on wetland SSSIs*. English Nature Freshwater Series No. 4. English Nature, Peterborough.

Berendse, F., Oomes, M.J.M., Altena, H.J. and Elberse, W.T. (1992) Experiments on the restoration of species rich meadows in The Netherlands. *Biological Conservation*, **62**, 59–65.

Bignal, E. and McCracken, D. (1992) *Prospects for nature conservation in European pastoral farming systems*. Joint Nature Conservation Committee, Peterborough.

Bignal, E.M. and McCracken, D.I. (1996) Low-intensity farming systems in the conservation of the countryside. *Journal of Applied Ecology*, **33**, 413–424.

BirdLife International European Agriculture Task Force (1996) *Nature conservation benefits of plans under the Agri-environment Regulation (EEC 2078/92)*. BirdLife International, Cambridge.

Blake, S. and Foster, G.N. (1998) The influence of grassland management on body size in Carabidae (ground beetles) and its bearing on the conservation of wading birds. In: Joyce, C.B. and Wade, P.M. (eds), *European wet grasslands: biodivesity, management and restoration*, pp. 163–169. John Wiley, Chichester.

Brinson, M.M., Bradshaw, H.D. and Kane, E.S. (1984) Nutrient assimilative capacity of an alluvial floodplain swamp. *Journal of Applied Ecology*, **21**, 1041–1057.

Burgess, N.D., Evans, C.E. and Thomas, G.J. (1990) Vegetation changes on the Ouse Washes wetland, England, 1972–88 and effects on their conservation importance. *Biological Conservation*, **53**, 173–189.

Campbell, B.D., Stafford Smith, D.M. and McKean, G.M. (1997) Elevated CO_2 and water supply interactions in grasslands: a pastures and rangelands management perspective. *Global Change Biology*, **3**, 177–187.

Council of the European Communities (1992) Council Directive 92/43/EEC of 2 April 1992 on the conservation of natural habitats and of wild fauna and flora. *Official Journal of the European Communities*, No. L206.

Dargie, T.C. (1993) *The distribution of lowland wet grassland in England*. English Nature Research Report No. 49. English Nature, Peterborough.

Dister, E., Gomer, D., Obrdlik, P., Petermann, P. and Schneider, E. (1990) Water management and ecological perspectives of the Upper Rhine's floodplains. *Regulated Rivers: Research and Management*, **5**, 1–5.

Drake, M. (1998) The important habitats and characteristic rare invertebrates of lowland wet grassland in England. In Joyce, C.B. and Wade, P.M. (eds), *European wet grasslands: biodiversity, management and restoration*, pp. 137–149. John Wiley, Chichester.

Dziewulska, A. (1990) The spatial differentiation of grasslands in Europe. In: Breymeyer, A.I. (ed.), *Managed grasslands. Ecosystems of the world 17A*, pp. 1–13. Elsevier, Amsterdam.

Ellenberg, H. (1988) *Vegetation ecology of central Europe*. Cambridge University Press, Cambridge.

Ernst, W.H.O. (1990) Ecophysiology of plants in waterlogged and flooded environments. *Aquatic Botany*, **38**, 73–90.

Fuller, R.M. (1987) The changing extent and conservation interests of lowland grasslands in England and Wales: a review of grassland surveys 1930–84. *Biological Conservation*, **40**, 281–300.

Glaves, D.J. (1998) Environmental monitoring of grassland management in the Somerset Levels and Moors Environmentally Sensitive Area, England. In Joyce, C.B. and Wade, P.M. (eds). *European wet grasslands: biodiversity, management and restoration*, pp. 74–94. John Wiley, Chichester.

Guth, J. and Prach, K. (1996) Scenarios of possible future floodplain development. In: Prach, K., Jenik, J. and Large, A.R.G. (eds), *Floodplain ecology and management. The Lužnice River in the Třeboň Biosphere Reserve, central Europe*, pp. 237–243. SPB Academic, Amsterdam.

Haury, J., Merot, P. and Riviere, J-M. (1990) Under-utilization and intensification in two Brittany wetlands. *Bulletin d'Ecologie*, **21**, 61–64.

Higler, L.W.G. (1976) Observations on the macrofauna of a Dutch ditch. *Hydrobiological Bulletin*, **10**, 66–73.

Hill, D., Yates, T., Treweek, J. and Pienkowski, M. (eds) (1996) *Actions for biodiversity in the UK: approaches in UK to implementing the Convention on Biological Diversity.* Ecological Issue No. 6. British Ecological Society and Field Studies Council, Shrewsbury.

Hingley, M.R. (1979) The colonisation of newly-dredged drainage channels on the Pevensey Levels (East Sussex), with special reference to gastropods. *Journal of Conchology,* **30**, 105–122.

Hötker, H. (ed.) (1991a) *Waders breeding on wet grasslands.* Wader Study Group Bulletin No. 61. Wader Study Group, Tring.

Hötker, H. (1991b) Waders breeding on wet grasslands in the countries of the European Community – a brief summary of current knowledge on population densities and population trends. In: Hötker, H. (ed.), *Waders breeding on wet grasslands.* Wader Study Group Bulletin No. 61. Wader Study Group, Tring.

International Union for Conservation of Nature and Natural Resources (1990) *Environmental status reports: 1988/1989*, Vol. 1, *Czechoslovakia, Hungary, Poland.* International Union for Conservation of Nature and Natural Resources, Gland, Switzerland.

International Union for Conservation of Nature and Natural Resources (1991) *The lowland grasslands of central and eastern Europe.* International Union for Conservation of Nature and Natural Resources, Gland, Switzerland.

International Union for Conservation of Nature and Natural Resources (1993) *The wetlands of central and eastern Europe.* International Union for Conservation of Nature and Natural Resources, Gland, Switzerland.

Jefferson, R.G. and Grice, P.V. (1998) The conservation of lowland wet grassland in England. In: Joyce, C.B. and Wade, P.M. (eds), *European wet grasslands: biodiversity, management and restoration,* pp. 31–48. John Wiley, Chichester.

Joyce, C.B. (1998) Plant community dynamics of managed and unmanaged analysis. In Joyce, C.B. and Wade, P.M. (eds), *European wet grasslands: biodiversity, management and restoration,* pp. 173–191. John Wiley, Chichester.

Kirby, P. (1992) *Habitat management for invertebrates: a practical handbook.* Joint Nature Conservation Committee, Peterborough.

Kminiak, M. (1994) Amphibians in the alluvium of the Morava River. *Ecology (Bratislava),* **13**, 77–88.

Kvet, J. (1996) Obecné ekologiecké funkce nivnich luk. (Ecological functioning of floodplain meadows.) *Příroda, Praha,* **4**, 21–23.

Leibak, E. and Lutsar, L. (eds) (1996) *Estonian coastal and floodplain meadows.* Estonian Fund for Nature and WorldWide Fund for Nature Denmark, Tallinn.

Manchester, S., Treweek, J., Mountford, O., Pywell, R. and Sparks, T. (1998) Restoration of a target and wet grassland community on ex-arable land. In: Joyce, C.B. and Wade, P.M. (eds), *European wet grasslands: biodiversity, management and restoration,* pp. 277–294. John Wiley, Chichester.

Ministry of Agriculture, Czech Republic (1994) *Basic principles of the agricultural policy of the Government of the Czech Republic up to 1995 and for a further period.* Adore, Prague.

Mountford, J.O., Lakhani, K.H. and Kirkham, F.W. (1993) Experimental assessment of the effects of nitrogen addition under hay-cutting and aftermath grazing on the vegetation of meadows on a Somerset peat moor. *Journal of Applied Ecology,* **30**, 321–332.

Newbold, C. and Mountford, O. (1997) *Water level requirements of wetland plants and animals.* English Nature Freshwater Series No. 5. English Nature, Peterborough.

Norriss, D.W. and Wilson, H.J. (1993) Seasonal and long-term changes in habitat selection by Greenland white-fronted geese *Anser albifrons flavirostris* in Ireland. *Wildfowl,* **44**, 7–18.

Oomes, M.J.M. and Mooi, H. (1981) The effect of cutting and fertilizing on the floristic composition and production of an *Arrhenatherion elatioris* grassland. *Vegetatio,* **47**, 233–239.

Oomes, M.J.M., Olff, H. and Altena, H.J. (1996) Effects of vegetation management and raising the water table on nutrient dynamics and vegetation change in a wet grassland. *Journal of Applied Ecology,* **33**, 576–588.

Pinay, G., Décamps, H., Chauvet, E. and Fustec, E. (1990) Functions of ecotones in fluvial systems. In: Naiman, R.J. and Décamps, H. (eds), *The ecology and management of aquatic–terrestrial ecotones,* pp. 141–169. Parthenon, Carnforth.

Prach, K. (1994) Succession of woody species in derelict sites in central Europe. *Ecological*

Engineering, **3**, 49–56.

Prach, K. and Straškrabová, J. (1996) Louky v nivěřeky Lužnice v biosférické reservaci Třeboňsko – moznosti obnovy. (Meadows in the Luznice River floodplain in the Třeboň Biosphere Reserve – potential for restoration.) *Příroda, Praha*, **4**, 163–168.

Prach, K., Jeník, J. and Large, A.R.G. (eds) (1996) *Floodplain ecology and management. The Lužnice River in the Třeboň Biosphere Reserve, central Europe*. SPB Academic, Amsterdam.

Puerto, A., Rico, M., Matias, M.D. and Garcia, J.A. (1990) Variation in structure and diversity in Mediterranean grasslands related to trophic status and grazing intensity. *Journal of Vegetation Science*, **1**, 445–452.

Regnéll, G. (1980) A numerical study of successions in an abandoned, damp calcareous meadow in S. Sweden. *Vegetatio*, **43**, 123–130.

Rimes, C. (1992) *Freshwater acidification of SSSIs in Great Britain. I. Overview*. English Nature Science No.1. English Nature, Peterborough.

Rodwell, J.S. (ed.) (1991) *British plant communities*, Vol. 2, *Mires and heaths*. Cambridge University Press, Cambridge.

Rodwell, J.S. (ed.) (1992) *British plant communities*, Vol. 3, *Grasslands and montane communities*. Cambridge University Press, Cambridge.

Royal Society for the Protection of Birds (1993) *Wet grasslands – what future?* Royal Society for the Protection of Birds, Sandy.

Royal Society for the Protection of Birds, English Nature and Institute of Terrestrial Ecology (1997) *The wet grassland guide: managing floodplain and coastal wet grasslands for wildlife*. Royal Society for the Protection of Birds, Sandy.

Ružička, M. (ed.) (1994) Ecological potential of floodplain area of the River Morava. *Ecology (Bratislava)*, **13**, 1–216.

Rychnovská, M. (1993) Temperate semi-natural grasslands of Eurasia. In: Coupland, R.T. (ed.), *Natural grasslands. Ecosystems of the world 8B*, pp. 125–166. Elsevier, Amsterdam.

Rychnovská, M., Blažková, D. and Hrabé, F. (1994) Conservation and development of floristically diverse grasslands in central Europe. In: 't Mannetje, L. and Frame, J. (eds), *Grassland and society*, pp. 266–277. Wageningen Pers, Wageningen.

Self, M., O'Brien, M. and Hirons, G. (1994) Hydrological management for waterfowl on RSPB lowland wet grassland reserves. *RSPB Conservation Review*, **8**, 45–56.

Sheail, J. (1971) Formation and maintenance of water-meadows in Hampshire, England. *Biological Conservation*, **3**, 101–106.

Spieksma, J.F.M., Schouwenaars, J.M. and van Diggelen, R. (1995) Assessing the impact of water management options upon vegetation development in drained lakeside wetlands. *Wetland Ecology and Management*, **3**, 249–262.

Straškrabová, J. and Prach, K. (1998) Five years of restoration of alluvial meadows: a case study from central Europe. In: Joyce, C.B. and Wade, P.M. (eds), *European wet grasslands: biodiversity, management and restoration*, pp. 295–303. John Wiley, Chichester.

Straškrabová, J., Prach, K., Joyce, C. and Wade, M. (eds) (1996) Aluviální louky – jejich současý stav a možnosti obnovy. (Floodplain meadows – ecological functioning, contemporary state and possibilities for restoration.) *Příroda, Praha*, **4**, 1–176.

Thomas, G.J. (1982) The ecology of breeding waterfowl at the Ouse Washes, England. *Wildfowl*, **31**, 73–88.

Thomas, J. and Lewington, R. (1991) *The butterflies of Britain and Ireland*. Dorling Kindersley, London.

Tickner, M.B. and Evans, C.E. (1991) *The management of lowland wet grasslands on RSPB reserves*. Royal Society for the Protection of Birds, Sandy.

Traczyk, T., Traczyk, H. and Pasternak, D. (1976) The influence of intensive mineral fertilization on the yield and floral composition of meadows. *Polish Ecological Studies*, **2**, 39–47.

Traczyk, T., Traczyk, H. and Pasternak-Kusmierska, D. (1984) Reaction of meadow vegetation after seven years of intensive inorganic fertilization. *Ekologia Polska*, **4**, 581–596.

UK Steering Group, The (1995) *Biodiversity: The UK Steering Group report*, Vol. 1, Meeting the Rio challenge; Vol. 2, *Action Plans*. HMSO, London.

van der Hoek, D. (1987) The input of nutrients from arable lands on nutrient poor grasslands and their impact on the hydrological aspects of nature conservation. *Ecology (CSSR)*, **6**, 313–323.

van Diggelen, R., Grootjans, A. and Burkunk, R. (1994) Assessing restoration perspectives of disturbed brook valleys: the Gorecht area, The Netherlands. *Restoration Ecology*, **2**, 87–96.

van Dijk, G. (1991) The status of semi-natural grasslands in Europe. In: Goriup, P.D., Batten, L.A. and Norton, J.A. (eds), *The conservation of lowland dry grassland birds in Europe*, pp. 15–36. Joint Nature Conservation Committee, Peterborough.

Verdonschot, P.F.M. and Higler, L.W.G. (1989) Macroinvertebrates in Dutch ditches: a typological characterization and the status of the Demmerik ditches. *Hydrobiological Bulletin*, **23**, 135–142.

Vermeer, J.G. (1986) The effect of nutrient addition and lowering of the water table on shoot biomass and species composition of a wet grassland community (*Cirsio–Molinetum* Siss. et de Vries, 1942). *Oecologia Plantarum*, **7**, 145–155.

Ward, D. (1994) Management of lowland wet grasslands for breeding waders. *British Wildlife*, **6**, 89–98.

Wascher, D.M. (1998) European legislation and strategies for the conservation of lowland wet grasslands. In: Joyce, C.B. and Wade, P.M. (eds), *European wet grasslands: biodiversity, management and restoration*, pp. 15–29. John Wiley, Chichester.

Wells, T.C.E. and Sheail, J. (1988) The effects of agricultural change on the wildlife interest of lowland grasslands. In: Park, J.R. (ed.), *Environmental management in agriculture: European perspectives*, pp. 186–201. Belhaven Press, London.

Part One

STATUS

2 European Legislation and Strategies for the Conservation of Lowland Wet Grasslands

DIRK M. WASCHER

European Centre for Nature Conservation, Tilburg, The Netherlands

INTRODUCTION

In 1935, A.G. Tansley introduced the term 'ecosystem' to refer to a community of organisms in the context of, and including, their physical environment (Tansley 1935). In recent years, Tansley's definition of an ecosystem is probably one of the most quoted concepts finding its way into contemporary international journals and books on subjects such as biodiversity, landscape ecology and conservation biology. The early date of his scientific observation certainly deserves a great amount of respect and admiration. However, at the same time it is a reminder that the concept of ecosystems is anything but new and it has evolved to include an ever-extending branch of research with numerous widely recognized achievements.

Given the long history of ecosystem research and the bulk of resulting knowledge, the consequent adaptation of national and international nature conservation policies in response to this concept has taken a long time: an incubation period of more than 50 years was necessary to start legislative initiatives that focus on key habitat types of international conservation concern. Habitats such as lowland wet grasslands do not equal ecosystems, but they do form significant components of them. Traditionally, habitat conservation was thought to follow automatically the protection of individual species, and the implementation of species protection programmes has resulted in many practical management schemes for improving or restoring the functions of habitats at the regional, national and international level. Although difficult to quantify, the European Union (EU) Birds Directive (Council of the European Communities 1979) has had a significant knock-on effect on wetland ecosystems, including habitat types such as lowland wet grasslands. However, the concentration on only one species group, such as birds, or on other rare or endangered species has left out a wide range of ecotopes that are thought to contribute substantially to the continent's biodiversity. The existing highly fragmented and largely disintegrated patchwork of protected sites and management agreements has failed to stop or reverse the negative trends of habitat loss and species decline. Despite the increase of nature reserves and national parks (Figure 2.1), the overall loss and degeneration of many ecologically valuable habitat types, and the associated decline of species, have continued. The environmental impact on habitats

See Glossary, p. 305, for explanation of technical terms. Scientific names of vascular plants follow Tutin, T. G. *et al.* (1964–80) *Flora Europaea* Volumes 1–5. Cambridge University Press. See p. 319.

European Wet Grasslands: Biodiversity, Management and Restoration. Edited by Chris B. Joyce and P. Max Wade.
© 1998 John Wiley & Sons Ltd.

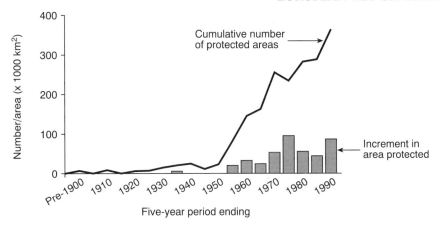

Figure 2.1   Growth of protected area network in Europe 1900–1990 (From Groombridge 1992)

outside or at the fringe of protected areas has turned out to be too large to be compensated by traditional nature conservation strategies. This is particularly true of lowland wet grasslands (Jefferson and Grice 1998).

The approval of the EU Habitats Directive (Council of the European Communities 1992) under the NATURA 2000 initiative and the adoption of the United Nations (UN) Convention on Biological Diversity (CBD) in 1992 (United Nations Environmental Programme 1992) have brought about substantial improvements at the legal and institutional level. Though still criticized for being incomplete, the inclusion of more species from all species groups and especially of explicitly listed habitat types gives the new EU Habitats Directive a policy format that is unprecedented at many national levels and which is armed with potentially powerful legal tools. This, in combination with the less binding but even more complex and offensive approaches laid out by the CBD, means that European nature conservation is now facing the challenge to provide definitions, concepts and methodologies that can bridge the gaps between regional and international actions as well as between law and reality.

After a brief review of those examples of European legislation on nature conservation which have been in place for many years, the potential role of new international policies and their institutional frameworks will be portrayed and critically examined. The analysis will have a special focus on European lowland wet grasslands.

EUROPEAN NATURE CONSERVATION LEGISLATION

THE BERN CONVENTION

The Convention on the Conservation of European Wildlife and Natural Habitats (Bern Convention) came into force in 1982. The Convention was promoted and developed by the Council of Europe and now has 31 contracting parties, including the 27 Member States of the Council of Europe, four non-Member States (Moldova, Monaco, Senegal and Burkina Faso), and the European Community. An increasing number of central and eastern European countries are in membership negotiation with the Council of Europe and are likely to join in the near future.

There is a governmental standing committee that monitors the enforcement of the Convention. This committee can organize international seminars and special studies to promote understanding of the issues, and adopts recommendations to the governments of contracting parties designed to further protect specific habitats and species.

The principal aim is to protect flora and fauna and their habitats, and to promote international co-operation among the contracting parties in resolving trans-frontier issues, with emphasis on the protection of endangered and vulnerable species and their habitats, and migratory species in particular. The Convention includes three annexes, listing more than 500 species of plants and more than 600 species of fauna. The implementation of the legal objectives is mainly carried out through expert groups on different species groups. On the basis of scientific information and inventories, the expert groups (e.g. on birds or invertebrates) provide recommendations to be followed by the Member States. While these have been relatively successful, for example in the case of the invertebrate group in Spain, Greece, Norway, Switzerland and France, some countries do not always follow these recommendations (Gruttke 1996).

Despite the increasing co-operation of Member States to improve their national legislation to meet the standards of the Convention, and despite the excellent scientific work of the expert groups, a number of shortcomings can be observed (Fernández-Galiano 1995):

- proposals and suggestions are in a relatively weak legal form
- passages concerning the legal protection of threatened natural habitats lack precision
- the legislation does not affect agricultural, forestry or fishery policies (of particular significance to lowland wet grasslands)
- no solution is provided for some severe and long-lasting problems concerning endangered species (e.g. marine turtles in Greece)
- the recommendations of the expert groups often do not reach, or are simply ignored by, the wider nature conservation community and public

The lack of specifications for conserving endangered natural habitats must be considered as especially critical. The large majority of the species listed in these annexes are extremely rare, endemic or geographically restricted to only certain regions. Consequently many sites that for other reasons have a high nature conservation value but do not support such species are, in this context, not considered to be of international importance and are ignored. As with other grassland habitats, lowland wet grasslands can support species and communities that are becoming regionally or nationally rare and even though such sites are frequently declining in nature conservation value they do not necessarily appear on international lists. For the same reason, the Convention's explicitly mentioned protection of breeding, staging, wintering, moulting and feeding areas of migratory species is beneficial for certain wetland ecotopes but applies only randomly (Haapanen 1995). The Bern Convention's definition of and emphasis on European-wide endangered species of fauna and flora provide only limited legal support for conserving lowland wet grasslands.

On the occasion of the Monaco Symposium in 1995, it was declared that the Bern Convention was considered to act as the European instrument for implementing the CBD. However, at the same time it has also been recognized that the scope of the Bern Convention is clearly narrower than that of the UN Convention. The differences between the two conventions have recently been analysed (Council of Europe 1993) and it has been found that with respect to the CBD the Bern Convention lacks reference to the following:

- genetic material or genetic resources *per se*
- *ex situ* conservation
- processes for identification of problems and monitoring of biological diversity
- environmental impact assessment
- control over genetically modified or living modified organisms
- conservation incentives
- domesticated or cultivated species

In addition to these formal differences, the Bern Convention lacks the political dimension of the CBD which identifies the need for 'fair and equitable sharing of the benefits of biodiversity', dealing *inter alia* with access to genetic resources, and access to and transfer of technologies and funding (Koester 1994).

The observed problems when implementing objectives derived from a traditional interpretation of the Bern Convention's legal framework and its narrow scope compared to the CBD, set certain limits to the expectations that have been raised in the Monaco Declaration. The CBD and the EU Habitats Directive have been created because existing legal instruments revealed shortcomings in their political and strategic dimensions. The Bern Convention will continue to play a key role for the future of Europe's nature conservation. However, it can only do so if it is efficiently integrated into the newly evolving legislative and institutional structures. One important aspect is a closer co-operation with environmental information centres such as the European Environment Agency and the World Conservation Monitoring Centre (Haapanen 1995). Only if all relevant information on the distribution and conservation status of the annex species is managed in a harmonized way, and if conceptual links are established to habitat monitoring and inventories, can the goals of the Bern Convention be addressed at several levels and play the complementary role that has been anticipated.

THE BIRDS DIRECTIVE

The European Community Directive on the conservation of wild birds (Council of the European Comunities 1981) is a piece of EU legislation which as such is legally binding on all Member States. All Member States are duty bound to 'preserve, maintain or re-establish a sufficient diversity and area of habitats' for all wild birds. They must ensure that a sufficient habitat is maintained to enable the survival of all migratory species and 175 specifically mentioned endangered species (Annex I of the Directive).

Sites set aside for this must be classified as Special Protection Areas (SPAs). The European Court of Justice has ruled that Member States have no power to modify or reduce the extent of such areas once they have been defined. So far 1195 sites have been designated, covering a total of approximately 7 million ha (European Topic Centre for Nature Conservation 1995). However, the International Council for Bird Preservation (ICBP, now BirdLife International) and the International Wetland and Waterfowl Research Bureau (IWWRB, now Wetland International) claim that twice as many sites qualify as SPAs, covering an area of 16 million ha, and that their protection is mandatory if the Directive is to succeed in its aim (Stanners and Bourdeau 1995). Sites classified under the Birds Directive as SPAs will form part of the NATURA 2000 ecological network currently being established (Figure 2.2).

By implementing its conservation goals through the establishment of SPAs as designated sites, the Birds Directive has a conceptual advantage over, for instance, the Bern Convention, by making its implementation more programmatic, transparent and measurable. Furthermore, the protection of bird habitats (especially migration sites, see Article 4.2) makes this Directive also a conservation instrument for wetlands, including many lowland wet grasslands. A number of bird species listed in the Directive's annex such as *Circus pygargus* (Montagu's harrier), *Pluvialis apricaria* (golden plover), *Anthus campestris* (tawny pipit), *Philomachus pugnax* (ruff), *Grus grus* (crane), *Ciconia nigra* (black stork), *Ciconia ciconia* (white stork) and *Numenius tenuirostris* (slender-billed curlew) can be associated more or less directly with lowland wet grassland habitat types.

Unfortunately, the Birds Directive SPAs are not inventoried in a very habitat-specific way. The newly developed database of the European Commission Ornithological Information System (ORNIS) (Institute Royal des Sciences Naturelles Belge 1993), however, provides information on population sizes and trends, temporal and spatial aspects of reproduction, migration and wintering, on hunting and also on habitats.

A Commission paper (Commission of the European Communities 1993) on the implementation of the Birds Directive notes: 'Up to now Member States have not supplied the Commission with adequate data on populations of Annex I and other migratory species within individual SPAs allowing evaluation of the protected area network.' The relevant bird population data sections of the recording form SFF4 which Member States are requested to complete when notifying the Commission of newly classified sites are in many cases poorly filled in and there is not yet any updating at regular time intervals to enable the reassessment of sites. There is no commonly accepted form for the information on the sites that need to be designated so that Member States can fulfil the obligations of Articles 4.1 and 4.2 of the Directive. Most Member States do not officially accept the list held by the Commission, which is based largely on information compiled by the ICBP (now BirdLife International) (the SFF3 list). Yet several of these same States have still to supply the Commission with an alternative list of sites.

The habitat information for ORNIS is provided on the basis of the CORINE (Co-ordination of Information on the Environment) habitat classification, terminology and coding system, which is analogous to that of the Habitats and Species Directive (see following section). However, the geographic reference base of ORNIS are the EU's Nomenclature des Unités Territoriales Statistiques (NUTS) regions and though basic information on species habitat requirements is given, further data gathering and transfer are necessary in order to identify locations and the condition of lowland wet grasslands. Only the results of the CORINE biotopes programme (see below) allow a habitat-specific analysis at the European level.

The Birds Directive is an important tool to use in the conservation of lowland wet grasslands but it must be remembered that birds are only one component of this complex habitat and other instruments are needed to provide overall conservation and protection.

THE EU HABITATS DIRECTIVE

The EU Habitats Directive (Council of the European Communities 1992) is a recent EU legislative instrument following on from the Birds Directive. It establishes a common

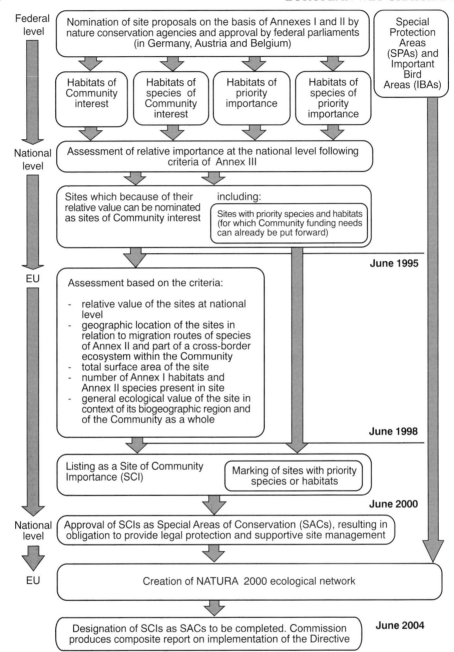

Figure 2.2   Implementation process for the EU's NATURA 2000 network

framework for 'the conservation of natural and semi-natural habitats and of wild fauna and flora' and constitutes the development of a network of Special Areas of Conservation (SACs) that will contribute to the ecological network NATURA 2000 (Figure 2.2). The Habitats Directive aims to combine the concern to protect endangered species with a wider concern to protect and enhance habitats of interest in their own right. Apart from the Bern Convention, the Habitats Directive is the only other instrument that furthers conserving natural habitats for their own sake and not only because they are home to certain species. It is intended that measures should be taken to allow species to move between sites through the identification of habitat corridors in order to increase their range or regain lost territory.

For the purpose of the Directive, 'habitat' means a natural or semi-natural area with particular, unique biogeographical characteristics. Annex I of the Directive lists 169 habitat types including rare and small habitats, such as alpine lakes and shifting dunes, as well as habitats known for their high biological diversity, such as calcareous grasslands. Other examples are estuaries (important for migratory species), traditionally managed agricultural lands (e.g. dehesas – Rey Benayas et al. 1998) and certain continental broad-leaved forests. Under Annex II the Directive lists 134 vertebrates (no birds), 59 invertebrates and 278 species of plants that require the protection of their habitats to ensure their survival. Annex IV provides strict protection for another 173 species of plants, 71 species of invertebrates and 160 species of vertebrates, placing a ban on the deliberate capture and killing of the animal species concerned and on the disturbance of these species, particularly during critical stages of their life cycle (breeding, rearing, hibernation and migration); deliberate picking, collecting, uprooting and destruction of the plants are prohibited.

In the first implementation phase, Member States are required to provide site proposals based on the criteria established in Annex III of the Directive (Figure 2.2). The national criteria for selecting habitats listed under Annex I include representativeness, the size of a site in relation to the total area of the habitat type in question, the global importance, the condition of the site and restoration possibilities. The national selection criteria for Annex II sites are based on size, density and isolation of a species population in relation to the national total, the condition of the site and its importance in the global context. The full description of the procedure for the implementation of the Habitats Directive is far more complex and goes beyond the scope and objective of this chapter.

In the second phase, the European Commission will apply another set of criteria to evaluate these proposals and to identify a European list of Sites of Community Importance (SCIs) ready for designation as SACs (Figure 2.2). In this evaluation process, the Commission will be actively assisted by the European Topic Centre for Nature Conservation (ETC/NC, see below) whose task it is to manage all relevant information on Europe's nature and biodiversity. One of the information sources of the ETC/NC is going to be data from the CORINE biotopes programme (see below), which holds descriptive files for more than 7000 sites.

At a technical seminar on the quality of the Habitats Directive, organized in San Lucar, Spain in Autumn 1993, it was observed that the implementation of this Directive was very inconsistent throughout the EU. The initial list of site proposals of each Member State was supposed to be delivered by June 1995, a date that has only been met by two countries. At the time of writing this chapter (nine months after the official

Table 2.1 Implementation of the EU Birds Directive and Habitats Directive under NATURA 2000. Situation as of 1 April 1996 on the basis of information supplied officially by the Member States

Member State	Birds Directive Special Protection Area (SPA) classification			Habitats Directive Special Area of Conservation (SAC) designation (stage 1)				
	No. of SPAs	Total area (km²)	Progress	National list	No. of sites	Total area (km²)	Site maps	NATURA 2000 forms
Austria	n/a	n/a	+	✓✓	94	±3 620	✓✓	✓✓
Belgium	36	4 313	+++	–	–	–	–	–
Denmark	111	9 601	+++	✓✓	175	±9 000	✓✓	–
Finland	15	n/a	+	✓✓	370	24 726	✓✓	–
France	99	7 069	+	–	–	–	–	–
Germany	494	8 537	++	–	–	–	–	–
Greece	26	1 916	+	–	–	–	–	–
Ireland	75	1 579	++	–	–	–	–	–
Italy	80	3 164	+	✓	±2 800	?	–	–
Luxembourg	6	14	+	–	–	–	–	–
Netherlands	23	3 276	++	–	–	–	–	–
Portugal	36	3 323	++	✓	30*	414	✓✓✓	✓✓
Spain	149	25 338	++	–	–	–	–	–
Sweden	75	1 460	+	✓✓	563	40 498	✓✓	✓✓
UK	126	4 396	++	✓✓	211	7 429	✓✓	–

Birds Directive: n/a, not available; +++, classification complete. Habitats Directive: –, no progress; ✓✓✓, list/maps/forms complete. *Madeira and Azores only.

Note on SPAs: Some Member States, especially Denmark and The Netherlands, have designated significant parts of their coastal waters (i.e. non-land area). Certain SPAs in Germany have been classified for nature conservation values other than their importance for birds.

Source: *NATURA 2000 Newsletter*, May 1996.

deadline), not even half of the Member States have complied with the requirements. Site lists have been put forward by Finland, Sweden, Denmark, the UK, Austria, Italy and Portugal (only for Madeira and the Azores). Much of the information reported on these sites was incomplete and did not meet the formal requirements. Among those countries who first forwarded initial site lists, Denmark was clearly leading in terms of quantity (more than 20% of its national territory) and the quality of the data supplied. For further details of the implementation of both the Birds and the Habitats Directives, see Table 2.1.

The Member States' implementation differs in terms of political and technical procedures. While Germany is largely developing existing biotope inventories and data, some Mediterranean Member States such as Italy and especially Spain have started a new data-gathering process for identifying SCIs. In Spain, a comprehensive project involving more than 200 scientists for mapping the vegetation at the scale 1:50 000 has recently been launched. By taking advantage of existing satellite data from the CORINE land cover project, the teams are working towards a precise boundary identification for all SCIs.

Large differences can be observed in the political and educational approaches of the Member States (Ssymank 1994). Most countries have decided strictly to separate the technical nature conservation-related selection procedure from the parallel active political–administrative working groups. France and the UK in particular have started from the beginning to support the implementation with a comprehensive public education campaign.

EUROPEAN LOWLAND WET GRASSLANDS IN CORINE AND THE HABITATS DIRECTIVE

The emphasis of the Annex I list of the Habitats Directive is on habitats that are naturally rare, biologically diverse or traditionally managed. One source for establishing this Annex I list has been the habitat classification developed for the CORINE biotopes programme of the European Commission. CORINE has the objective of improving the environmental information at a European level by gathering, harmonizing and compiling data in a structured and technically advanced way by the use of Geographic Information Systems. Within the CORINE biotopes programme, more than 7000 biotope sites have been identified throughout the former EU (12 Member States) and the associated habitat classification comprises more than 2200 different habitat types.

Among the 169 habitat types listed under Annex I of the Directive are semi-natural tall-herb humid meadows and mesophile grasslands with four types of plant communities that potentially resemble components of lowland wet grasslands:

C37.31   *Molinia* meadows on chalk and clay (*Eu–Molinion*)
C37.4   Mediterranean tall-herb and rush meadows (*Molinia–Holoschoenion*)
C37.7   Eutrophic tall herbs (*Calystegiao–Alliarietalia*)
C38.2   Lowland hay meadows (*Alopecurus pratensis–Sanguisorba officinalis*)

Table 2.2 shows the results of an analysis of the presence of the above Annex I habitat types as records within the CORINE biotopes database (Commission of the European Communities 1994). Of the EU Member States examined, neither The Netherlands nor the new Member States (Austria, Sweden and Finland) were included due to data revisions, only eight countries having supplied information on these habitat types. When interpreting the table, it is important to keep in mind that the resulting figures reflect to a certain extent the quality and completeness of the reporting mechanisms. Many countries in which CORINE biotopes have been identified did not complete those passages of the form that asked for specifications for habitat types. Even where habitat information had been provided, such as in the case of the eight countries, data on the surface area were in most cases given for the whole site, but not for the habitat type in question. The number of hectares in the last column of Table 2.2 represents the total extent of all sites where these habitats have been reported to be present. Hence, the actual size of lowland wet grassland habitats must be considered to be much smaller. Nevertheless, the results listed in Table 2.2 can be regarded as indicative of the overall situation. It does not come as a surprise that lowland wet grasslands (especially habitat types 37.7 and 38.2) are more concentrated in northwestern regions and that large countries such as France and Germany are holding the larger share. While the total number of sites in France and Germany are almost equal, the French sites cover an area that is more than four times larger than those of Germany. Because the database was only partially complete at the time of analysis, and given their biogeographic location, countries such as the UK,

Table 2.2 Potential presence of lowland wet grassland habitat types in eight EU countries according to the CORINE biotopes database (number of CORINE biotope sites where habitat types have been recorded and total surface area in which they can occur)

Member State	Habitat type				Total no. of sites	Area (ha)
	37.31	37.4	37.7	38.2		
Germany	12		114	197	323	500 000
France	2		34	298	334	2 380 000
Belgium	10		2	53	65	400 000
Luxembourg	3		1		4	180
Ireland	2		1		3	12 000
Denmark	2				2	1 700
Greece			1		1	8 000
Spain	2	27	8	6	43	1 370 000

CORINE habitat types: C37.31, *Molinia* meadows on chalk and clay (*Eu–Molinion*); C37.4, Mediterranean tall herb and rush meadows (*Molinia–Holoschoenion*); C37.7, Eutrophic tall herbs (*Calystegiao–Alliarietalia*); C38.2, Lowland hay meadows (*Alopecurus pratensis–Sanguisorba officinalis*).
Sites for the UK, Portugal and Italy did not show any entries for the above habitat types due to incomplete response to the CORINE form. Information on sites in the Netherlands was lacking due to revision at the time the assessment was undertaken.

Denmark and The Netherlands are likely to have more lowland wet grassland than is currently identifiable within CORINE biotopes. Among the UK's total number of 350 identified CORINE biotope sites based on the 1992 inventory, less than 40% have been described according to their habitat qualities. A comparison with the German CORINE contribution shows that for a total number of 1200 selected sites, information on habitats had been provided for almost 90%, allowing the identification of sites of the four habitat types associated with lowland wet grasslands. The UK's shortage of information and reporting on CORINE biotope sites, and consequently on lowland wet grasslands, is also reflected in the first approach in implementing the Habitats Directive (see previous section).

Among the Mediterranean countries, Spain has performed the most comprehensive and complete implementation of the CORINE biotopes programme, identifying 43 lowland wet grassland sites with a total of more than 1.3 million ha of grassland habitats. The largest proportion (27 sites) is accounted for by Mediterranean tall-herb and rush meadows (habitat code C37.4), indicating future conservation needs.

As indicated above, the role of the CORINE biotopes database is currently being reviewed. However, the example of lowland wet grassland habitats demonstrates that the EU's potential information base on the presence of sites can differ greatly from country to country (Table 2.2).

While the CORINE habitat classification has been frequently criticized for not being complete and systematically correct (WorldWide Fund for Nature 1995), its quality is above all dictated by the level of data that the Member States input. Whatever role the CORINE habitat classification is going to play within the implementation of the Habitats Directive, it remains clear that the lack of data for any given Member State on habitats in general and on lowland wet grasslands in particular will limit its potential use for assessing that State's site proposals.

Taking as an example the recent list the UK Government presented in the framework of the Habitats Directive, this comprises 211 sites due to be updated in the near future. This is 140 less than the total number of sites inventoried for the CORINE biotopes programme. National nature conservation groups analysed these site proposals and came to the conclusion that this list would in terms of quantity and quality not constitute an adequate UK contribution to NATURA 2000 (Friends of the Earth 1995). A more detailed analysis of the proposed sites (Joint Nature Conservation Committee 1995) concluded that the nine sites of Eutrophic tall herbs (C37.7 and C37.8) and the three sites of Lowland hay meadows (C38.2) must be considered as largely under-representative of these habitat types in the UK. This analysis does not include the site proposals under Annex II for which those species that are indicative of lowland wet grasslands need to receive special attention.

THE EUROPEAN TOPIC CENTRE FOR NATURE CONSERVATION

From 1985 to 1993, and especially through the establishment of the European Environment Agency Task Force in 1991, CORINE has been the initial instrument for preparing and testing improved data management. Since its establishment in 1994, the European Environment Agency (EEA) has been the institutional body of the EU officially responsible for the management of environmental data and in charge of setting up a European Information and Observation Network (EIONET). The EEA itself is carrying out this work by co-ordinating other expert organizations in the various environmental media, the so-called European Topic Centres. The European Topic Centre for Nature Conservation (ETC/NC) is located in Paris and has three main tasks:

1. General Approach: assessing information needs and data availability related to nature conservation policies.
2. State and Trends of Biodiversity: setting up a methodology for a permanent monitoring of biodiversity on the basis of a preliminary statement of current state and principal trends in the different biogeographic regions of Europe.
3. Support for the NATURA 2000 Network: providing information and tools to support the implementation of the Habitats Directive and the Birds Directive.

As part of the third task, the ETC/NC is presently revising and extending CORINE biotope databases and classification systems to ensure full compatibility with the requirements of the NATURA 2000 legislation. Other activities include the review of existing national and international databases on species and habitats in order to prepare for harmonization and integration into a European database. The ETC's work programme is designed to build up a reference and information base that improves the EU's capacities in environmental assessment and reporting as the base for better policy implementation.

THE PAN-EUROPEAN BIOLOGICAL AND LANDSCAPE DIVERSITY STRATEGY

The Pan-European Biological and Landscape Diversity Strategy (Council of Europe 1995) is a European response to support the implementation of the CBD. The Strategy was proposed in the Maastricht Declaration 'Conserving Europe's Natural Heritage' 1993, and builds upon the Bern Convention, the European Conservation Strategy, the ministerial conferences within the 'Environment for Europe' process (Dobris, Lucerne and Sofia) and the UN Conference for Environment and Development (UNCED, Rio de Janeiro, 1995). One of the most prominent sources for the Strategy's Action Plan is the report *Europe's environment – the Dobris assessment* (Stanners and Bourdeau 1995). The Strategy aims to strengthen the application of the Bern Convention in relation to the CBD following the Monaco Declaration by introducing a co-ordinating and unifying framework and by building on existing initiatives. It does not aim to introduce new legislation or programmes, but to fill gaps where initiatives are not implemented to their full potential or fail to achieve desired objectives. Furthermore, the Strategy seeks to integrate more effectively ecological considerations into all relevant socio-economic sectors, and will increase public participation in, and awareness and acceptance of, conservation interests.

The Strategy's vision for the future is to achieve the conservation and sustainable use of biological and landscape diversity for the whole continent of Europe and all its regions within 20 years. It will seek to ensure that the threats to Europe's biological and landscape diversity are reduced substantially, or where possible removed; the resilience of European biological and landscape diversity is increased and the ecological coherence of Europe as a whole is strengthened; and public involvement and awareness concerning biological and landscape diversity issues are increased considerably. These aims should be achieved within 20 years and through a series of five-year Action Plans. The Action Plan 1996–2000 identifies the following Action Themes:

1. Establishment of a Pan-European Ecological Network
2. Integration of biological and landscape diversity considerations into sectors
3. Raising awareness and support with policy makers and the public
4. Conservation of landscapes
5. Conservation of coastal and marine ecosystems
6. Conservation of river ecosystems and related wetlands
7. Conservation of inland wetland ecosystems
8. Conservation of grassland ecosystems
9. Conservation of forest ecosystems
10. Conservation of mountain ecosystems
11. Action for threatened species

A first international workshop under the guidance of the European Centre for Nature Conservation and the World Conservation Union's European Programme was convened in October 1996 to identify an adequate implementation framework and suitable projects. The Action Theme most relevant to lowland wet grasslands is Action Theme 8 along with aspects of Action Themes 5, 6 and 7 with which it overlaps on changes in land-use policy and practices throughout Europe with special emphasis on agricultural policy reform as well as on privatization of land and regional development policies. Within the agricultural sector there are a number of legislative opportunities that are of actual or potential support for grassland ecosystems: the Common Agricultural Policy reform (1992) including Agri-environment Measures (EC Regulation 2078/92/EEC), the European Agricultural Guidance and Guaranteed Fund (EAGGF) instrument, and the organic farming regulation (2092/91/EEC).

Despite all of those legal instruments which are already in existence, the Strategy is considered to provide stronger and more co-operative support to the identified priority Action Themes. During 1996, the UN Environmental Programme, Council of Europe, International Union for the Conservation of Nature and Natural Resources, and the European Centre for Nature Conservation prepared the establishment of a Strategy secretariat as well as holding a working conference on nature conservation funding and investments. This conference identified existing sources and new opportunities for supporting practical implementation of the Pan-European Biological and Landscape Diversity Strategy and the CBD in Europe.

CONCLUSIONS

The lack of complete and reliable information has been viewed as one of the key obstacles in the implementation of EU directives and international conventions. The current establishment of new European institutes and mechanisms offers potentially strong technical and administrative facilities to improve this situation. However, the process of defining ecological standards and compiling data throughout Europe involves a large number of administrative bodies and semi-political procedures. In this situation, existing bodies of expertise in the non-governmental field at the level of universities and working groups face the challenge of making contributions as well as critically observing these procedures. Only a full understanding of the newly established mechanisms by the different expert groups will allow efficient participation and co-operation within international nature conservation policy implementation. In order to meet the challenge to produce internationally harmonized information on species, habitats and landscapes, regional and national experts should actively support the ETC/NC, which is responsible for assessing the contributions of Member States implementing NATURA 2000. At the same time, the Pan-European Biological and Landscape Diversity Strategy is designed to improve the situation for specific ecosystem types including grasslands, by channelling funding to Action Theme related initiatives and by promoting an integration of ecological considerations into socio-economic sectors.

SUMMARY

As with many other European habitat types, lowland wet grasslands have declined in many regions whilst remaining sites are highly fragmented and face increasing pressures from competing human land uses and pollution. Due to the traditional focus on species protection and to the lack of support for implementation by Member States, existing nature conservation policies such as the Bern Convention and the European Union (EU) Birds Directive are likely to improve the situation only marginally. With the Convention on Biological Diversity and the EU Habitats Directive in place since 1992, this new set of legal and formal objectives requires more holistic and integrative approaches than previously. Success will also depend on the ability and willingness of the Member States to comply adequately with these requirements and on the methodologies developed to accompany and assess this process. The EU CORINE programme, although only experimental in nature, has shown how non-compliance to technical requirements can result in inadequate implementation procedures and incomplete survey results. While the current state of the implementation of the EU Habitats Directive is now facing similar obstacles, the EU has established the European Environment Agency and the European Topic Centre for Nature Conservation to provide better technical support for these legal instruments. In recognition of the problem in harmonizing existing initiatives, the Council of Europe is preparing a Pan-European Biological and Landscape Diversity Strategy, endorsed at the Sofia conference of environmental ministers in 1995. The Strategy's five-year action programme also includes a plan to maintain and restore Europe's grassland ecosystems by means of nature conservation as well as by integrating these goals into all relevant socio-economic sectors.

Keywords: Bern Convention, Biological diversity, Birds Directive, CORINE, Habitats Directive, International co-operation, ORNIS, Protected areas, Sustainability.

REFERENCES

Commission of the European Communities (1993) *State and perspectives of the classification of Special Protected Areas by Member States.* ORNIS 93/7. Commission of the European Communities, internal document.

Commission of the European Communities (1994) *Use of the CORINE biotopes database for the implementation of the Directive on the conservation of the natural habitats and of wild fauna and flora (92/43/EEC)*, Vol. I. List of CORINE sites for which habitats, listed under Annex I of the Directive, have been reported. Prepared for the Habitats Committee of 3–4 February 1994. Commission of the European Communities, Brussels.

Council of Europe (1993) *Connections and comparisons between the Bern Convention, the Convention on Biological Diversity and Council Directive 92/43/EEC.* T-PVS (93) 21. Council of Europe, Strasbourg.

Council of Europe (1995) *Pan-European Biological and Landscape Diversity Strategy.* ECE/CEP/23. Submitted by the Council of Europe at the Ministerial Conference Environment for Europe, Sofia, Bulgaria, 23–25 October 1995.

Council of the European Communities (1979) Council Directive 79/409/EEC on the conservation of wild birds. *Official Journal of the European Communities*, No. L291, 29 October 1979.

Council of the European Communities (1992) Council Directive 92/43/EEC of 2 April 1992 on the conservation of natural habitats and wild fauna and flora. *Official Journal of the European Communities*, No. L206.

European Topic Centre for Nature Conservation (1995) *Biodiversity and nature conservation: a European general approach.* European Topic Centre for Nature Conservation, Paris.

Fernández-Galiano, E. (1995) A personal view. *Naturopa*, **77**, 5.

Friends of the Earth (1995) *Council Directive on the conservation of natural habitats and of wild fauna and flora (92/43/EEC) – the Habitats Directive. A list of possible Special Areas of Conservation in the UK. Friends of the Earth's, Plantlife's and Derek Ratcliffe's joint response to the Consultation Document.* Friends of the Earth, London.

Groombridge, B. (1992) *Global biodiversity. Status of the earth's living resources.* Chapman & Hall, London.

Gruttke, H. (1996) Bern Convention on invertebrates. *Natur und Landschaft*, **71**, 7–11.

Haapanen, A. (1995) The Bern Convention: its potential and objectives. *Naturopa*, **77**, 6–7.

Institute Royal des Sciences Naturelles Belge (1993) *ORNIS database.* Report XI/743/93. Institute Royal des Sciences Naturelles Belge, Brussels.

Jefferson, R.G. and Grice, P.V. (1998) The conservation of lowland wet grassland in England. In: Joyce, C.B. and Wade, P.M. (eds), *European wet grasslands: biodiversity, management and restoration*, pp. 31–48. John Wiley, Chichester.

Joint Nature Conservation Committee (1995) *Council Directive on the conservation of natural habitats and of wild fauna and flora (92/43/EEC) – the Habitats Directive. A list of possible Special Areas of Conservation in the UK.* Joint Nature Conservation Committee, Peterborough.

Koester, V. (1994) The Bern Convention: prospects for action with a view to applying UNCED instruments, in particular the Convention on Biological Diversity, at the regional level. *Environmental Encounters,* **22**, 73–78.

Ray Benayas, J.M., Sánchez-Colomer, M.G., Levassor, C. and Vásques-Dodero, I. (1998) The role of wet grasslands in biological conservation in Mediterranean landscapes. In: Joyce, C.B. and Wade, P.M. (eds), *European wet grasslands: biodiversity, management and restoration,* pp. 61–72. John Wiley, Chichester.

Ssymank, A. (1994) New challenges for nature conservation in Europe. The network NATURA 2000 and the Habitats Directive of the European Union. (In German.) *Natur und Landschaft,* **69**, 395–406.

Stanners, D. and Bourdeau, P. (eds) (1995) *Europe's environment – the Dobris assessment.* European Environment Agency, Copenhagen.

Tansley, A.G. (1935) The use and abuse of vegetational concepts and terms. *Ecology,* **16**, 284–307.

United Nations (1992) *Report of the United Nations Conference on Environment and Development,* Rio de Janeiro, 3–14 June 1992, 1, Resolution adopted by the Conference, Sales no. E93, New York.

WorldWide Fund for Nature (1995) *CORINE: databases and nature conservation – the new politics in the European Union.* A report supported by the WorldWide Fund for Nature UK compiled by Waterton, C., Grove-White, R., Rodwell, J. and Wynne, B., Lancaster University, Lancaster.

3 The Conservation of Lowland Wet Grassland in England

RICHARD G. JEFFERSON and PHILIP V. GRICE
English Nature, Peterborough, UK

INTRODUCTION

Lowland wet grassland consists of land managed as pasture or hay meadow occurring in areas with a high water table or subject to periodic flooding. It normally occurs at less than 200 m above sea level in river valleys, behind sea defences or in areas with impeded drainage. Lowland wet grassland has high nature conservation value, supporting a wide range of indigenous species of plants, invertebrates and birds (Ratcliffe 1977; Fuller 1982; Drake 1998). A substantial proportion of its total area has been lost or undergone a marked decline in its biodiversity. This has principally been due to drainage and agricultural improvement over the last 50 years (Williams *et al.* 1983; Nature Conservancy Council 1984; Ratcliffe 1984; Williams and Bowers 1987; Williams and Hall 1987; O'Brien and Smith 1992; Mountford 1994; O'Brien and Self 1994). UK conservation organizations in both the statutory and voluntary sector, including English Nature and the Royal Society for the Protection of Birds (RSPB), unanimously view the conservation of wet grassland as one of the highest priorities (Buisson and Williams 1991; Housden *et al.* 1991; Brown and Grice 1993; Moffat 1994). The UK Biodiversity Action Plan (The UK Steering Group 1995) has identified 38 key habitats requiring conservation action. Costed action plans have been produced for 14 of these, to date, of which two, coastal and floodplain grazing marsh and *Molinia caerulea* and *Juncus* pasture, comprise lowland wet grassland.

DEFINITIONS

The term 'lowland wet grassland', which is now in wide use, appears to have been originally introduced by ornithologists. It is a generic term and most definitions are fairly broad and ill defined but indicate that the habitat principally consists of permanent grassland that is periodically waterlogged. Lowland wet grassland is normally considered by ornithologists to encompass grazing marshes, flood meadows and man-made washlands and water meadows, which are descriptions based on management and hydrology rather than on species composition. Plant community ecologists, in contrast, have tended to describe and classify wet grasslands and related communities in terms of their plant

See Glossary, p. 305, for explanation of technical terms. Scientific names of vascular plants follow Tutin, T. G. *et al.* (1964–80) *Flora Europaea* Volumes 1–5. Cambridge University Press. See p. 319.

European Wet Grasslands: Biodiversity, Management and Restoration. Edited by Chris B. Joyce and P. Max Wade.
© 1998 John Wiley & Sons Ltd.

Table 3.1 National Vegetation Classification (NVC) plant communities (Rodwell, 1991, 1992, 1995) considered as lowland wet grassland. Equivalent European phytosociological alliances and CORINE biotopes are also provided (Commission of the European Communities 1991)

NVC plant communities		European phytosociological alliances	CORINE biotopes	
(a) Neutral grassland				
*MG4	*Alopecurus pratensis–Sanguisorba officinalis*	Cynosurion	38.2	Lowland hay meadows
MG6	*Lolium perenne–Cynosurus cristatus* (*Alopecurus geniculatus, Deschampsia cespitosa* and *Iris pseudacorus* variants of the typical subcommunity)	Cynosurion	38.111	*Lolium perenne* pastures
MG7	*Lolium perenne* (*L. perenne–A. pratensis* flood pasture and *L. perenne–A. pratensis* grassland)	Lolio–Plantaginion	38.111	*Lolium perenne* pastures
*MG8	*Cynosurus cristatus–Caltha palustris*†	Calthion	37.214	*Senecio aquaticus* meadows
MG9	*Holcus lanatus–Deschampsia cespitosa*	Calthion	37.213	*Deschampsia cespitosa* meadows
MG10	*Holcus lanatus–Juncus effusus*	Calthion	37.217	*Juncus effusus* meadows
*MG11	*Festuca rubra–Agrostis stolonifera–Potentilla anserina*	Elymo–Rumicion	37.242	*Agrostis stolonifera* and *Festuca arundinacea* swards
*MG12	*Festuca arundinacea*	Elymo–Rumicion	37.242	*Agrostis stolonifera* and *Festuca arundinacea* swards
*MG13	*Agrostis stolonifera–Alopecurus geniculatus*	Elymo–Rumicion	37.242	*Agrostis stolonifera* and *Festuca arundinacea* swards
(b) Fen meadow				
*M22	*Juncus subnodulosus–Cirsium palustre*	Calthion	37.218	*Juncus subnodulosus* meadows
*M23	*Juncus effusus/acutiflorus–Galium palustre*	Juncion acutiflori	37.22	*Juncus acutiflorus* meadows
*M24	*Molinia caerulea–Cirsium dissectum*	Junco conglomerati–Molinion	37.312	Acid purple moor grass meadows (*Junco-Molinion*)
*M25	*Molinia caerulea–Potentilla erecta*	Junco conglomerati–Molinion	37.312	Acid *Molinia caerulea* (*Junco-Molinion*)
(c) Swamp				
*S5	*Glyceria maxima*	Phragmition	53.15	*Glyceria maxima* beds
*S22	*Glyceria fluitans*	Spargario–Glycerion	53.4	Small reed beds of fast flowing waters
*S28	*Phalaris arundinacea*	Magnacaricion	53.16	*Phalaris arundinacea* beds

*Communities considered to be agriculturally unimproved and semi-natural in character
†Included in this category is an *Agrostis stolonifera–Carex nigra–Senecio aquaticus* wet grassland described from the Somerset Levels. This community is related to MG8 but is more closely related to the *Senecioni–Brometum racemosi* described from The Netherlands (Rodwell. pers. comm.)

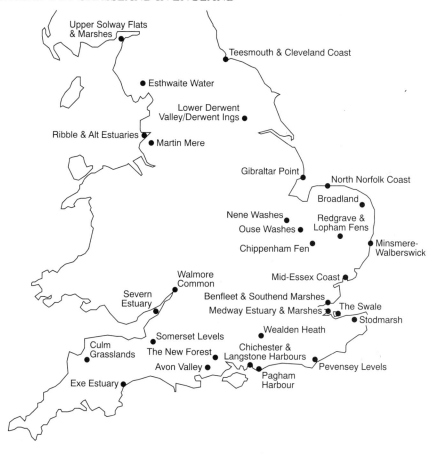

Figure 3.1 The location of lowland wet grassland sites mentioned in the text

species composition (Tansley 1939; Ratcliffe 1977; Rodwell 1991, 1992, 1995). The plant communities considered as lowland wet grassland in this chapter are given in Table 3.1.

Table 3.1 includes those communities which: (i) normally contain a mixture of grasses, broad-leaved herbs, sedges and rushes in varying proportions; (ii) occur in lowland situations (i.e. <200 m above sea level); (iii) are regularly managed as pasture or meadow; and (iv) have a high water table (mean field level ranges from −60 to +10 cm) or are periodically inundated with fresh or brackish water. In addition to the communities listed in Table 3.1, stands of vegetation in grazing marshes behind sea walls may, in some situations, show affinities with saltmarsh communities such as National Vegetation Classification (NVC) SM16 *Festuca rubra* and SM28 *Elymus repens* (Rodwell, in press). The latter community has particularly close phytosociological affinities with neutral grassland types MG11, MG12 and MG13 (Rodwell 1992), which can all occur in coastal localities. Many lowland wet grasslands, particularly coastal grazing marshes and flood-plain grasslands, are dissected by ditches containing standing water with their component plant and animal communities; examples of this are the Somerset Levels and the Pevensey Levels. (Figure 3.1 shows the location of sites mentioned in the text.) In many localities, wet grassland often forms part of a mosaic with 'non-grassland' wetland communities including swamps and tall-herb fens (for example, NVC S25 *Phragmites australis–Eupatorium cannabinum* tall-herb fen; Rodwell 1995), wet heath (for example, NVC M16 *Erica tetralix–Sphagnum compactum* wet heath; Rodwell 1991), saltmarsh (see above) and open water. These habitat mosaics are particularly prevalent in coastal areas and on river floodplains.

DISTRIBUTION AND EXTENT

At present there is no definitive figure for the extent of lowland wet grassland in England. However, Dargie (1993) estimated that there is approximately 220 000 ha of land which is 'probable' lowland wet grassland. Not all of this area, however, will support wildlife communities of high nature conservation value. This figure underestimates the full extent of lowland wet grassland as it does not include any blocks of less than 10 ha. Only a small proportion of this resource consists of agriculturally unimproved semi-natural grassland communities and again, at present, it is not possible to give an accurate figure for the extent of these communities. Unimproved wet neutral grassland (NVC MG4, MG8, MG11, MG12 and MG13) is likely to occupy an area of less than 10 000 ha in England (Jefferson and Robertson 1996). There are currently no national estimates for the extent of the other semi-natural fen-meadow, rush pasture and swamp communities listed in Table 3.1 (marked with an asterisk). However, we suggest a rough estimate of less than 20 000 ha. The spatial distribution of lowland wet grassland in England is reasonably well documented (Dargie 1993), with the largest extent in south-west England (i.e. the Somerset Levels) and the smallest in the north-east. As far as individual vegetation communities are concerned, the maps in Rodwell (1991, 1992, 1995) provide a general guide. However, these should be used with caution as they represent the location of quadrat samples taken for the NVC that have subsequently been ascribed to community types following numerical analysis and are unlikely to reflect the complete distribution of these plant communities. Some field surveys of neutral grassland and fen meadows have been undertaken, principally by English Nature and the former Nature Conservancy Council (see, for example, Palmer and Blake 1991) but the extent of knowledge varies geographically.

BIODIVERSITY AND NATURE CONSERVATION VALUE

Lowland wet grassland supports a wide range of plants and animals and thus has high biodiversity. Biodiversity can be defined as the variety of organisms considered at all levels (subspecies, species, genera etc.) as well as the variety of communities or biotopes (Wilson 1992). It follows that lowland wet grassland as a resource is highly valued for nature conservation. The principal interests are plant communities, vascular plants, breeding and wintering birds, and invertebrates. Ditches can also support important assemblages of aquatic plants (Ratcliffe 1977; Palmer and Newbold 1983; Evans 1991) and invertebrates (Drake, in press).

PLANT COMMUNITIES

Of the plant communities that constitute lowland wet grassland (Table 3.1), a number are considered to be of special nature conservation value as they represent 'ancient' semi-natural vegetation communities composed of native plant species that have not been adversely affected by agricultural treatments such as ploughing, reseeding or the application of fertilizers and herbicides (Nature Conservancy Council 1989). Many are now rare and continue to suffer losses and declines in their biodiversity, particularly outside of protected areas (see Introduction to this chapter; and Fuller 1987; Hopkins and Hopkins 1994; Jefferson and Robertson 1996).

Some types of wet neutral grassland and fen meadow, such as NVC communities MG4, MG8, M22, M23 and M24, are also particularly species-rich, with the mean number of plant species per 4 m^2 sample ranging from 19 to 28 (Rodwell 1991, 1992).

VASCULAR PLANTS

Lowland wet grassland supports a number of species of vascular plants that are rare or scarce both nationally and internationally. Table 3.2 lists nationally rare and scarce species associated with this habitat as defined in Table 3.1. Of the communities that comprise wet grassland, the highest proportion of the rare and scarce species appears to be associated with fen-meadow vegetation. England's lowland wet grassland is of lesser significance for the conservation of rare vascular plants than other lowland grassland types, particularly calcareous grassland (Stewart *et al.* 1994; Jefferson and Robertson 1996). However, when all habitats are analysed, wetland plant species comprise 20% of all nationally rare and 40% of nationally scarce vascular plant species (Evans 1991).

A further seven nationally rare and seven nationally scarce vascular plant species of open ground and ditches occur within areas of lowland wet grassland. The nationally rare species are as follows: *Carex vulpina, Chenopodium botryodes, Cyperus fuscus, Leersia oryzoides, Lythrum hyssopifolia, Mentha pulegium* and *Pulicaria vulgaris.* All of these except *Chenopodium botryodes* are also globally threatened or declining as defined by the Biodiversity Action Plan (The UK Steering Group 1995). The nationally scarce species are as follows: *Hordeum marinum, Polygonum minus, P. mite, Polypogon monspeliensis, Puccinellia fasciculata, P. rupestris* and *Sium latifolium.*

BIRDS

A number of England's rare, threatened and most important bird species are either wholly or partly dependent on lowland wet grasslands (Buisson and Williams 1991; Brown and Grice 1993). These include several rare species that are almost entirely restricted to this habitat during the breeding season, such as *Anas querquedula* (garganey), *Philomachus pugnax* (ruff) and *Limosa limosa* (black-tailed godwit). Several other species breed in large numbers and at high densities on lowland wet grassland although are not confined to it. They include *Anas clypeata* (shoveler), *Vanellus vanellus* (lapwing), *Numenius arquata* (curlew), *Gallinago gallinago* (snipe) and *Tringa totanus* (redshank). Other important breeding species include *Tyto alba* (barn owl) and *Motacilla flava* (yellow wagtail).

English lowland wet grasslands also support the majority of Britain's internationally

Table 3.2 Nationally rare and scarce species associated with lowland wet grassland. Nationally rare, occurrence in 15 or fewer 10-km squares; nationally scarce, occurrence in 16–100 10-km squares. Nationally rare species are given in bold

---

(a) Coastal grazing marsh
    Alopecurus bulbosus
    Althaea officinalis
    Bupleureum tenuissimum
    Carex divisa
    Cyperus longus
    Lepidium latifolium
    Trifolium squamosum
(b) Fen meadows and *Juncus* pasture
    Erica vagans (RDB)
    Gentiana pneumonanthe
    Hypericum undulatum
    Lathyrus palustris
    Lobelia urens (RDB, GT)
    Peucedanum palustre
    Scorzonera humilis (S.8, RDB, GT)
    Selinum carvifolia (S.8, RDB, GT)
(c) Inland wet neutral grassland
    Apium repens (RDB, GT)
    Carex elongata
    Carex tomentosa (RDB)
    Fritillaria meleagris
    Oenanthe silaifolia

S.8, listed on Schedule 8 of the Wildlife and Countryside Act, 1981 or added subsequently. RDB, listed in the British Red Data Book (Perring and Farrell 1983). GT, globally threatened and declining species as defined by the Biodiversity Action Plan (The UK Steering Group 1995).
Coastal grazing marsh includes NVC communities MG6, MG7, MG11, MG12 and MG13 (Rodwell 1992) (Table 3.1). Fen meadows and *Juncus* pasture include NVC communities M22, M23, M24 and M25 (Rodwell 1991) (Table 3.1). Inland wet neutral grassland includes all the NVC communities listed under headings (a) and (c) in Table 3.1 except MG12.

important wintering population of *Cygnus bewickii* (Bewick's swan), and most of Britain's *Anser fabalis* (bean goose), a localized wintering species largely confined to grazing marshes in the Yare Valley, Norfolk, in eastern England. A number of other wintering waterfowl species are supported in large numbers including *Cygnus cygnus* (whooper swan), *Branta leucopsis* (barnacle goose), *Anas penelope* (wigeon), *A. strepera* (gadwall), *A. crecca* (teal), *A. acuta* (pintail), *A. clypeata* (shoveler), *Aythya ferina* (pochard), *Pluvialis apricaria* (golden plover), *V. vanellus* (lapwing) and *N. arquata* (curlew). Wet grasslands also support large numbers of passage waders and wildfowl during the spring and autumn, being of particular significance for passage *Numenius phaeopus* (whimbrel).

Many of the breeding and non-breeding species associated with England's lowland wet grasslands are recognized as of high conservation priority in the UK and as Species of European Conservation Concern (SPECs) (Tucker and Heath 1994; Royal Society for the Protection of Birds 1996).

The results of a series of surveys of breeding wading birds (waders) (Smith 1983; O'Brien and Smith 1992; and summarized in O'Brien and Self 1994) and the new atlas of breeding birds (Gibbons *et al.* 1993) have shown that both the numbers and distribution of the key wet grassland wader species have been greatly reduced in many areas of

England in recent decades. In view of these declines and the overall ornithological significance of the habitat, both statutory and voluntary conservation bodies have recognized lowland wet grasslands as a high priority habitat for bird conservation (Housden *et al.* 1991; Brown and Grice 1993). This has prompted much research by the RSPB into the habitat requirements and therefore management needs of the key breeding and wintering species (Thomas 1982; Green 1986; Allport 1989; Burgess and Hirons 1990; Tickner and Evans 1991; Self *et al.* 1994).

It should be emphasized that the breeding and wintering bird populations associated with lowland wet grassland are not confined to areas of agriculturally unimproved (semi-natural) wet grassland. Factors such as substrate, hydrology (water-table height, duration and extent of inundation), vegetation structure, grassland management and freedom from disturbance are often of greater significance than is botanical composition (Green 1986), and therefore nationally or internationally significant numbers of birds can be supported in areas of limited botanical interest.

INVERTEBRATES

Research to date suggests that the areas of greatest importance for invertebrates within the wet grassland habitat are ditches and their margins, with the intervening fields often being of relatively low value, particularly if these are managed as hay meadows (Drake 1998; Drake, in press; R. Crossley, pers. comm.). It is thought that the low diversity of invertebrates in hay meadows is due to their structural uniformity and the periodic disturbance caused by cutting (Kirby 1992). Wet fen meadows and *Juncus* pastures that are summer-grazed are, however, an exception to this overall pattern (Drake, in press). A more detailed overview is given in Drake 1998.

MANAGEMENT

Maintenance of the nature conservation value of lowland wet grassland, as with the majority of semi-natural plagioclimax habitats, is dependent on the continuation of extensive or low-intensity agricultural management, i.e. grazing and hay-cutting (Crofts and Jefferson 1994), and the maintenance of high water tables or seasonal inundation. The latter are not always controllable by conservation bodies especially if influenced by surrounding land uses or climate.

On some wet grassland nature reserves owned and managed by conservation bodies such as English Nature and RSPB, the nature conservation value has been enhanced by obtaining more precise control of site hydrology (Self *et al.* 1994). This has been achieved by the use of pipes, sluices, bunds and pumps. Examples of sites where hydrology has been manipulated include the Lower Derwent Valley and Ouse Washes, although conditions at both sites are still strongly influenced by the water budgets of the respective river catchments.

It is not proposed to elaborate on the management of wet grassland as more detailed overviews are available elsewhere (Green 1986; José and Self 1994; Ward 1994).

Table 3.3 Designated Special Protection Areas (SPAs) and Ramsar sites containing areas of lowland wet grassland

Site name	County	Grassland type
Benfleet and Southend Marshes	Essex	CGM
Broadland	Norfolk/Suffolk	FM
Chichester and Langstone Harbours	Hants/West Sussex	CGM
Chippenham Fen (R)	Cambs	FM
Esthwaite Water (R)	Cumbria	FM
Exe Estuary	Devon	CGM
Gibraltar Point	Lincolnshire	CGM
Lower Derwent Valley/Derwent Ings	Humberside/North Yorks	IWNG
Martin Mere	Lancashire	IWNG
Medway Estuary and Marshes	Kent	CGM
Mid-Essex Coast	Essex	CGM
Minsmere-Walberswick	Suffolk	CGM
Nene Washes	Cambs	IWNG
New Forest	Hampshire	FM
North Norfolk Coast	Norfolk	CGM
Ouse Washes	Cambs/Norfolk	IWNG
Pagham Harbour	West Sussex	CGM
Redgrave and Lopham Fens (R)	Norfolk/Suffolk	FM
Ribble and Alt Estuaries	Lancashire	CGM
Severn Estuary	Somerset/Gloucs/Avon	CGM
Stodmarsh	Kent	IWNG
Teesmouth and Cleveland Coast	Cleveland	CGM
The Swale	Kent	CGM
Wealden Heath (Phase 1): Thursley, Hankley and Frensham Commons (SPA)	Surrey/Hants	FM
Upper Solway Flats and Marshes	Cumbria	CGM
Walmore Common	Gloucs	IWNG

(CGM, coastal grazing marsh; FM, fen meadow; IWNG, inland wet neutral grassland, R, Ramsar site only, SPA, Special Protection Area only

REPRESENTATION OF LOWLAND WET GRASSLAND WITHIN DESIGNATED AREAS

Table 3.3 lists the sites designated as Special Protection Areas (SPAs) under the European Union (EU) Birds Directive and as Wetlands of International Importance under the Ramsar Convention that contained significant areas of lowland wet grassland as at 30 June 1996 (see Figure 3.1 for site locations). These sites constitute 46% of the total number of designated SPA/Ramsar sites, emphasizing the importance of this habitat for breeding and wintering birds. A number of other sites containing important concentrations of lowland wet grassland are candidates for these international designations. These include the Somerset Levels and Moors, the Avon Valley and the Pevensey Levels.

Forty-one of the 168 (24%) National Nature Reserves declared by 30 September 1995 contain significant areas of lowland wet grassland. Of the 3065 biological sites notified by 1 July 1996 as Sites of Special Scientific Interest (SSSIs) under Section 28 of the Wildlife and Countryside Act 1981, approximately 47% have the presence of lowland grassland as a principal reason for notification. Approximately 18% have been notified with lowland wet grassland being a criterion for SSSI selection. Of those sites notified with grassland as a key interest, approximately 39% have wet grassland as a key component. It should be remembered when interpreting these figures that SSSIs can be selected and notified on the basis of more than one interest or feature. It is not practical, at present, to provide figures of the areal extent of lowland wet grassland within the various types of designated site due to inadequacies in the methods of data collection and presentation.

Annex I of the EU Habitats Directive (Council of the European Communities 1992) lists biotopes of European importance, and Member States are required to select and designate sites conforming to these as Special Areas of Conservation (SACs) by 2004 (see Chapter 2). The Annex I biotopes that fall within the current definition of lowland wet grassland are MG4 (*Alopecurus pratensis–Sanguisorba officinalis*) grassland (Lowland hay meadows) and M24 (*Molinia caerulea–Cirsium dissectum*) fen meadow (*Molinia* meadows on chalk and clay) (Table 3.1). Examples of sites selected for these biotopes and which will be designated as SACs include the Lower Derwent Valley and the Culm grasslands complex respectively.

KEY CONSERVATION ISSUES

In this section, we consider the key issues that need to be addressed to ensure the maintenance and enhancement of wet grassland biodiversity in the foreseeable future.

CHANGES IN AGRICULTURAL MANAGEMENT

Until very recently, the key threat to lowland wet grassland was drainage followed by conversion to arable or improved grassland, by ploughing and reseeding with high-yielding varieties of *Lolium perenne*. For example, between 1935 and 1982 there was a 48% loss of coastal grazing marsh on the North Kent Marshes (Williams *et al.* 1983) and between 1984 and 1990 a 10% loss of Culm grasslands (wet acid grassland/fen meadow) in Devon and Cornwall, primarily due to agricultural improvement (Devon Wildlife Trust 1990). In the 1990s, the situation changed somewhat, with inappropriate agricultural management being a greater threat to the integrity and biodiversity of wet grassland than drainage and improvement now (Royal Society for the Protection of Birds 1993; Grice *et al.* 1994). Table 3.4 summarizes the likely impact of such changes on the plant and animal communities of lowland wet grassland.

HYDROLOGY

Lowering of water tables in the vicinity of wet grassland and a reduction or cessation of winter flooding as a result of water abstraction, mineral extraction and flood alleviation continue to pose significant threats to the resource (Royal Society for the Protection of Birds 1993; Grice *et al.* 1994). An increase in the soil moisture deficit can lead to vegetation change with, in some situations, the replacement of wetland plant communities with drier grassland communities. Drying of sites also results in reduced suitability for breeding and wintering waders and wildfowl, due to the loss of, or loss of access to,

Table 3.4   Impacts of agricultural practices on the biota of lowland wet grassland

Management practice	Impact on biota
Use of artificial fertilizers and/or slurry	Reduction in plant species richness (e.g. Mountford *et al.* 1993). Fertilizer use often proceeds in tandem with a shift from hay to silage production
Change from hay to silage production	Earlier cutting results in reduced breeding success for birds, especially waders (Green 1986; José and Self 1994) Reduction in seed return as a result of more rapid cutting/baling with potential for long-term impact on seedling recruitment in semi-natural swards (R. Smith, pers. comm.)
Cessation of grazing/cutting, or undergrazing	Vegetation change often resulting in a decline in plant species richness and decline in nature conservation value (Duffey *et al.* 1974) Change in vegetation structure reduces habitat suitability for breeding waders (Green, 1986; Buisson and Williams 1991) Cessation of aftermath grazing in alluvial flood meadows may result in a reduction in botanical diversity if results from northern hay meadows can be extrapolated (Smith and Rushton 1994)
Overgrazing	Nest trampling causes reduced breeding success for waders (Green 1986; José and Self 1994) Overgrazing may result in reduced species-richness of unimproved wet grassland communities. For example, alluvial flood meadow (NVC MG4) may be converted to impoverished *Lolio–Cynosuretum* (NVC MG6) with heavy grazing (Rodwell 1992)

invertebrate prey, the loss of aquatic vegetation used for breeding sites and food, and the loss of secure water roosts (Green and Robins 1993).

Conversely, increased summer flooding and increased wetness can result in a reduction in the plant species richness and nature conservation value, as neutral grassland communities are replaced by more species-poor swamp communities (Burgess *et al.* 1990). Summer inundation can also lead to reduced breeding success of waders due to flooding of nests (Green *et al.* 1987). It is likely that the increased frequency of summer flooding events in some areas is principally the result of land drainage improvements over the last 25 years (Green *et al.* 1987). This is a particular problem where the grassland forms part of catchment flood storage networks, such as washlands.

HABITAT RESTORATION AND CREATION

Existing UK countryside policy and incentive measures including Environmentally Sensitive Areas (ESAs), long-term set-aside and the Countryside Stewardship Scheme (CSS), and the Wildlife Enhancement Scheme (WES) offer opportunities for both the restoration and creation of lowland wet grassland.

It is likely that such opportunities will increase in the future on the assumption that the above-mentioned schemes become more widely available and agricultural policies will increasingly integrate environmental benefits into the commodity support mechanisms of the Common Agricultural Policy (CAP), with the UK Biodiversity Action Plan (Department of the Environment 1994; The UK Steering Group 1995) providing an underlying stimulus.

Targeted wet grassland creation should link habitat fragments and thus facilitate the expansion of species and communities that were formerly more widespread. It may also assist in the management of existing grassland areas by increasing the area of land available for grazing or hay-making, thus enhancing its viability for low-intensity agricultural management. It should be stressed that the creation of exact replicas of existing semi-natural wet grassland is unlikely ever to be feasible (Hopkins 1989). Creation should not be seen as providing a substitute for this resource but rather as an additional benefit for assisting the conservation of species associated with wet grassland. It is estimated that in the UK, 1.2 million ha of land has potential for the creation of wet grassland (Self *et al.* 1994). This gives an indication of the opportunity that exists for an expansion of the habitat to benefit nature conservation, particularly for fauna.

OTHER ISSUES

There are a number of other issues relating to the conservation of wet grassland that are not, as yet, as important as those discussed above, although may ultimately prove to be so.

Climate change will undoubtedly have some impact on the distributions and populations of the component species of lowland wet grassland, including rare species, depending on the nature of the change (Elmes and Free 1994). Related to this, relative sea-level rise may result in plant and bird community change mediated through physical habitat changes in some coastal grazing marshes. Relative sea-level rise poses both threats to coastal grazing marshes and opportunities for their creation in a number of areas in England, particularly on the coasts of East Anglia and the south-east (Brown *et al.* 1994; Norris and Buisson 1994). In areas where important coastal habitats, including freshwater wetlands, are unable to retreat fully due to the terrain or hard sea defences, loss of grazing marshes will inevitably occur. There is thus a need to create replacement areas elsewhere within the coastal zone as compensation. However, in areas where intensively managed farmland abuts the coast, there may be opportunities for the creation of grazing marsh along with other coastal habitats should the coastline undergo managed retreat.

Nest predation by both mammalian and avian predators can be an important factor determining the breeding success of waders on lowland wet grassland (Green 1986). Whilst nest predation is known to be the most important cause of reproductive failure in birds, its role in the long-term declines of bird populations is far from clear and predator removal programmes have not usually provided an effective long-term solution to predation problems, other than on islands (Côté and Sutherland 1995). Despite this, many land managers are convinced that predation is an important factor determining breeding bird populations, especially on small, highly fragmented areas of semi-natural habitat, such as most of the remaining lowland wet grassland sites. A number of options are available to land managers to reduce the effect of nest predation on waders. Most important is to ensure that suitable nesting and chick-rearing conditions are maintained for as long as possible through the summer so waders can lay replacement clutches if nest predation occurs early in the breeding season (Green 1986). Where a small or vulnerable population is under extreme predation pressure and habitat improvements and/or creation are not options in the short term, predator removal or deterrence using lethal and/or non-lethal methods may be appropriate. Non-lethal methods can include the removal of trees and bushes used as nest sites and watch-points for avian predators such as corvids. More work is needed fully to assess the effectiveness of both predator removal and alternative methods of predator control.

Recreational disturbance may also be an important factor reducing breeding success of waders at some wet grassland sites. A recent review concluded that human-induced disturbance can have a significant negative effect on bird breeding success and reduce the use of sites by birds outside the breeding season. In particular, public and vehicular access to open landscapes has been shown to have a detrimental affect on breeding waders and wintering geese (Hockin *et al.* 1992).

Wildfowl shooting is an important recreational use of many wet grassland sites and the waterfowl that they support. Disturbance due to shooting can, however, cause waterfowl to change their behaviour, distribution and habitat utilization either within a site or over a larger area (Bell and Fox 1991). Such disturbance can be particularly important to herbivorous wildfowl, such as *Anas penelope* (wigeon), which need to spend long periods of the day feeding in order to satisfy their energy requirements (Mayhew 1988). The impact of shooting disturbance on populations is unclear. However, it is important to recognize the crucial role played by disturbance-free refuges in the management of waterfowl populations at both coastal and inland wetland sites. An appropriately design-ed refuge network can increase the carrying capacity of a site or site complex, such that the quality of the site for both conservation and wildfowling can be enhanced (Bell and Fox 1991). Wildfowling clubs can also play an important role in site safeguard, through land purchase and long-term land management (Laws 1984).

KEY ACTIONS

In this section we briefly review the key actions that need to be taken to address the issues identified above. Much action has already been instigated or is planned to take effect in the near future.

DESIGNATIONS AND INCENTIVE MECHANISMS

In recent years, a great deal of effort has been expended by conservation organizations in using the full range of statutory designations such as SSSIs, Ramsar sites and SPAs to protect and conserve important areas of lowland wet grassland in England. Incentive schemes such as the CSS and ESAs have also been implemented in the last few years. While these have their limitations in terms of the areas over which they apply or the habitats that are targeted and the nature of the management prescribed, they have undoubtedly played a part in conserving and enhancing the wet grassland resource (Land Use Consultants 1994; Russell 1994; Whitby 1994). It is essential that remaining lowland wet grasslands are safeguarded for the future using a combination of the appropriate designations and incentive mechanisms. It is equally important that the latter are used, where possible, to create new areas of wet grassland. Creation should be targeted ideally at areas adjacent to existing grassland or where fragmented sites can be linked, in order to maximize colonization opportunities and to establish appropriate management. Opportunities to target incentive schemes more closely to wet grassland should also be sought.

DEVELOPMENT CONTROL AND FACILITATING POSITIVE MANAGEMENT

Site designation does not, in itself, always guarantee favourable management to fulfil nature conservation objectives (Brotherton 1990). It is, therefore, imperative that any proposals that will result in significant loss or damage to lowland wet grassland are resisted or, if loss or damage is unavoidable, then mitigating measures should be secured as part compensation.

Acquisition of sites by conservation bodies, particularly those in the voluntary sector, has been seen as one solution to ensuring the protection and appropriate management of wet grassland sites. It is likely that acquisition will continue to have a role to play in the conservation of the resource in the future but it is likely that there will always be a need for a variety of other mechanisms as discussed here.

Sustainable management of designated wet grassland needs to be encouraged through the appropriate use of positive management agreements and enhancement schemes such as English Nature's Wildlife Enhancement Scheme (WES) and Reserves Enhancement Scheme (RES), which provide financial support for management of SSSIs. The former provides flat-rate payments to SSSI owners and occupiers to secure positive management, such as grazing, on selected habitats. One scheme, for example, covers the so-called 'Culm grasslands' in Devon and Cornwall, which consist principally of wet acid grassland (NVC fen-meadow communities M22, M23, M24 and M25). As at 1 January 1996, of a total of 1125 ha of Culm grassland notified as SSSI, approximately 70% was covered by 44 WES agreements. One of the main objectives of the RES is to maintain and enhance the nature conservation value of SSSI nature reserves managed by Wildlife Trusts and other Voluntary Conservation Organizations (VCOs). It provides standard hectarage payments scaled according to the habitat type. As at 30 June 1996, 30 VCOs were involved in the scheme covering 524 individual SSSI reserves. It is not possible to separate out the hectarage covering lowland wet grassland due to the method of data collection, but the scheme currently covers 16 378 ha in total (all habitats), of which 11% relate to all types of grassland.

It is considered essential that the management of lowland wet grassland sites of high nature conservation value takes full account of the specific needs of the rare, threatened and vulnerable species that they support. The preparation of species action plans for Red Data Book and globally threatened/declining species that are closely associated with wet grassland remains a high priority for conservation bodies. These plans will highlight the particular habitat requirements of species and the key actions that are required to deliver the agreed population targets. The Biodiversity Action Plan (The UK Steering Group 1995) has produced action plans for 116 key species. Responsibility for taking forward the actions will rest with the statutory conservation agencies and other government departments and agencies in partnership with VCOs. This may include the potential opportunities for recovery-type projects that will seek to maintain and enhance populations of a given species, for example, through English Nature's Species Recovery Programme.

INFLUENCING AGRICULTURAL AND OTHER LAND-USE POLICY

The achievement of nature conservation objectives in areas of wet grassland in England is affected by policies formulated by the EU and central and local government. It is thus essential that conservation objectives are fully integrated into policies and plans formulated by relevant organizations, which in addition to those cited above, include the Ministry of Agriculture, Fisheries and Food (MAFF), the Environment Agency and the Internal Drainage Boards.

The Environment Agency is preparing Local Environment Agency Plans (LEAPs) (formerly known as Catchment Management Plans) for 163 of the principal river catchments in England and Wales. These are intended to provide a framework for decision-making on a catchment-wide basis and will involve examining the interactions between the air and water environment and the land uses and activities that relate to it. The Environment Agency is seeking to reconcile conflicts and protect and improve environmental quality through the production of action plans that form part of the LEAPs. It is anticipated that these plans should help to achieve the conservation and enhancement of wet grassland in river catchments (English Nature 1995).

In a European context it is essential to ensure closer integration of environmental objectives within the CAP to secure more appropriate agricultural management (Dixon 1994). Support measures for low-intensity farming systems that benefit nature conservation in Europe should ideally adopt a holistic approach to rural development, taking into account the broad spectrum of social, cultural and educational needs of the farming community (Beaufoy *et al.* 1994).

For wet grassland in England, there is a need to encourage extensive livestock systems targeted at areas with significant concentrations of this resource or at sites where there is scope for habitat creation. Extensive beef production systems are probably the most compatible with lowland wet grassland, particularly for grazing the wetter, semi-natural types such as the fen-meadow communities (Table 3.1). Grazing rates of 100–250 livestock unit days ha^{-1} are generally suitable for maintaining the nature conservation value of lowland wet grassland, the precise rate depending on particular conservation objectives (Ward 1994).

Losses of coastal grazing marsh and other freshwater wetlands as a result of managed retreat should, as far as possible, be compensated by the provision of replacement areas to ensure that there is no net loss within each management zone. This requires a strategic approach to the conservation of all coastal habitats, involving the assessment of the likely losses and potential gains, the identification of replacement areas and a review of the possible mechanisms that are available. Such an approach is proceeding in north Norfolk (Brown *et al.* 1994).

IMPROVING THE KNOWLEDGE BASE

The acquisition of high-quality data and information relating to the wet grassland resource is essential to ensure its effective conservation and enhancement. The types of information required include:

- knowledge of biodiversity
- the extent, distribution and condition/status of communities and species

- the identification of areas with potential for creation of wet grassland
- the impact of management practices on communities and species
- the relationship between hydrology and vegetation
- autecology of rare and vulnerable species
- research into effective techniques of grassland creation

Much progress has been made on the first three listed (Ratcliffe 1977; Fuller 1982; Smith 1983; Owen *et al.* 1986; Palmer and Blake 1991; Dargie 1993) and on the ecological requirements and therefore management needs of some key species, notably the breeding waders, but there is still a need, for example, to research aspects relating to the long-term impact of differing hay-cutting dates on the botanical composition of wet meadows and the hydrological conditions that support the differing semi-natural communities which constitute wet grassland (Mountford and Chapman 1993). Further work on the techniques of grassland creation and collation of data from actual field projects would be instructive and would build on existing knowledge (Firbank *et al.* 1992; Parker 1995).

ACKNOWLEDGEMENTS

We thank Dr Martin Drake, Mike Edwards, Gerry Hamersley, Maurice Massey, Dr Heather Robertson and Martin Wiggington for their advice and provision of information.

We are grateful to Dr Andy Brown, and two anonymous reviewers for comments on the text.

SUMMARY

Lowland wet grassland consists of land managed as pasture or meadow occurring in areas with a high water table or subject to periodic inundation. It normally occurs at less than 200 m above sea level in river floodplains, behind sea defences or in areas with impeded drainage. It has been estimated that there are approximately 220 000 ha of probable lowland wet grassland in England. Lowland wet grassland has high biodiversity and is highly valued for nature conservation. The key nature conservation interests are unimproved species-rich grassland, breeding and wintering waterfowl, and rare and scarce vascular plants and invertebrates. Many remaining areas of lowland wet grassland are covered by statutory nature conservation designations. For example 46% (26 sites) of designated Special Protection Areas and Wetlands of International Importance contain areas of wet grassland. Key conservation issues affecting this habitat are inappropriate agricultural management, hydrological changes and the need for targeted expansion of the resource by restoration and re-creation.

Keywords: Agriculture, Biodiversity, Birds, Conservation, Creation, Hydrology, Lowland wet grassland, Plant communities, Restoration, Vascular plants.

REFERENCES

Allport, G. (1989) Norfolk's bean geese and their management. *RSPB Conservation Review*, **3**, 59–60.

Beaufoy, G., Baldock, D. and Clark, J. (1994) *The nature of farming. Low intensity farming systems in nine European countries*. Institute for European Environmental Policy, London.

Bell, D.V. and Fox, P.J.A. (1991) *Shooting disturbance: an assessment of the impacts and effects on overwintering waterfowl populations and their distribution in the United Kingdom.* Unpublished report to the Nature Conservancy Council by the Wildfowl and Wetlands Trust. Wildfowl and Wetlands Trust, Slimbridge.

Brotherton, I. (1990) On loopholes, plugs and inevitable leaks – a theory of SSSI protection in Great Britain. *Biological Conservation*, **52**, 187–204.

Brown, A.F. and Grice, P.V. (1993) *Birds in England: context and priorities.* English Nature Research Report No. 62. English Nature, Peterborough.

Brown, A.F., Grice, P.V., Radley, G.P., Leafe, R.N. and Lambley, P. (1994) *Towards a strategy for the conservation of coastal habitats in north Norfolk.* English Nature Research Report No. 74. English Nature, Peterborough.

Buisson, R. and Williams, G. (1991) RSPB action for lowland wet grasslands. *RSPB Conservation Review*, **5**, 60–64.

Burgess, N.D. and Hirons, G.J.M. (1990) *Techniques of hydrological management at coastal lagoons and lowland wet grasslands on RSPB reserves. Management case study.* Royal Society for the Protection of Birds, Sandy.

Burgess, N.D., Evans, C.E. and Thomas, G.J. (1990) Vegetation change on the Ouse Washes wetland, England, 1972–88 and effects on their conservation importance. *Biological Conservation*, **53**,173–189.

Commission of the European Communities (1991) *CORINE biotopes manual: habitats of the European Community.* Commission of the European Communities, Luxembourg.

Council of the European Communities (1992) Council Directive 92/43/EEC of 2 April 1992 on the conservation of natural habitats and of wild fauna and flora. *Official Journal of the European Communities*, No. L206, 7–50.

Côté, I.M. and Sutherland, W.J. (1995) *The scientific basis for predator control for bird conservation.* English Nature Research Report No. 144. English Nature, Peterborough.

Crofts, A. and Jefferson, R.G. (1994) *The lowland grassland management handbook.* English Nature/The Wildlife Trusts, Peterborough.

Dargie, T.C. (1993) *The distribution of lowland wet grassland in England.* English Nature Research Report No. 49. English Nature, Peterborough.

Department of the Environment (1994) *Biodiversity. The UK action plan.* HMSO, London.

Devon Wildlife Trust (1990) *Survey of Culm grasslands in Torridge District.* Devon Wildlife Trust, Exeter.

Dixon, J. (1994) The potential role of agricultural policy for achieving nature conservation on farmland. In: Bignal, E.M., McCracken, D.I. and Curtis, D.J. (eds), *Nature conservation and pastoralism in Europe*, pp. 110–116. Joint Nature Conservation Committee, Peterborough.

Drake, C.M. (1998) The important habitats and characteristic rare invertebrates of lowland wet grassland in England. In: Joyce, C.B. and Wade, P.M. (eds), *European wet grasslands: biodiversity, management and restoration*, pp. 137–149. John Wiley, Chichester.

Drake, C.M. (in press) Factors influencing the species richness of aquatic invertebrates in grazing marsh ditches. In: Harpley, J. (ed.), *Nature conservation and the management of drainage system habitat.* Wiley, Chichester.

Duffey, E., Morris, M.G., Sheail, J., Ward, L.K., Wells, D.A. and Wells, T.C.E. (1974) *Grassland ecology and wildlife management.* Chapman & Hall, London.

Elmes, G.W. and Free, A. (eds) (1994) *Climate change and rare species in Britain.* Institute for Terrestrial Ecology Research Publication No. 8. HMSO, London.

English Nature (1995) *Conservation in catchment management planning – a handbook.* English Nature, Peterborough.

Evans, C.E. (1991) The conservation importance of the ditch flora on RSPB reserves. *RSPB Conservation Review*, **5**, 65–71.

Firbank, L.G., Arnold, H.R., Eversham, B.C., Mountford, J.O., Radford, G.L., Telfer, M.G., Treweek, J.R., Webb, N.R.C. and Wells, T.C.E. (1992) *The potential uses of set-aside land to benefit wildlife.* Institute of Terrestrial Ecology, Abbots Ripton.

Fuller, R.J. (1982) *Bird habitats in Britain.* T. & A.D. Poyser, London.

Fuller, R.M. (1987) The changing extent and conservation interest of lowland grasslands in England and Wales: a review of grassland surveys 1930–84. *Biological Conservation*, **40**,

281–300.

Gibbons, D.W., Reid, J.B. and Chapman, R.A. (1993) *The new atlas of breeding birds in Britain and Ireland: 1988–1991*. British Trust for Ornithology/Scottish Ornithologists' Club/Irish Wildbird Conservancy. T. & A.D. Poyser, London.

Green, R.E. (1986) *The management of lowland wet grassland for breeding waders*. Nature Conservancy Council Chief Scientists Directorate Report No. 626. Nature Conservancy Council, Peterborough.

Green, R.E. and Robins, M. (1993) The decline of the ornithological importance of the Somerset Levels and Moors, England and changes in the management of water levels. *Biological Conservation*, **66**, 95–106.

Green, R.E., Cadbury, C. and Williams, G. (1987) Floods threaten black-tailed godwits breeding at the Ouse Washes. *RSPB Conservation Review*, **1**, 14–16.

Grice, P.V., Brown, A.F., Carter, I.C. and Rankine, C.A. (1994) *Birds in England: a natural areas approach*. English Nature Research Report No. 114. English Nature, Peterborough.

Hockin, D., Ounsted, M., Gorman, M., Hill, D., Keller, V. and Barker, M.A. (1992) Examination of the effects of disturbance on birds with reference to its importance in environmental assessments. *Journal of Environmental Management*, **36**, 253–286.

Hopkins, A. and Hopkins, J.J. (1994) UK grasslands now: agricultural production and nature conservation. In: Haggar, R.J. and Peel, S. (eds), *Grassland management and nature conservation*, pp. 10–19. Occasional Symposium No. 28. British Grassland Society, Reading.

Hopkins, J.J. (1989) Prospects for habitat creation. *Landscape Design*, **179**, 19–23.

Housden, S., Thomas, G., Bibby, C. and Porter, R. (1991) Towards a nature conservation strategy for bird habitats in Britain. *RSPB Conservation Review*, **5**, 9–16.

Jefferson, R.G. and Robertson, H.J. (1996) *Lowland grassland: wildlife value and conservation status*. English Nature Research Report No. 169. English Nature, Peterborough.

José, P. and Self, M. (1994) The management of lowland wet grassland for birds. In: Crofts, A. and Jefferson, R.G. (eds), *The lowland grassland management handbook*, pp. 10:1–10:13. English Nature/The Wildlife Trusts, Peterborough.

Kirby, P. (1992) *Habitat management for invertebrates: a practical handbook*. Royal Society for the Protection of Birds, Sandy.

Land Use Consultants (1994) *Countryside schemes and nature conservation*. Unpublished report to English Nature. Land Use Consultants, London.

Laws, A.R. (1984) *The conservation of wildlife habitats in England by BASC members*. British Association for Shooting and Conservation, Rossett.

Mayhew, P.W. (1988) The daily energy intake of European wigeon in winter. *Ornis Scandinavia*, **19**, 217–223.

Moffat, A. (ed.) (1994) *Priorities for habitat conservation in England*. English Nature Research Report No. 97. English Nature, Peterborough.

Mountford, J.O. (1994) Floristic change in English grazing marshes: the impact of 150 years of drainage and land use change. *Watsonia*, **20**, 3–24.

Mountford, J.O. and Chapman, J.M. (1993) Water regime requirements of British wetland vegetation using the moisture classifications of Ellenberg and Londo. *Journal of Environmental Management*, **38**, 275–288.

Mountford, J.O., Lakhani, K.H. and Kirkham, F.W. (1993) Experimental assessment of the effects of nitrogen addition under hay cutting and aftermath grazing on the vegetation of meadows on a Somerset peat moor. *Journal of Applied Ecology*, **30**, 321–332.

Nature Conservancy Council (1984) *Nature conservation in Great Britain*. Nature Conservancy Council, Peterborough.

Nature Conservancy Council (1989) *Guidelines for selection of biological SSSIs*. Nature Conservancy Council, Peterborough.

Norris, K. and Buisson, R. (1994) Sea-level rise and its impact upon coastal birds in the UK. *RSPB Conservation Review*, **8**, 63–71.

O'Brien, M. and Self, M. (1994) Changes in the numbers of breeding waders on lowland wet grasslands in the UK. *RSPB Conservation Review*, **8**, 38–44.

O'Brien, M. and Smith, K.W. (1992) Changes in the status of waders breeding on wet lowland grasslands in England and Wales between 1982 and 1989. *Bird Study*, **39**, 165–176.

Owen, M., Atkinson-Willes, G.L. and Salmon, D.G. (1986) *Wildfowl in Great Britain*, 2nd edn. Cambridge University Press, Cambridge.

Palmer, M. and Blake, C. (1991) *Review of the extent of grassland survey in England*. England Field Unit Project No. 101. Nature Conservancy Council, Peterborough.

Palmer, M. and Newbold, C. (1983) *Wetland and riparian plants in Great Britain*. Focus on Nature Conservation, No. 1. Nature Conservancy Council, London.

Parker, D.M. (1995) *Habitat creation – a critical guide.* English Nature Science, No. 21. English Nature, Peterborough.

Perring, F.H. and Farrell, L. (1983) *British Red Data Books: 1. Vascular plants*, 2nd edn. Royal Society for Nature Conservation, Lincoln.

Ratcliffe, D.A. (ed.) (1977) *A nature conservation review*. Cambridge University Press, Cambridge.

Ratcliffe, D.A. (1984) Post-medieval and recent changes in British vegetation: the culmination of human influence. *New Phytologist*, **98**, 73–100.

Rodwell, J.S. (ed.) (1991) *British plant communities*, Vol. 2, *Mires and heaths.* Cambridge University Press, Cambridge.

Rodwell, J.S. (ed.) (1992) *British plant communities*, Vol. 3, *Grasslands and montane Communities.* Cambridge University Press, Cambridge.

Rodwell, J.S. (ed.) (1995) *British plant communities*, Vol. 4, *Aquatic communities, swamps and tall-herb fens.* Cambridge University Press, Cambridge.

Rodwell, J.S. (ed.) (in press) *British plant communities*, Vol. 5, *Maritime and weed communities.* Cambridge University Press, Cambridge.

Russell, N. (1994) Grassland conservation in an arable area: the case of the Suffolk river valleys. In: Whitby, M. (ed.), *Incentives for countryside management. The case of Environmentally Sensitive Areas*, 25–39. CAB International, Wallingford.

Royal Society for the Protection of Birds. (1993) *Wet grasslands – what future? An account of wet grassland loss in the UK*. Royal Society for the Protection of Birds, Sandy.

Royal Society for the Protection of Birds (1996) *Birds of conservation concern in the United Kingdom, Channel Islands and Isle of Man*. Royal Society for the Protection of Birds, Sandy.

Self, M., O'Brien, M. and Hirons, G. (1994) Hydrological management for waterfowl on RSPB lowland wet grassland reserves. *RSPB Conservation Review*, **8**, 45–56.

Smith, K.W. (1983) The status and distribution of waders breeding on wet lowland grassland in England and Wales. *Bird Study*, **30**, 177–192.

Smith, R.S. and Rushton, S.P. (1994) The effects of grazing management on the vegetation of mesotrophic (meadow) grassland in northern England. *Journal of Applied Ecology*, **31**, 13–24.

Stewart, A., Pearman, D.A. and Preston, C.D. (eds) (1994) *Scarce plants in Britain*. Joint Nature Conservation Committee, Peterborough.

Tansley, A.G. (1939) *The British Islands and their vegetation.* Cambridge University Press, London.

Thomas, G.J. (1982) Autumn and winter feeding ecology of waterfowl at the Ouse Washes, England. *Journal of Zoology, London*, **197**, 131–172.

Tickner, M. and Evans, C.E. (1991) *The management of lowland wet grassland on RSPB reserves. Management case study*. Royal Society for the Protection of Birds, Sandy.

Tucker, G.M. and Heath, M.F. (1994) *Birds in Europe: their conservation status*. BirdLife International, Cambridge.

UK Steering Group, The (1995) *Biodiversity: The UK Steering Group report*. HMSO, London.

Ward, D. (1994) Management of lowland wet grassland for breeding waders. *British Wildlife*, **6**, 89–98.

Whitby, M. (ed.) (1994) *Incentives for countryside management. The case of Environmentally Sensitive Areas*. CAB International, Wallingford.

Williams, G. and Bowers, J. (1987) Land drainage and birds in England and Wales. *RSPB Conservation Review*, **1**, 25–30.

Williams, G. and Hall, M. (1987) The loss of coastal grazing marsh in south and east England, with special reference to east Essex, England. *Biological Conservation*, **39**, 243–253.

Williams, G., Henderson, A., Goldsmith, L. and Spreadborough, A. (1983) The effects on birds of land drainage improvements in the North Kent Marshes. *Wildfowl*, **34**, 33–47.

Wilson, E.O. (1992) *The diversity of life.* Allen Lane Penguin Press, London.

4 The Ecology of Floodplain Grasslands in Estonia

LAIMI TRUUS[1] **and ANDRES TÕNISSON**[2]
[1]*Institute of Ecology, Tallinn, Estonia*
[2]*Estonian Institute of Meteorology and Hydrology, Saku, Estonia*

HISTORICAL OVERVIEW

Phytogeographically, Estonia belongs to the northern part of the forest subzone of the north temperate zone. Under natural plant cover, coniferous and mixed forests dominate. Mires have also occurred abundantly in the past, and nowadays they cover 21% of the Estonian territory. In the forest zone, primary grassland communities can occur only in places that are unfavourable for tree growth, such as in those areas which are either too dry (dry limestone grasslands), too wet (paludified grasslands), or have recently risen from the sea (sea-shore grasslands; see Puurmann and Ratas 1998). Cattle breeders and crop growers settled the river valleys of Estonia in 2000 BC (Jaanits *et al.* 1982), floodplain forests were felled around the settlements and grasslands were formed. These floodplain grasslands can be regarded as the oldest anthropogenic vegetation type in Estonia (Pork 1964).

Exploitation of semi-natural grasslands for hay and pasture was of great significance in the earlier land use of Estonia. The history of pastoral usage reaches back over a thousand years and that of hay-mowing to the first centuries AD. Floodplain grasslands have been managed in a consistent manner for centuries. Consequently plant communities have developed that are in equilibrium with this human influence. Most floodplain grasslands have been cut for hay, and grazing has been negligible. Management for hay increased rapidly when the scythe came into use, and became the prevalent land use about 1000 years ago (Jaanits *et al.* 1982). After that time the area of semi-natural grasslands continued to expand and reached its maximum at the beginning of the twentieth century, when it accounted for 11 084 km², i.e. 24.4% of the territory of Estonia (Laasimer 1965). The area of floodplain grasslands has never been large, relative to other types of semi-natural grasslands in Estonia. At the time when the management of semi-natural grassland was at its maximum, floodplain grasslands covered only 7.5% of the total grassland area, i.e. 830 km² (Laasimer 1965).

In the 1950s, mechanized hay-mowing began to replace mowing by hand. Since then the area of semi-natural grasslands has been decreasing rapidly: some grasslands have been afforested, some cultivated, but most have been abandoned. Hay meadows were changed to pastures or began to overgrow with shrubs and trees. At present, natural grasslands cover only about 6% of the Estonian territory, and floodplain grasslands constitute about one-tenth of these. Aug and Kokk (1983), reporting the results of an

See Glossary, p. 305, for explanation of technical terms. Scientific names of vascular plants follow Tutin, T. G. *et al.* (1964–80) *Flora Europaea* Volumes 1–5. Cambridge University Press. See p. 319.

European Wet Grasslands: Biodiversity, Management and Restoration. Edited by Chris B. Joyce and P. Max Wade.
© 1998 John Wiley & Sons Ltd.

inventory of grasslands made in 1978–81, suggest that the area of floodplain grasslands on farms maintained by the State accounted for 270 km². In addition to this, floodplain grasslands occurred on land maintained by forestry organizations and in areas protected for nature conservation.

CLASSIFICATION AND DISTRIBUTION

PLANT COMMUNITIES

Due to the long history of regular mowing (or grazing) a variety of floodplain grassland community types have developed in Estonia. In general, six different community types are recognized (Krall *et al.* 1980), based on moisture and trophic conditions and species composition (Table 4.1). These basic types have been divided into numerous associations and subassociations (Krall *et al.* 1980).

1.   *Dry floodplain grasslands* are a restricted type in Estonia, occupying about 4.7% of the total area of floodplain grasslands. They are situated on higher landforms of floodplains, where present-day sedimentation is not substantial. Therefore the soils are of low fertility, water availability is not particularly good, and plant cover is somewhat sparse. Dominant plant species are psychromesophytes with low nutrient demands. On the carbonaceous alluvial sediments of west Estonia, these grasslands are species-rich; on the sandy sediments of east and south-east Estonia, they are species-poor. In species-rich grasslands the two most frequent plant associations are *Seslerio–Festucetum ovinae* and *Seslerio–Nardetum.* In species-poor grasslands four associations are often encountered: *Thymo–Festucetum, Danthonio–Nardetum, Anthoxantho–Agrostetum* and *Galio–Agrostetum capillaris.*

2.   *Moderately moist floodplain grasslands* are distributed on fluvisols that are fertile because of substantial sedimentation. They are located near the river channel or in the central part of the floodplain and form 10.9% of the total area of floodplain grasslands. Plant cover is dense and luxuriant consisting mainly of mesophytes, many of them being nitrophytes. The most agriculturally valuable natural hay meadows of Estonia belong to this type of grassland. They are represented by two main plant associations, *Agrostetum giganteae* and *Deschampsieto–Festucetum rubrae*, distributed mostly in south Estonia.

3.   *Moist floodplain grasslands* are located near the river channel of small rivers and streams and in the central part of the floodplain of larger rivers. Sedimentation is less substantial than in the previous type, but the availability of water is favourable and plant cover is luxuriant, consisting mainly of hygromesophytes. This type of grassland occupies 30.3% of the total area of floodplain grasslands. Four plant associations are encountered: *Cirsio–Polygonetum bistortae, Filipendulo–Geranietum palustris, Deschampsieto–Caricetum cespitosae* and *Elymo–Alopecuretum arundinacei.* This type of grassland is frequent in all regions of Estonia.

4.   *Wet floodplain grasslands with tall grasses* occupy 9.0% of the total area of floodplain grasslands. They are located in the central part of the floodplain or near the terrace. Plant cover is luxuriant and dense and the dominant plants are hygromesophytes. Two main

Table 4.1 Plant community types of floodplain grasslands in Estonia (after Krall *et al.* 1980; Aug and Kokk 1983)

Plant community type	Flooding and moisture conditions	Characteristic plant species	Productivity $(\mathrm{kg\,ha^{-1}\,yr^{-1}})$	Area (ha)
1. Dry floodplain grasslands	Dry – occasional flooding	*Sesleria caerulea, Festuca ovina, Nardus stricta, Agrostis capillaris*	300–1500	1287
2. Moderately moist floodplain grasslands	Regularly flooded, well drained	*Agrostis gigantea, Festuca rubra, Alopecurus pratensis, Deschampsia cespitosa*	1000–4000	2986
3. Moist floodplain grasslands	Regularly flooded, well drained	*Cirsium palustre, Filipendula ulmaria, Deschampsia cespitosa, Calamagrostis stricta*	1000–4000	8311
4. Wet floodplain grasslands with tall grasses	Poorly drained to saturated	*Phalaris arundinacea, Deschampsia cespitosa, Calamagrostis stricta*	1000–5000 (exceptionally 7000)	2454
5. Wet floodplain grasslands with tall sedges	Permanently saturated	*Carex cespitosa, C. acuta*	1000–4500	2626
6. Floodplain marshes	Permanently inundated	*Sesleria caerulea, Carex panicea, C. nigra, C. elata, C. cespitosa, C. lasiocarpa*	400–3000	9748

plant associations are recognized: *Stellario–Deschampsietum* and *Phalaroidetum.* These are most frequently encountered in east and south Estonia.

5.   *Wet floodplain grasslands with tall sedges* occupy 9.6% of the total area of floodplain grasslands. They occur in similar parts of the floodplain to the previous type, but they are found most frequently in meanders and deltas of rivers where sedimentation is plentiful, and plant cover is often dense and tall. In areas that receive less sediments, however, plant cover tends to be sparse. Three plant associations are frequently encountered: *Caricetum distichae, Caricetum acutae* and *Caricetum rostrato–vesicariae.* This type of community is located mainly in east and south Estonia.

6.   *Floodplain marshes* are the most common type, occupying 35.6% of the total area of floodplain grasslands. They occur in permanently inundated areas where sedimentation is negligible and peat is present. The thickness of the peat layer ranges from 30 cm to 1 m and more. Plant cover consists mainly of sedge, and the main plant associations are: *Seslerio–Caricetum paniceae, Caricetum paniceo–nigrae, Caricetum diandro–nigrae, Caricetum cespitoso–appropinquatae, Caricetum elatae* and *Drepanoclado–Caricetum*

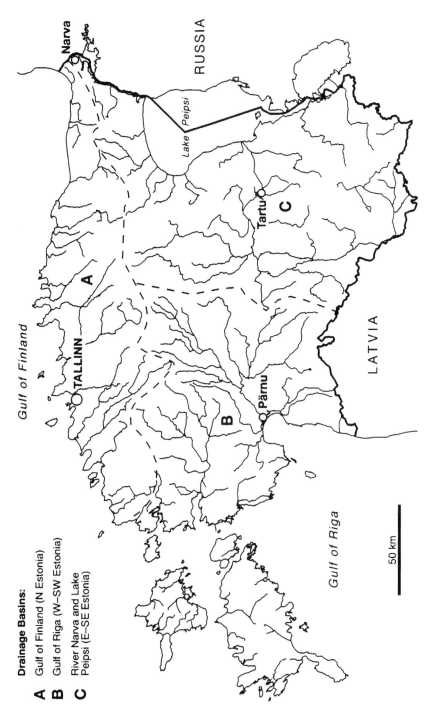

Figure 4.1 Drainage basins of Estonian rivers

Drainage Basins:

A Gulf of Finland (N Estonia)
B Gulf of Riga (W–SW Estonia)
C River Narva and Lake Peipsi (E–SE Estonia)

Gulf of Finland

Gulf of Riga

50 km

TALLINN

Narva

Tartu

Pärnu

Lake Peipsi

RUSSIA

LATVIA

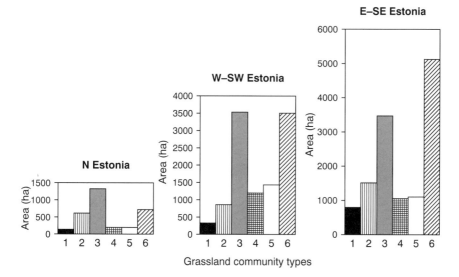

Figure 4.2 Area of floodplain grassland community types in different river drainage basins. Grassland community types: 1, dry floodplain grasslands; 2, moderately moist floodplain grasslands; 3, moist floodplain grasslands; 4, wet floodplain grasslands with tall grasses; 5, wet floodplain grasslands with tall sedges; 6, floodplain marshes. See Table 4.1 for further details of plant community types and Figure 4.1 for location of river drainage basins

lasiocaarpa. This type of community is frequently found in south and west Estonia and tends to dominate the lower courses of rivers.

DISTRIBUTION

Estonia is characterized by numerous small streams and a flat topography. The relative heights of landforms do not exceed 20 m as a rule and only seldom are 50 m or more; the highest point in Estonia is Suur-Munamägi Hill at 318 m above sea level. Due to abundant precipitation (mean 700 mm yr^{-1}), Estonia is rich in wetlands. There are more than 7300 rivers, brooks and main drains in the country, with a total length of 31 150 km. However, only 10 rivers are longer than 100 km, and only 13 rivers have a mean annual discharge over 10 m^{3}s^{-1}. Most major Estonian rivers are 15–20 m wide and 2–5 m deep in their lower reaches. Estonian rivers are divided by their watersheds into three drainage basins: (west and the Gulf of Finland (north Estonia), the Gulf of Riga south-west Estonia), and the River Narva and Lake Peipsi (east and south-east Estonia) (Figure 4.1). These drainage basins have contrasting landscape structure and geological development.

The majority of the north Estonian rivers (drainage basin of the Gulf of Finland) flow northwards along ancient valleys. Their basins are narrow and the number of tributaries is usually small. In their lower course they flow over the north Estonian glint, where rapids and waterfalls are formed. Flowing on limestone bedrock, they carry little sediment and floods are not usually extensive. Floodplain grasslands occur sparsely, their area is not large, and plant community type 3 predominates (Figure 4.2).

South-west Estonian rivers (drainage basin of the Gulf of Riga) have an even and gently dipping transverse profile. On the west coast they flow over limnoglacial and marine plains. The rivers of this drainage basin are, in terms of discharge, the largest

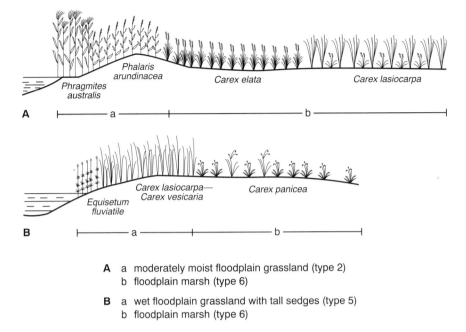

A a moderately moist floodplain grassland (type 2)
   b floodplain marsh (type 6)

B a wet floodplain grassland with tall sedges (type 5)
   b floodplain marsh (type 6)

Figure 4.3 Distribution of plant communities on different shore types on the upper course of the River Suur Emajõgi (after Laasimer 1965). See Table 4.1 for details of plant community types

running waters in Estonia and have numerous tributaries. Due to the relatively imperme-able catchments (underlain by clays) and slow flows, floods are common in this region. The amount of sediment carried is greater than for the previous drainage basin. The largest area of floodplain grassland in Estonia, on the floodplain of the River Kasari (about 40 km²) is located in the northern part of this drainage basin. Mean depth of floods there is 30–40 cm (Pork 1973). Due to the level topography, floods in this region have a long duration. Therefore, the ground is saturated and floodplain marshes (type 6) are abundant (Figure 4.2). Community types 4 and 5 are also relatively common. Thus plant associations with *Deschampsia cespitosa* and different species of *Carex* (Figure 4.3) are usual. Along the River Halliste in the eastern part of this drainage basin, wooded grasslands of community type 3 are typical.

In the drainage basin of the River Narva and Lake Peipsi (east and south-east Estonia) transverse profiles of rivers are concave, being characterized by a variable but relatively steep gradient in the upper courses and a shallow gradient in the lower courses. The prevailing sandstone bedrock creates more favourable conditions for erosion in compari-son with the north Estonian carbonate bedrock. A higher proportion of grasslands in this area are floodplain grasslands, compared to other drainage basins. Floodplain grass-lands occur as small patches along the upper courses of rivers, are common on middle courses (types 2 and 3), and make extensive floodplain marshes (type 6) along lower courses (Figure 4.2).

HYDROLOGY AND ECOLOGY

FLOODS

The persistence of floodplain grasslands as semi-natural plant communities depends upon human influence. Besides this, the main ecological factors that determine the soil and vegetation types on floodplains are the quantity and physical and chemical character of deposited sediment and the duration of floods. Spring floods in Estonia are usually due to melting snow. They normally start in March and reach their peak in April. The peak autumn runoff is in November. Floodplain width is typically 30 m to 0.5 km, exceptionally up to 1 km, and the depth of floodwater on the floodplains ranges from 0.3 to 3 m. Floods remain longer on the lower courses of rivers. For example, in the floodplain of the River Pedja in central Estonia the duration of floods ranges from three to 45 days. On the marshy floodplain at the mouth of the River Emajõgi, floods have lasted as long as 104 days.

In recent years, the runoff into many rivers has been much greater than the long-term average. Due to cultivation, the seasonal inequalities of the flow rates have increased. Spring floods are usually more extensive but of shorter duration than in earlier periods. Intensive drainage and cultivation have caused accelerated runoff of nutrients from catchment areas, and compounded another reason for floods, namely the overgrowing of river channels mostly as a result of eutrophication.

OTHER HYDROLOGICAL ASPECTS

The distribution and character of alluvial deposits in Estonian river valleys have been little studied so far. However, the differences in river turbidity are better understood. Estonian rivers can carry a remarkable amount of suspended matter, especially in the south-east where the soils are more susceptible to erosion. The river sediments, mostly fine particles, are derived from their surroundings and their composition is diverse. Suspended matter (turbidity) as high as $343 \, \text{g m}^{-3}$ has been recorded occasionally in south-eastern Estonian rivers. The mean values for north Estonia are usually below $10 \, \text{g m}^{-3}$, and for south Estonia below $25 \, \text{g m}^{-3}$ (Anon 1972). In central Estonia the mean annual layer of particulate matter deposited on the floodplain is estimated to be between 0.5 and 25 mm (Pork 1984).

In recent years, researchers have quantified the buffering capacity of riparian wetlands. Different studies demonstrate the importance of riparian buffer stripes in nutrient removal. Observations made by Mander *et al.* (1995) indicate that the plant biomass of riparian wetland communities accumulates up to $70 \, \text{g N m}^{-2}$ during the growing season and may remove 20–30% of nutrient input. A high retention capacity has been found in a study of the floodplain of the River Valejõgi. This area (18 ha) has received primary waste-water effluent for 25 years. The retention values for total inorganic nitrogen and phosphorus and the biological oxygen demand (BOD_5) have been measured as 0.58, 0.30 and $9.3 \, \text{g m}^{-2} \, \text{day}^{-1}$ respectively (Mander *et al.* 1991).

Grassland plant communities on Estonian floodplains form narrow strips according to relief. An example of this sequence in relation to moisture and trophic conditions, which are greatly influenced by relief, is illustrated in Figure 4.4, which shows the ecological amplitudes of the communities in the Kasari floodplain, west Estonia. It indicates that

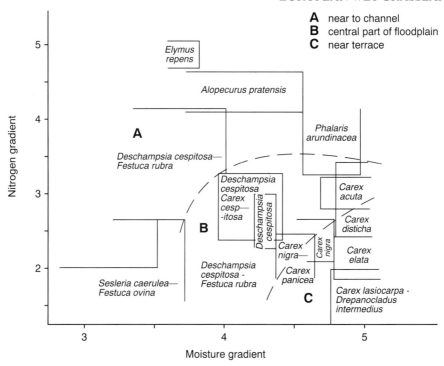

Figure 4.4   Distribution of plant communities of the Kasari floodplain, west Estonia, relative to nitrogen and moisture gradients according to the ecological indicator values of Ellenberg (after Pork 1973)

floodplain grassland communities are very sensitive to environmental conditions. Even annual hydrological conditions can be reflected in biomass and species dominance (Tetter *et al.* 1988).

NATURE CONSERVATION VALUES

Floodplain grasslands do not belong to the most species-rich grassland communities in Estonia. This distinction belongs to the calcareous dry wooded meadows of western Estonia, with up to 63 vascular plant species per square metre (Kull and Zobel 1991). Among floodplain grasslands, the most diverse communities are also those associated with drier conditions and calcareous sediments, where grasses are not tall and light availability is favourable for a wide variety of herb species (type 1). No complete survey has been carried out on wet meadows, but some of their plant communities are of coenogeographical interest, being on the boundaries of their international distributions in Estonia (e.g. *Iris pseudacorus–Carex acuta, Carex paniculata, Cnidium dubium* and *Molinia caerulea*) or common in the Baltic region but rare elsewhere (e.g. *Carex disticha* association).

Out of 155 species of vascular plants included in the Red Data Book of Estonia (Kask and Kuusk 1981) at least 20% are found on floodplain grasslands or in the shallow water of rivers, streams and ditches. Among them are species classified as endangered (e.g. *Angelica palustris, Carex ligerica, Isoetes setacea, Silene tatarica*), vulnerable (e.g. *Carex pediformis* subsp. *rhizodes, Crepis mollis, Lepidotis inundata, Ligularia sibirica, Nuphar pumila, Sanguisorba officinalis, Selaginella selaginoides*) or rare (e.g. *Elatine hydropiper, Gladiolus imbricatus*). Numerous orchid species, many included in the Red Data Book, also occur on floodplain grasslands.

Many species that are common in Estonian floodplains, but threatened in neighbouring countries, have been included in the Red Data Book of the Baltic region (Ingelörf *et al.* 1993). Among them are representatives of such genera as *Carex, Eleocharis, Equisetum, Eriophorum, Potamogeton, Ranunculus, Rumex, Sparganium, Stellaria, Thalictrum, Veronica* and *Viola,* as well as species such as *Angelica sylvestris, Blysmus compressus, B. rufus, Calla palustris, Cyperus fuscus, Iris sibirica, Pinguicula vulgaris, Saussurea alpina* subsp. *esthonica, Schoenus ferrugineus, Scorzonera humilis, Succisa pratensis* and *Valeriana officinalis.*

CHANGES IN THE LAST HALF-CENTURY

Floodplain grassland maintenance had been undertaken in a similar manner through the centuries until the last 50 years or so. Recent impacts have resulted in notable vegetation changes, often including reductions in species diversity.

Dredging and straightening of rivers took place in the 1920s and 1930s. As a result the duration of floods decreased by at least half and less nutrients were deposited on floodplains. Floodplain grasslands became drier and easier to mow so that plant biomass was frequently removed through regular mowing for hay. Hence the nutrient balance was disturbed and grasslands became less productive. During a campaign of land 'improvement' in the 1960s, the channelization of rivers was accompanied by converting floodplain grasslands into arable fields and cultivated grasslands. In this way, for example, the largest and most floristically interesting Korva floodplain on the central course of the River Väike-Emajõgi in south Estonia was destroyed.

Since the 1950s hand-mowing has been gradually replaced by mechanized mowing and large parts of the formerly managed floodplain grasslands have been abandoned. This is because they were too wet and soft for heavy tractors, too far from large centralized farms, and cultivated grasslands were found to provide sufficient production.

Agriculture became more intensive in the last decades of the Soviet period (in the 1970s and 1980s). Increased use of fertilizers and waste from livestock concentrated into huge farms caused an increase in plant nutrients, particularly NO_3-N and PO_4-P, in rivers (Järvekülg, 1994). This led to eutrophication of the rivers and a greater input of nutrients into floodplain grassland ecosystems. Concentrations of NO_3-N and PO_4-P in the water remained considerably higher than the ecologically recommended maximum concentration (1200 $mg N m^{-3}$ and 30 $mg P m^{-3}$ respectively). Indeed NO_3-N pollution was recorded in 38% and PO_4-P pollution in 31% of river reaches studied (Järvelkülg 1994). High nitrogen loads (500–7000 $mg N m^{-3}$) were more common in north and central Estonian rivers, whereas there were no regional differences in phosphorus load. The increased nutrient input into floodplain grasslands by agricultural pollution of surface water has been reflected in changes in plant cover, for example by a greater coverage of *Phragmites australis* in the estuary of the River Kasari (Ksenofontova 1985) and of *Filipendula ulmaria* on floodplains of smaller rivers (Aasalo *et al.* 1990).

Political change in 1991 caused an extensive transformation in rural life. The system of

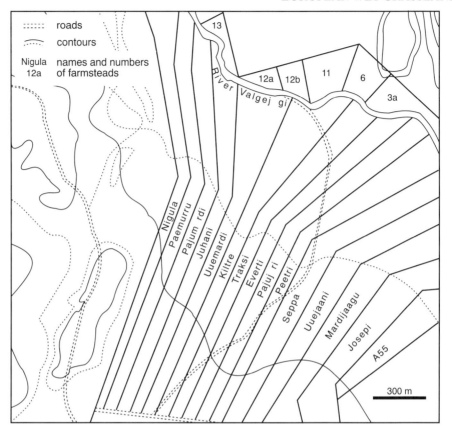

Figure 4.5 Hay meadows of the farmsteads on the floodplain of the River Valgejõgi. Land-use map from 1935

collective farms collapsed, and agricultural production and the use of mineral fertilizers rapidly declined. This was followed by a decrease in agricultural pollution and an improvement in river water quality. For example, in 1992 in the catchment area of Lake Võrtsjärv, agricultural pollution was a quarter to a third lower than in 1988–90 (Järvet and Nôges 1994). However, the decrease in agricultural activity left many grasslands unmanaged. Cessation of mowing has induced a great change in plant community composition (Pork 1981). Species with higher moisture and nutrient demands, those of later phenological development and those which were previously suppressed by cutting have become prevalent. Floodplain grasslands that are unmown for long periods are characterized by the dominance of tall grasses and sedges. Replacement of mowing by grazing can also bring about impoverishment of the flora. Grasslands that have been abandoned for decades are found to be overgrown with *Alnus, Betula* and *Salix* and, as a rule, species diversity decreases.

At present, land reform in Estonia is in progress. Restoration of private ownership of floodplain grasslands is being undertaken, but it is a process with several uncertainties. Traditionally the land ownership pattern in the countryside has been intricate, as indicated by Figure 4.5, but now it is not known how many former owners want the parcels of meadows to be returned to them and how they will choose to manage these overgrown portions. It seems that the majority of owners are not interested in reclaiming their former meadows and probably these parcels will be left unmanaged.

Floodplain grasslands are mowed and grazed in very few places in Estonia nowadays. This is due to the use of cultivated grasslands for fodder production as well as a decrease in cattle breeding. Only some of the new landowners are interested in, and are able to continue, traditional land-use practices. Such traditional management of floodplain grasslands in large areas can only be expected in nature conservation areas through special financial support, such as incentives.

ACKNOWLEDGEMENTS

We are indebted to E. Nillson from the Institute of Ecology, Tallinn for reading the manuscript and useful contribution.

SUMMARY

Estonian floodplain grasslands are ancient landscapes of human origin that have been managed in a similar manner through the centuries. Although they are not the most widespread grassland type, a wide variety of floodplain grassland communities has developed. Estonian floodplain grasslands have been divided into six main types (with numerous subassociations), which are described in this chapter. As well as regular mowing and grazing the main ecological factors that determine the type of vegetation on Estonian floodplains are hydrological (e.g. physical and chemical characteristics and quantity of sediments; duration of floods).

The area of natural and semi-natural grasslands in Estonia reached its maximum at the beginning of the twentieth century, when they covered $11\,084\,km^2$, i.e. 24.5% of the whole territory of Estonia. Of this, floodplain grasslands covered $830\,km^2$ i.e. 7.5%. However, since the beginning of the 1950s, the use of grasslands and their area have decreased rapidly. An inventory of grasslands made in 1978–81, for example, indicated that the area of floodplain grasslands was $270\,km^2$, although this included only agricultural land and not other land uses.

Estonian floodplain grasslands are of considerable nature conservation value, supporting plant communities that are notable from an international perspective as well as many plant species that are rare in Estonia and the Baltic region.

During the last half-century, human impact on floodplain grasslands has changed dramatically. Mowing has reduced significantly, and only very few places are presently mown or grazed. River and land 'improvement' has diminished the duration of floods and the quantity of sediments. An increase in agricultural pollution of floodwaters in the 1970s and 1980s, and a subsequent reduction in the 1990s, also influenced floodplain grasslands. These factors have induced marked changes in plant cover. At present, floodplain grasslands have lost their agricultural economic value and it is not known whether this will be restored in the future.

Keywords: Classification, Distribution, Floodplain grasslands, Hydrology, Management.

REFERENCES

Aasalo, L., Elvisto, T. and Samuel, H. (1990) *Lääne-Eesti roostike lammi-ja puisnitude püsivus, nende mõju infiltreeruvale ja vooluveele.* Unpublished manuscript, Institute of Zoology and Botany, Tartu.
Anon (1972) Ressursy Poverhostnyh vod SSR. Pribaltiiskii raion. Estonia. *Leningrad,* **4**, 1–554.

Aug, H. and Kokk, R. (1983) *Eesti NSV looduslike rohumaade levik ja saagikus.* ENSV ATK IJV, Tallinn.

Ingelörf, T., Andersson, R. and Tjörnberg, M. (eds) (1993) *Red Data Book of the Baltic region,* Part 1. Swedish Threatened Species Unit, Uppsala, and Institute of Biology, Riga.

Jaanits, L., Laul, S., Lõugas, V. and Tonisson, R. (1982) *Eesti esiajalugu.* Eesti Raamat, Tallinn.

Järvekülg, A. (1994) Eesti jõgede vee troofsusaste ja algproduktsiooni tase suvel. Eesti jõgede ja jävade seisund ning kaitse. *Teaduste Akadeemia Kirjastus, Tallinn,* 148–165.

Järvet, A. and Nôges, P. (1994) Võrtsjärve osa maastiku aineringes. Eesti jõgede ja järvede seisund ning kaitse. *Teaduste Akadeemia Kirjastus, Tallinn,* 16–31.

Kask, M. and Kuusk, V. (1981) Plant species in the 'Red Data Book of the Estonian SSR'. In: Laasimer, L. (ed.), *Anthropogenous changes in the plant cover of Estonia,* pp. 5–17. Publications Advisory Committee of the Academy of Sciences of the Estonian SSR, Tartu.

Krall, H., Pork, K., Aug, H., Püss, O., Rooma, I. and Teras, T. (1980) *Eesti NSV looduslike rohumaade tüübid ja tähtsamad taimekooslused.* ENSV Põllumajandusministeerium IJV, Tallinn.

Ksenofontova, T. (1985) Matsalu lahe pilliroog ja roostikud. In: *Matsalu-rahvusvahelise tähtsusega märgala,* pp. 113–125. Valgus, Tallinn.

Kull, K. and Zobel, M. (1991) High species richness in an Estonian wooded meadow. *Journal of Vegetation Science,* **2,** 715–718.

Laasimer, L. (1965) *Eesti NSV taimkate.* Valgus, Tallinn.

Mander, Ü., Matt, O. and Nugin, U. (1991) Perspectives on vegetated shoals, ponds and ditches as extensive outdoor systems of wastewater treatment in Estonia. In: Etnier, C. and Guterstom, B. (eds), *Ecological engineering for wastewater treatment,* pp. 271–282. Proceedings of an International Conference, Stemsund Folk College, Sweden, 24–28 March.

Mander, Ü., Kuusemets, V. and Ivask, M. (1995) Nutrient dynamics of riparian ecotones: a case study from the Porijogi River catchment, Estonia. *Landscape and Urban Planning,* **31,** 333–348.

Pork, K. (1964) Taimkatte genees ja antropogeensed suksessioonid luhtadel. Eesti NDSV TA LUŚ-i Aastaraamat. *Valgus, Tallinn,* **56,** 97–111.

Pork, K. (1973) Kasari joe alamjooksu luha taimkate. Matsalu maastik ja linnud. *Valgus, Tallinn,* 40–59.

Pork, K. (1981) Anthropogenous dynamics of meadows in recent decades. Protection of meadow communities. In: Laasimer, L. (ed.), *Anthropogenous changes in the plant cover of Estonia,* pp. 46–63. Publications Advisory Committee of the Academy of Sciences of the Estonian SSR, Tartu.

Pork, K. (1984) Joeluhtade looduslikus seisundis säilitamisest. Looduskaitse ja pollumajandus. *ENSV TA Looduskaitse Momision, Tartu,* 58–70.

Puurmann, E. and Ratas, U. (1998) The formation, vegetation and management of sea-shore grasslands in West Estonia. In: Joyce, C.B. and Wade, P.M. (eds), *European wet grasslands: biodiversity, management and restoration,* pp. 97–110. John Wiley, Chichester.

Tetter, M., Květ, J., Suchý, K. and Dvořáková, H. (1988) Produkční potanciál travních společenstev v nivě horního toku Lužnice (in Czech with English summary). *Sborník visové Školy Zemědělské v Praze Agronokické fakulty v Českých Budějovích,* **5,** 119–129.

5 The Role of Wet Grasslands in Biological Conservation in Mediterranean Landscapes

JOSÉ M. REY BENAYAS[1], MANUEL G. SÁNCHEZ-COLOMER[2],
CATHERINE LEVASSOR[3] and IÑIGO VÁZQUEZ-DODERO[1]

[1] *Universidad de Alcalá, Spain*
[2] *Paseo Bajo de la Virgen del Puerto, Madrid, Spain*
[3] *Universidad Autónoma de Madrid, Cantoblanco, Spain*

INTRODUCTION

An overview is provided of plant community diversity in wet Mediterranean grasslands: the patterns and causes of diversity, the threats to these ecosystems, and the implications for their conservation and management are considered. In environments characterized by a warm, dry period lasting several months, most wetlands are hypogenic, i.e. fed by the discharge of groundwater flows or seepages. The proximity of groundwater to the surface gives rise to a system of grasslands and *Carex* meadows, including those related to river systems, and different types of small water bodies (Bernáldez *et al.* 1989). These ecosystems are wet islands in dry climatic regions dominated by either croplands or xerophytic vegetation. In intensively cultivated areas, the presence of uncultivated enclaves is of great importance given their positive effects on landscape and biological diversity. Outcropping of groundwater results in the local interruption of the cultivated areas and provides habitats that are mostly used as pastures. These pastures are virtually the only species-rich plant communities in Mediterranean agricultural landscapes and are worthy of protection in order to conserve species diversity.

There is widespread international concern for the conservation of these wet grasslands, mostly within a European context, since there has been a marked reduction in the extent and biodiversity of natural and semi-natural grasslands (Bernáldez *et al.* 1993). The perception of grasslands has shifted from one of traditional productivity value to the acknowledgement of their ecological and environmental value (Mannetje and Frame 1994). From a qualitative perspective, the preservation of wetland species diversity is particularly important due to the scarcity of such wetlands in these arid and semi-arid areas within the Mediterranean region (International Wetland Research Bureau 1991).

In this chapter, we first define the types and measures of diversity used here to assess wet grassland plant communities. Annual species are not considered in this analysis as they are less abundant and relatively unimportant in this type of ecosystem. This is particularly so in Mediterranean grasslands where the contribution of annual species to plant diversity is only important in the driest zones (Montalvo *et al.* 1991). We then explain the patterns and causes of this diversity in representative areas of Mediterranean

See Glossary, p. 305, for explanation of technical terms. Scientific names of vascular plants follow Tutin, T. G. *et al.* (1964–80) *Flora Europaea* Volumes 1–5. Cambridge University Press. See p. 319.

European Wet Grasslands: Biodiversity, Management and Restoration. Edited by Chris B. Joyce and P. Max Wade.
© 1998 John Wiley & Sons Ltd.

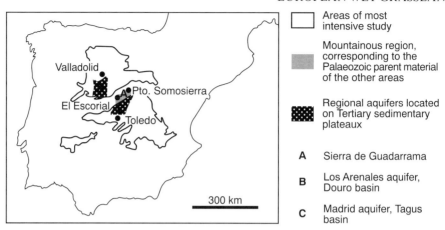

Figure 5.1 Location of three areas in Spain where wet meadow ecology has been studied intensively. Area A (Sierra de Guadarrama) is a mountainous region, corresponding to the Palaeozoic parent material of the other areas. Areas B (Los Arenales aquifer, Douro basin) and C (Madrid aquifer, Tagus basin) correspond to regional aquifers located on Tertiary sedimentary plateaux

landscapes (Figure 5.1), with special attention being paid to the dynamics of regional aquifers in sedimentary basins, and altitude in mountainous regions. Further information is provided through the comparison of wet grassland diversity with that of the surrounding xerophytic plant communities. Finally, we discuss the threats to wet grasslands, their conservation value and the implications for their management.

TYPES AND MEASURES OF DIVERSITY

Biodiversity means the variety of organisms in ecological units (e.g. areas). The study of various levels of biodiversity is important because the overall species richness in a landscape is a product of both within- and between-community diversity (Whittaker, 1977) and because biodiversity is a complex synthetic structural parameter of communities. Three types of biodiversity are examined:

1. *Inventory diversity* or the variety of species within ecological units. Two scales are considered: within-community species richness (α diversity) and total species richness across all communities (γ diversity).
2. *Differentiation diversity* or the distribution of species of an ecological unit between its subunits (similarity between communities or β diversity). A high similarity indicates that the set of communities is not very diverse and has a high species overlap; a low similarity indicates a low species overlap.
3. *Mosaic diversity* (μ diversity or compositional pattern diversity). Mosaic diversity measures the complexity of community arrangements (Scheiner 1992) such that a low value of mosaic diversity is indicative of a uniform landscape with one or a few underlying environmental gradients and dominated by a few species; a high value is indicative of a complex landscape with many environmental gradients and no ubiquitous species.

THE ROLE OF REGIONAL AQUIFER DYNAMICS IN WET GRASSLAND DIVERSITY

The regional aquifers associated with the sedimentary plateaux in Spain, the latter being formed of Tertiary deposits, are systems with relatively homogeneous lithological conditions and comparatively simple groundwater flow patterns. Chemical changes in groundwater occur along flow paths through a process of groundwater evolution, although they take a long time in silicate substrates where the residence time of water before reaching the surface can be thousands of years (Freeze and Cherry 1979). The chemical composition of groundwater where it reaches the surface depends on the age of the water in relation to both the permeability and the length of the path, and on the lithological composition of the saturated zone. The most common chemical change is an increment in salinity. This gradient is continued in those soils under the influence of groundwater, i.e. wetlands, assisted by considerable evapotranspiration, and often by a flat topography (Bernáldez and Rey Benayas 1992). However, some independent aquifers can be found above the regional aquifers (Figure 5.2), and they produce a high landscape heterogeneity with regard to the wetland characteristics.

Rey Benayas and Scheiner (1993) and Montalvo and Herrera (1993) have studied the diversity patterns of plant communities in wet grasslands linked to regional aquifers (Figure 5.1). Geochemical patterns in these aquifer wetlands affect both community composition (Rey Benayas *et al.* 1990) and diversity. Species richness is highly variable, ranging from only one to 37 perennial plant species in 100-m^2 plots, averaging approximately 15 species per plot. There is an indication that soil toxicity, rather than nutrient levels, is the key factor explaining diversity patterns of wet grasslands, with the most species-poor plots being located in salt pans.

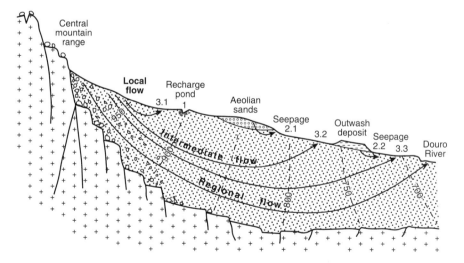

Figure 5.2 Conceptual hydrogeological sketch of the Los Arenales aquifer (central Spain), with different types of wetlands superimposed. Wetlands linked to the main aquifer (Tertiary arkose sediments) arise from the discharge of local (3.1), intermediate (3.2) or regional (3.3) flow systems. Wetlands not connected with the main aquifer can be ponds in recharge areas (1), seepages from post-Tertiary sediments such as aeolian sands (2.1) or outwash deposits (2.2). (From Bernáldez *et al.* 1993)

Table 5.1 Multiple regression analysis of species richness on (a) composite soil variables and (b) selected nutrients and toxins

	b'	SE	$p<$
(a) Composite variables ($R^2 = 0.35$)			
Halite concentration	−3.68	0.77	0.0001
Salinity	−3.23	0.82	0.0001
Alkalinity	−2.86	0.67	0.0001
(b) Nutrients and toxins ($R^2 = 0.27$)			
Conductivity	−0.65	0.13	0.0001
pH	−0.26	0.07	0.0006
Ca^{2+}	0.32	0.13	0.02

The magnitudes of the standardized regression coefficients (b') indicate the relative importance of each variable. These results derive from 156 100-m^2 plots sampled in wet grasslands located in the Los Arenales aquifer (Area B, Figure 5.1). (From Rey Benayas and Scheiner 1993.)

   Three composite soil stress variables (salinity, alkalinity and halite concentration) explained the greatest variation of species richness in the different plant communities observed across 156 plots located in the Douro basin (Table 5.1a). The multiple regression analysis for individual geochemical variables showed a positive effect of Ca^{2+} concentration after correcting for the negative effect of conductivity and pH (Table 5.1b). These plant communities were classified into six types based on characteristic species: halocalcicole, alkalinophyte, subalkalinophyte, moderately tolerant, subglycophyte and glycophyte. The tolerance to salinity of these community types decreases from the first to the last type. Halocalcicole species are those adapted to salty groundwater and soils (electrical conductivity is > 2000 $\mu S\,cm^{-1}$) and with a relatively high proportion of calcium. Alkalinophytes have their highest occurrence on soils with pH > 8.3, and subalkalinophytes on soils with a pH ranging from 7.7 to 8.3. Species classified as moderately tolerant resist intermediate salt concentrations (500–1000 $\mu S\,cm^{-1}$). Subglycophytes are most abundant on low salinity (< 500 $\mu S\,cm^{-1}$) and neutral soils, and glycophytes thrive on soils with the lowest salt concentrations (< 350 $\mu S\,cm^{-1}$) and pH < 7. Table 5.2 summarizes the measurements of diversity that were obtained for these six community types and lists respective plant indicators. Alkaline wetlands contained vascular plant communities that had very few species but the between-community diversity was high. In contrast, the overall richest community types (glycophytes and subglycophytes) tended to share the same species. In contrast to species richness, the complexity of wetland plant communities, as measured by mosaic diversity, was not related to environmental stress. Rather, the greatest effect was on community species richness, with less effect on community similarity, and very little effect on landscape complexity.
   Montalvo and Herrera (1993) attribute the diverse features of wet grasslands to abiotic and anthropogenic factors. These types of community are the historical result of the combined action of deforestation and grazing usage. They represent states with a relatively low level of ecological maturity, characterized by a low proportion of woody species. There are a small number of woody species adapted to the extreme alkaline and saline conditions but even for these species growth is severely limited. Only a few *Tamarix* species can be considered to have descended from the tree formations that originally populated the most saline wetlands. The decline in species diversity is most dramatic

Table 5.2 Diversity measures (means ±SD) for six community types sampled in the Los Arenales aquifer (Area **B**, Figure 5.1) and their plant indicators

Community type	Halocalcicole	Alkalinophyte	Subalkalinophyte	Moderately tolerant	Subglycophyte	Glycophyte
Number of species per plot	8.4 ± 4.5^c	8.4 ± 3.5^c	9.2 ± 4.0^c	17.4 ± 6.4^b	16.8 ± 8.0^b	21.5 ± 7.9
Mean similarity	0.18 ± 0.06^{ab}	0.17 ± 0.07^b	0.17 ± 0.06^b	0.19 ± 0.04^{ab}	0.20 ± 0.23^a	0.23 ± 0.05^a
Mosaic diversity	3.38 ± 0.07^b	3.30 ± 0.07^c	3.43 ± 0.07^b	4.10 ± 0.08^a	3.43 ± 0.05^b	4.04 ± 0.10^a
Number of species in (n) plots	40 (15)	58 (24)	55 (23)	87 (21)	110 (35)	123 (38)
Number of species in 1 ha	64	80	77	125	132	149
Indicator species	*Artemisia carulescens* *Juncus subulatus* *Limonium costae* *Suaeda vera*	*Juncus acutus* *Plantago maritima* *Puccinellia festuciformis* *Scirpus maritimus*	*Festuca arundinacea* *Hordeum secalinum* *Medicago sativa* *Trifolium fragiferum*	*Camphorosma monspeliaca* *Carex divisa* *Cynosurus cristatus* *Scirpus holoschoenus*	*Agrostis castellana* *Convolvulus arvensis* *Juncus inflexus* *Trifolium repens*	*Festuca ampla* *Juncus effusus* *Rumex angiocarpus* *Rumex papillaris*

Significantly different values ($P < 0.05$) are indicated by different superscripts. The number of species in 1 ha has been calculated from species–area curves. Number of species per plot is consistent with data gathered by Montalvo and Herrera (1993) in Madrid aquifer wet grasslands. (Modified from Bernáldez and Rey Benayas 1992 and Rey Benayas and Scheiner 1993.)

among the halophytic communities because abiotic factors can also reduce diversity by restricting ecological maturity.

DIVERSITY OF WET GRASSLAND COMMUNITIES IN MONTANE REGIONS

Most wet grasslands in montane regions are small, sometimes as little as 100 m^2, and scattered through the xerophytic vegetation. Nevertheless, vast *Fraxinus angustifolia* dehesas (savannahs formed by human activity) can be found with a carpet of hygrophytic grasses and herbs such as *Agrostis castellana* and *Potentilla reptans*. These small wet grasslands originate from groundwater seepages, often located on slopes.

In a recent survey of 92 100-m^2 plots in the Sierra de Guadarrama (altitude range 580–2400 m) (Figure 5.1), we found that the number of species in wet grassland communities (range 2–44 species per plot, mean 24) was mostly influenced by elevation. This variable alone explained 24% of species richness, which was maximal at intermediate altitudes (Sánchez-Colomer *et al.* 1995). Altitude is a complex environmental variable that is often linked to other environmental factors such as geomorphology, precipitation and temperature (Whittaker and Niering 1975). A multiple regression analysis revealed that geomorphology, topography, soil moisture, climate and soil ion content explained the greatest species richness across 90 of the 92 plots (Table 5.3). Species richness tended to be greater in areas with high soil moisture and relatively high soil ion content, and in which erosion was the dominant process rather than sedimentation. It decreased in areas with flooded soils and at high altitude, above 1500 m elevation, where higher precipitation and lower temperatures allowed for an excessive annual water balance. Other environmental factors that have been found to correlate individually and positively with species richness are slope and number of hygrotypes (both are measures of the environmental heterogeneity of a plot), soil texture and soil nitrate concentration.

COMPARATIVE DIVERSITY OF WET GRASSLANDS IN SEDIMENTARY BASINS AND MONTANE SYSTEMS

The comparison of the diversity of wet grasslands linked to regional aquifers in sedimentary basins and those linked to montane systems is interesting due to its evolutionary implications. Wet grasslands in sedimentary basins are much poorer in species than those in montane locations (170 species recorded in 16 500 m^2, giving 14.2 species 100 m^{-2}, versus 237 species recorded in 9200 m^2, giving 24 species 100 m^{-2} respectively). In the former, wet grassland plant communities have had to adapt to continuous increments in salinity, the evolution of groundwater and the concentration of salts in soils since the Tertiary period. In montane areas, groundwater evolution has not occurred, although plant communities have undergone more recent historical disturbances such as glaciation. Salinity has been highlighted as an environmental factor among the determinants of species richness due to the limitation it imposes on the evolution potential of tolerant genotypes (Grubb 1987). Increase in groundwater salinity decreases diversity at a local and a regional scale.

A comparison of the various measures of diversity of wet grassland communities with

Table 5.3 Stepwise multiple regression analysis of species richness over various environmental factors

Variable ($R^2 = 0.48$)	b'	SE	$P<$
Geomorphological and topographic position	−0.791	0.011	0.0001
Soil wetness	0.576	2.193	0.0009
Water balance and soil wetness interaction	−0.466	0.017	0.002
Water balance	−0.430	0.021	0.002
Soil ion content	0.181	0.383	0.02

Variables are those found by this analysis to be statistically significant. The magnitudes of the regression coefficients (b') have been standardized to indicate the relative importance of each variable. These results correspond to 90 of the 92 100-m^2 plots sampled in wet grasslands located in the Sierra de Guadarrama (Area A, Figure 5.1).

those of the surrounding, xerophytic plant communities revealed that wet grasslands are richer in species (within-community or α diversity, and total or γ diversity), but often have a similar between-community or β diversity (Table 5.4). Thus, the highest overall species richness is primarily due to the highest within-community number of species rather than the total distribution pattern of these species. Complexity as measured by mosaic diversity or μ, depends mostly upon the physiognomy of the ecosystem under consideration. *Pinus sylvestris* forest, *Quercus pyrenaica* forest and woodland ecosystems are more complex than the corresponding wet grasslands, but wet grasslands are more complex than the corresponding *Cytisus purgans* formation.

ENVIRONMENTAL IMPACT

In general, broad-scale changes that are likely to affect wetlands include climate change, land-use and land-cover change, water and air-borne pollution, a shift in disturbance/recovery regimes, and habitat fragmentation or destruction (Pearson 1994). The main threat to wet grasslands linked to regional aquifers is a fall in the water table due to pumping and drainage. Groundwater extraction has drastically decreased the area of wet grasslands in many Mediterranean regions (Bernáldez *et al.* 1993). Figure 5.3 illustrates one example. The desiccation of wetlands, the loss of economic value of pasture and the absence of flooding that prevents ploughing, especially in winter, are factors that facilitate tilling and land availability for irrigated crops. Glycophyte communities linked to local flow systems are usually little affected because of the lower usage of the areas for irrigation due to their steeper topography (Figure 5.2). However, there are a number of specific impacts on the tolerant, alkalinophyte and halophyte communities linked to intermediate and regional flow systems. These impacts can be: (i) an increase in non-phreatophyte annual species such as *Hordeum marinum, Trisetum paniceum* and *Bromus tectorum*, indicating dryness and an increasing nitrification due to the mineralization of unstable organic matter in the new and moist regime; (ii) disappearance of perennial plants, chiefly the most hygrophytic species such as *Carex* and *Juncus*; and (iii) an increase in the proportion of xerohalophyte species such as *Puccinellia festuciformis* and *Camphorosma monspeliaca*. Tilman and El Haddi (1992) have found that drought can limit species richness by causing the local extinction of rare species.

In contrast, wet grasslands located in montane regions are subject to a quite different

Table 5.4 Comparison of wet grassland diversity with the diversity of the surrounding xerophytic communities in the Sierra de Guadarrama (Area A, Figure 5.1)

	Xerophytic community type					Corresponding wet grassland				
	α	β	$\gamma(n)$	$\gamma*$	μ	α	β	$\gamma(n)$	$\gamma*$	μ
[1] Alpine grasslands	13.7	0.40	36(12)	56	4.31	—	—	—	—	—
[2] High mountain shrublands	9.7	0.22	65(23)	90	3.25	20.58	0.36	61(12)	94	4.07
[3] Pine forests	14.8	0.23	78(19)	115	4.27	28.71	0.29	114(21)	155	3.30
[4] Deciduous oak forests	28.3	0.23	138(21)	194	4.14	30.70	0.23	152(18)	225	3.94
[5] Mediterranean woodlands	15.4	0.22	76(16)	116	3.94	19.60	0.16	158(41)	189	3.57

The communities compared are within the same altitudinal range. α, number of species per plot; β, mean similarity; $\gamma(n)$, total number of species found in n 100-m^2 plots; $\gamma*$, number of species extrapolated for 10 000 m^2 based upon the species–area curves; μ, mosaic diversity.
Notes: Alpine grasslands are *Festuca indigesta* formations (mean altitude of the plots is 2195 m); high mountain shrublands are *Cytisus purgans* formations (1886 m); pine forests are *Pinus sylvestris* formations (1791 m); deciduous oak forests are *Quercus pyrenaica* formations (1453 m), and Mediterranean woodlands are formations of *Rosmarinus officinalis*, *Cistus ladanifer* and *Quercus rotundifolia* (990 m). The data sets corresponding to the xerophytic community types have been compiled from: (1) Rivas Martínez (1963); (2) Rivas Martínez (1963), Rivas Martínez and Cantó (1987), Rivas Martínez et al. (1987); (3) Rivas Martínez (1963), Rivas Martínez et al. (1987), Fernández-González (1991); (4) Fernández-González (1991); (5) Costa (1974), Fernández-González (1991), Rivas Martínez (1968), Rivas Martínez and Cantó (1987).

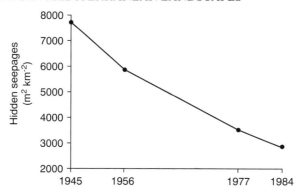

Figure 5.3 Changes in wet grasslands in representative areas of the Los Arenales aquifer between 1945 and 1984, expressed as rate of loss in area. This analysis is based on photointerpretation with wet meadows and *Carex* stands interpreted by hidden seepages. (After Bernáldez *et al.* 1993)

set of threats, most linked to changes in land use. Mountain regions in dry countries are in demand for their recreational value, and an increasing number of grasslands, mainly those close to towns and villages, are being sold for housing development. Traditional extensive livestock grazing is declining, and some grasslands are being invaded by woody, hydrophytic plants such as *Crataegus monogyna* and *Rubus* and *Rosa* species. In addition, lowland riverside grasslands are being planted with trees (*Populus* sp.) for wood production. It is likely that the impact on wet grasslands arising from groundwater extraction on sedimentary plateaux and land-use changes in mountain regions will be a reduction in diversity.

IMPLICATIONS FOR CONSERVATION AND MANAGEMENT

A number of ecological and environmental benefits have been recognized for wet grasslands in dry landscapes (Trettin *et al.* 1994). They include biological and landscape diversity, productivity, water cycle regulation, microclimate buffering, soil conservation and enhanced scientific value. We shall now focus on two areas of biological diversity: plant community diversity and the influence of wet grasslands on the conservation of other taxa such as birds.

As noted earlier, wet grasslands are almost the only species-rich communities in intensively cultivated areas. Groundwater discharge areas in agricultural regions are untilled enclaves, which act as refugia and sites of high biological diversity within vast expanses of croplands treated with biocides. Despite their small size, wet grasslands in non-agricultural areas are usually richer in species than the surrounding xerophytic natural or semi-natural vegetation, and such species are more persistent (Montalvo *et al.* 1993). Grazing is partly responsible for this and plays an important role in maintaining the diversity of Mediterranean grasslands (Naveh and Whittaker 1979; Hobbs and Mooney 1991).

In addition to the number of species, the attributes of each species are also important for the evaluation of diversity (Usher 1986). Thus, wet steppe grasslands have a considerable richness of organisms with highly specific adaptations to extreme conditions and a distribution area that, in western Europe, is restricted to the dry enclaves of Iberia and

Sardinia. Their isolation has fostered endemism and refugia, in some cases related to unique palaeoecosystems. Some plant examples are *Pholiurus pannonicus*, a halophyte species with a distribution area split between northwestern Spain and eastern Europe, *Hohenackeria polyodon*, a species endemic to Spain and Algeria, and *Carex lainzii*, an Iberian endemic. Other examples of species with highly restricted distribution areas are *Elymus curvifolius, Microcnemum coralloides* and *Sphenopus divaricatus* (Rey Benayas 1991).

These enclaves have also become reserves for fauna due to the microclimatic effects and the floristic diversity. The abundance of insects is fundamental for the survival of important endangered or game birds with an insectivorous diet, at least in their juvenile stage, for example *Otis tarda* (greater bustard), *Otis tetrax* (little bustard), *Alectoris rufa* (red-legged partridge) and others. The presence of invertebrates near the soil surface also makes these enclaves important for wintering species such as *Vanellus vanellus* (lapwing), *Pluvialis apricaria* (golden plover), *Anthus pratensis* (meadow pipit) and *Gallinago gallinago* (snipe). Small mammals, amphibians, reptiles and large insects encourage the presence of several endangered raptors including *Circus pygargus* (Montagu's harrier), and *C. cyaneus* (hen harrier), as well as *Falco tinnunculus* (kestrel) and *Ciconia ciconia* (white stork).

The Pan-European Biological and Landscape Diversity Strategy, which was approved at the Sofia conference in 1995, is a guide for the joint action of all European countries for the next 20 years (see Chapter 2). The first plan (1996–2000) includes 12 Action Themes. Action Theme 8 refers to the conservation and management of grasslands (Fernández-Galiano 1996). Special emphasis is placed on the conservation of hay meadows, alpine meadows, and arid and semi-arid steppes and pastures. The strategy aims to control irrigation, drainage, tilling and the use of biocides and fertilizers in natural grasslands, and also maintain traditional grazing practices. Apart from the ethical, environmental, natural and scientific justifications, the economic interest of correct husbandry of grasslands using native breeds of livestock, and the positive effects on certain game species, should be starting points and keystones on the way to fulfilling these goals.

ACKNOWLEDGEMENTS

The studies of vegetation under groundwater influence in Spain were fostered by the late Professor Fernando G. Bernáldez. Current research on wet grasslands in mountainous regions is being funded by the 'Humedales en áreas de descarga de acuíferos en territorios graníticos (Sierra de Guadarrama)' project under the auspices of the Madrid Regional Government.

SUMMARY

In Mediterranean landscapes most wet grasslands are fed by groundwater. They represent islands within dry climatic regions dominated by either croplands or xerophytic vegetation. Two main types of wet grasslands can be distinguished: those linked to regional aquifers and those in montane regions. Soil salinity is the key factor explaining diversity patterns of wet grasslands linked to regional, sedimentary aquifers. In mountainous regions, altitude is the variable most closely related to species richness, which it affects in a quadratic fashion, being maximal at intermediate altitudes. This complex environmental factor of altitude can be segregated into various related variables such as geomorphology and topography, soil wetness, climate and soil ion content. Wet mountain grasslands are more diverse than wet grasslands in sedimentary basins because change in ground-water chemistry along flow paths decreases species richness at the local and regional scales. The main threats to Mediterranean wet grasslands are groundwater extraction for irrigation, land-use change such as residential development, and the abandonment of traditional extensive grazing practices. These pastures are virtually the only species-rich plant communities in agricultural areas and are often richer than the local natural xerophytic vegetation. They also provide habitats for rare and endangered plants and animals. The Pan-European Biological and Landscape Diversity Strategy (1996–2000) includes an action theme for the conservation and management of grasslands with special attention being paid to wet grasslands. Besides their environmental and ecological value, the economic interest of wise pasture exploitation together with its positive effect on game species should be keystones for wet grassland conservation in Mediterranean landscapes.

Keywords: Altitude, Conservation, Diversity, Groundwater, Management, Salinity, Threats.

REFERENCES

Bernáldez, F.G. and Rey Benayas, J.M. (1992) Geochemical relationships between groundwater and wetland soils and their effects on vegetation in central Spain. *Geoderma*, **55**, 273–288.

Bernáldez, F.G., Rey Benayas, J.M., Levassor, C. and Peco, B. (1989) Landscape ecology of uncultivated lowlands in central Spain. *Landscape Ecology*, **3**, 3–18.

Bernáldez, F.G., Rey Benayas, J.M. and Martínez, A. (1993) Ecological impact of groundwater extraction on wetlands (Douro basin, Spain). *Journal of Hydrology*, **141**, 219–238.

Costa, M. (1974) Estudio fitosociológico de los matorrales de la provincia de Madrid. *Anales del Jardín Botánico de Madrid*, **31-I**, 225–315.

Fernández-Galiano, E. (1996) La estrategia paneuropea para la diversidad biológica y paisajística. *Quercus*, **126**, 42–46.

Fernández-González, F. (1991) La vegetación del Valle del Paular (Sierra de Guadarrama, Madrid). *Lazaroa*, **12**, 153–272.

Freeze, R.A. and Cherry, J.A. (1979) *Groundwater*. Prentice-Hall, Englewood Cliffs, New Jersey.

Grubb, P.J. (1987) Global trends in species richness in terrestrial vegetation: a view from the northern hemisphere. In: Gee, J.H.R. and Giller, P.S. (eds) *Organization of communities. Past and present*, pp. 99–118. Blackwell, Oxford.

Hobbs, R.J. and Mooney, H.A. (1991) Effects of rainfall variability and gopher disturbance on serpentine annual grassland dynamics. *Ecology*, **72**, 59–68.

International Wetland Research Bureau (1991) *A strategy to arrest and reverse wetland loss and degradation in the Mediterranean basin*. Managing Mediterranean wetlands and their birds for the year 2000 and beyond, IWRB Symposium. Grado, Italy.

Mannetje, L.'t and Frame, J. (eds) (1994) *Grassland and society*. Wageningen Pers, Wageningen.

Montalvo, J. and Herrera, P. (1993) Diversidad de especies de los humedales: criterios de conservación. *Ecología*, **7**, 215–231.

Montalvo, J., Casado, M.A., Levassor, C. and Pineda, F.D. (1991) Adaptation of ecological systems: compositional patterns of species and morphological and functional traits. *Journal of Vegetation Science*, **2**, 655–666.

Montalvo, J., Casado, M.A., Levassor, C. and Pineda, F.D. (1993) Species diversity patterns in Mediterranean grasslands. *Journal of Vegetation Science*, **4**, 213–222.

Naveh, Z. and Whittaker, R.H. (1979) Structural and floristic diversity of shrublands and woodlands in northern Israel and other Mediterranean areas. *Vegetatio*, **41**, 171–190.

Pearson, S.M. (1994) Landscape level processes and wetlands conservation in the Southern Appalachian Mountains. *Water, Air & Soil Pollution*, **77**, 125–136.

Rey Benayas, J.M. (1991) *Aguas subterráneas y ecología. Ecosistemas de descarga de acuíferos en Los Arenales*. ICONA, Madrid.

Rey Benayas, J.M. and Scheiner, S.M. (1993) Diversity patterns of wet meadows along geochemi-
    cal gradients in central Spain. *Journal of Vegetation Science*, **4**, 103–108.
Rey Benayas, J.M., Bernáldez, F.G., Levassor, C. and Peco, B. (1990) Vegetation of groundwater
    discharge sites in the Douro basin, central Spain. *Journal of Vegetation Science*, **1**, 461–466.
Rivas Martínez, S. (1963) Estudio de la vegetación y flora de las Sierras de Guadarrama y Gredos.
    Anales del Jardín Botánico de Madrid, **21-I**, 1–325.
Rivas Martínez, S. (1968) Los jarales de la Cordillera Central. *Collectanea Botanica,* **7**, 1033–1082.
Rivas Martínez, S. and Cantó, P. (1987) Datos sobre la vegetación de las Sierras de Guadarrama y
    Malagón. *Lazaroa*, **7**, 235–237.
Rivas Martínez, S., Belmonte, D., Cantó, P., Fernández-González, F., de la Fuente, V., Moreno,
    J.M., Sánchez-Mata, D. and Sancho, G.L. (1987) Piornales, enebrales y pinares oromediter-
    ráneos (*Pino–Cytison oromediterranei*) en el Sistema Central. *Lazaroa*, **7**, 93–124.
Sánchez-Colomer, M.G., Rey Benayas, J.M. and Levassor, C. (1995) Aspectos ecológicos de las
    áreas de descarga de agua subterránea en la Sierra de Guadarrama. *Hidrogeología y Recursos
    Hidrúlicos*, **20**, 353–367.
Scheiner, S.M. (1992) Measuring pattern diversity. *Ecology*, **73**, 1860–1867.
Tilman, D. and El Haddi, A. (1992) Drought and biodiversity in grasslands. *Oecologia*, **89**,
    257–264.
Trettin, C.C., Aust, W.M., Davis, M.M., Weakley, A.S. and Wisniewki, J. (1994) Wetlands of the
    interior south-eastern United States: conference summary statement. *Water, Air & Soil Pollu-
    tion*, **77**, 199–205.
Usher, M.B. (ed.) (1986) *Wildlife conservation evaluation*. Chapman & Hall, London.
Whittaker, R.H. (1977) Evolution of species diversity in land communities. *Evolutionary Biology*,
    10, 1–67.
Whittaker, R.H. and Niering, W.A. (1975) Vegetation of the Santa Catalina Mountains. V.
    Biomass, production and diversity along the elevational gradient. *Ecology*, **46**, 429–452.

6 Environmental Monitoring of Grassland Management in the Somerset Levels and Moors Environmentally Sensitive Area, England

DAVID J. GLAVES

Farming and Rural Conservation Agency, Exeter, UK

INTRODUCTION

The Somerset Levels and Moors in England was one of five 'first tranche' Environmentally Sensitive Areas (ESAs) launched by the Ministry of Agriculture, Fisheries and Food (MAFF) in 1987. It is the largest remaining grazing marsh in Britain, of outstanding value for its lowland wet grassland landscape, wildlife and well-preserved archaeological remains.

European Community (EC) Member States were authorized to introduce ESAs by the EC Council of Ministers in 1985, through Article 19 of the Structures Regulation (EC/797/85), subsequently modified by further regulations, notably EC Regulation 1760/87 and the European Union (EU) Agri-environment Regulation 2078/92.

The ESA scheme was introduced in England after the passage of enabling legislation (Section 18 of the Agriculture Act 1986), following the pilot Broads Grazing Marshes Conservation Scheme in 1985–86 (Ministry of Agriculture, Fisheries and Food 1989). Similar schemes have been introduced in a number of other EU Member States, notably in Germany, Denmark and The Netherlands, although to date not on the same scale as in the UK. Since 1987 the ESA scheme has been expanded in the UK and there are now 43 ESAs covering around 15% of agricultural land. Considerable experience has been gained from managing and monitoring the UK schemes, which may enable their use as a model for new schemes elsewhere in Europe.

The ESA scheme encourages farmers to help safeguard areas of the countryside where the landscape, wildlife and historical interest are of national importance, by offering five- or ten-year agreements to carry out beneficial agricultural practices in return for fixed annual payments. Participants in the scheme can also apply for a 'conservation plan' that provides grant aid for carrying out capital works to improve particular environmental features. The scheme is voluntary and is open to all managing suitable land within the designated areas. ESAs are administered by MAFF's regional organization, with profes-

---

See Glossary, p. 305, for explanation of technical terms. Scientific names of vascular plants follow Tutin, T. G. *et al.* (1964 80) *Flora Europaea* Volumes 1–5. Cambridge University Press. See p. 319.

European Wet Grasslands: Biodiversity, Management and Restoration. Edited by Chris B. Joyce and P. Max Wade.
© 1998 John Wiley & Sons Ltd.

sional management by the Farming and Rural Conservation Agency (formerly ADAS) Project Officers, who act as the main interface between farmers and MAFF.

Section 18(8) of the Agriculture Act 1986 requires that 'the Minister shall arrange for the effect on the area as a whole of the performance of agreements to be kept under review and shall from time to time publish such information as he considers appropriate about these effects'. To this end, environmental monitoring programmes have been established in individual ESAs. The monitoring activities are targeted according to the environment-al characteristics and objectives of each ESA. The general approach has been to monitor change by establishing a baseline record of conditions when the ESA was launched and to compare this with information from subsequent resurveys at intervals. The national strategy and methodologies adopted for environmental monitoring of ESAs are de-scribed in ADAS (1995).

This chapter reviews the impact of the Somerset Levels and Moors ESA. Each ESA is reviewed by MAFF on a five-year cycle. The first review of the Somerset Levels and Moors ESA took place in 1991 and the results of monitoring from 1987 to 1990 were described in Ministry of Agriculture, Fisheries and Food (1991). The second review of the scheme was undertaken during 1996 and the results of monitoring from 1987 to 1995 were described in a series of activity reports covering landscape, grassland, breeding and wintering birds, historical and water-level monitoring, and summarized in an overview report (ADAS 1996a). This forms the basis of the present chapter, which provides a brief description of the ESA, the scheme structure, the environmental aim and objectives of the scheme and the monitoring programme. The results and conclusions of the programme are summarized, with particular emphasis on the ecological monitoring activities.

GRAZING MARSH AND RIVER VALLEY GRASSLAND ESAs IN ENGLAND

There are 22 ESAs in England covering around 10% of the agricultural area (Figure 6.1). Within these designated areas there were 7479 agreements covering 409 962 ha up to 1995–96, representing payments to land managers totalling £29 million (Harrison, 1997). Lowland wet grassland associated with grazing marshes or river valley floodplains forms the main environmental interest in eight ESAs (Table 6.1) although small areas also occur in a number of others. In these eight ESAs grassland covers around 84 000 ha, of which it is estimated that at least 53 000 ha is wet grassland subject to periodic inunda-tion. This represents around 27% of the estimated total resource in England (Dargie 1993; The UK Steering Group 1995) (perhaps higher, as the national estimate is likely to represent the potential area rather than the true extent of grassland subject to regular inundation). Though most of this grassland is known to have been subject to agricultural improvement, important areas of floristically rich, 'unimproved' grassland occur in a number of ESAs, most notably around 1750 ha in the Somerset Levels and Moors (Glaves 1996a), plus smaller areas in several others (e.g. 273 ha in the original Test Valley ESA; Ministry of Agriculture, Fisheries and Food 1993b) (Figure 6.1). A number of these ESAs with significant arable areas have arable reversion tiers, which have resulted in the potential re-creation of wet grassland. Thus the ESA mechanism has an important role to play in the delivery of the coastal and floodplain grazing marsh habitat action plan

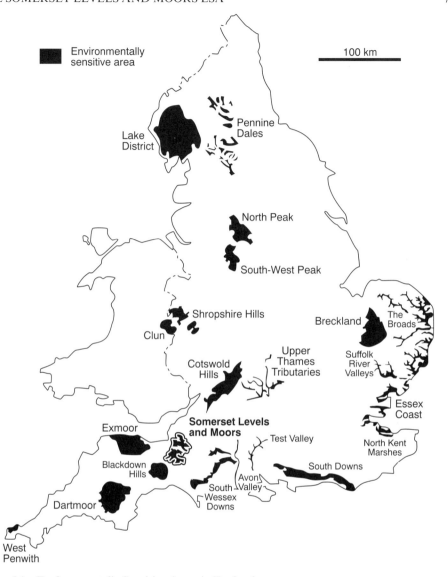

Figure 6.1 Environmentally Sensitive Areas in England

proposed by The UK Steering Group (1995). This sets objectives to maintain the existing habitat extent and sets targets for rehabilitation and creation of grazing marsh.

THE SOMERSET LEVELS AND MOORS ESA

ENVIRONMENTAL CHARACTER

The Somerset Levels and Moors ESA now extends over 29 300 ha of the central Somerset lowlands (Figure 6.1), bounded by the Mendip Hills to the north, low limestone escarp-

Table 6.1   Grazing marsh and river valley grassland Environmentally Sensitive Areas (ESAs) in England in 1996

ESA	Date designated	Total area (ha)[1]	Grassland area (ha)[1]	Estimated wet grassland area (ha)[1]
The Broads	1987	36 175	20 148[2]	[8 000+]
Somerset Levels and Moors	1987	27 678	23 695[3]	18 580
Suffolk River Valleys	1988	43 480	[10 500][4]	[7 150]
Test Valley	1988	4 776	[2 500][5]	[2 000]
Avon Valley	1993	5 220	3 257	2 630
North Kent Marshes	1993	14 683	7 349	[5 000]
Essex Coast	1994	28 599	[6 000]	[4 700]
Upper Thames Tributaries	1994	27 678	10 686	5 375
Total		187 827	84 135	[53 435+]

[1]Data from aerial photographic interpretation by ADAS as part of the ESA monitoring programme, mostly unpublished except [2]ADAS (1996a) and [3]ADAS (1996b).
[4,5]Estimates for revised extended schemes based on proportion of grassland in original schemes given in [4]Ministry of Agriculture, Fisheries and Food (1993a) (7770 ha excluding 'abandoned pasture') and [5] Ministry of Agriculture, Fisheries and Food (1993b) (1413 ha).
Figures in square brackets are imprecise estimates.

ments to the east, the Blackdown Hills to the south and the Quantock Hills to the west. The higher land surrounding the moors is excluded from the ESA, which is divided in two by the low Polden Hills ridge running east to west. The moors comprise an extensive area of very low-lying basin peat, with a few remnants of raised bog, surrounded by alluvial silt and clay. The peat is overlain in places by a varying thickness of riverine clay. Extending westwards from the moors to the coast lies an extensive area of slightly higher estuarine alluvium known as the Levels, most of which is excluded from the ESA.

The whole area forms the largest remaining grazing marsh in Britain and is consequently of outstanding environmental interest. The landscape value lies in the rectilinear pattern of traditionally managed fields and drainage channels within a low-lying, generally wet and open grassland landscape, containing scattered trees and scrub. The ecological interest is principally associated with wet, often floristically rich, pastures and meadows, which support overwintering and breeding birds, and the network of ditches, rhynes and drains, which are of special interest for aquatic flora and invertebrates. In archaeological and historical terms the area is internationally famous for the many prehistoric wooden trackways, preserved for millennia by the waterlogged ground conditions, along with evidence of past vegetation, human activity, settlement and technology unparalleled anywhere else in the country.

NATURE CONSERVATION INTEREST

The national importance of the Somerset Levels and Moors area for nature conservation is recognized by various statutory designations. National Nature Reserves within the ESA include the Somerset Levels, covering 300 ha (1% of the ESA). There are 13 biological Sites of Special Scientific Interest (SSSIs), covering 6104 ha (22% of the ESA). There is also a major reserve of the Royal Society for the Protection of Birds at West

Sedgemoor covering 510 ha (2% of the ESA) and several small Somerset Wildlife Trust and Vincent Wildlife Trust nature reserves.

The Somerset Levels and Moors is of international importance for wintering and passage birds and 6386 ha (some outside the ESA) are proposed as a Special Protection Area and Ramsar site. The site as a whole is also recognized as a nationally Important Bird Area (Grimmett and Jones 1989; Pritchard *et al.* 1992). Based on recent winter counts the area is internationally important for *Cygnus columbianus bewickii* (Bewick's swan), *Anas crecca* (teal) and *Vanellus vanellus* (lapwing) and nationally important for *Cygnus olor* (mute swan), *Anas penelope* (wigeon), *A. clypeata* (shoveler), *A. strepera* (gadwall) and *Pluvialis apricaria* (golden plover) (Quinn and Bell 1994). The area also qualifies as a wetland of international importance by regularly supporting over 20 000 waterfowl in winter, though some species are also attracted to nearby flooded former peat workings excluded from the ESA. The area is internationally important as a staging post for *Numenius phaeopus* (whimbrel) on spring passage (Ferns *et al.* 1979), although there is some evidence of a recent decline in numbers (Green and Robins 1993).

The wet grassland of the ESA supports an important assemblage of breeding wading birds (waders), in particular *Vanellus vanellus*, *Gallinago gallinago* (snipe), *Tringa totanus* (redshank) and *Numenius arquata* (curlew), and occasionally a few pairs of *Limosa limosa* (black-tailed godwit). The ESA is considered one of the five prime inland wet grassland areas for breeding waders in England and Wales, for example holding 4.4% of the total number of pairs counted in the lowlands in 1982–83 (Green and Cadbury 1987). There has been some decline in the ESA over the last two decades and none of the populations of individual species alone attain national importance, although all are regionally important.

The ESA also holds important breeding populations of a number of other grassland birds. These include *Motacilla flava flavissima* (yellow wagtail) and *Saxicola rubetra* (whinchat), which occur in regionally important numbers, although both species have shown recent declines (Glaves 1996b). Other species include relatively large populations of *Alauda arvensis* (skylark) and *Anthus pratensis* (meadow pipit), both nationally declining species in the lowlands (Gibbons *et al.* 1993). The ditches hold high densities of a number of wetland species, notably *Acrocephalus schoenobaenus* (sedge warbler) and *Emberiza schoeniclus* (reed bunting), probably in regionally important numbers. Small numbers of wildfowl breed, including *Cygnus olor* in nationally important numbers. Other nationally notable birds breeding include an increasing population of *Cettia cetti* (Cetti's warbler), *Coturnix coturnix* (quail) in small numbers, if somewhat erratically, and recently *Circus aeruginosus* (marsh harrier).

The pastures and meadows comprise a wide range of grassland, mire and swamp plant communities, including a number of damp mesotrophic grassland and fen-meadow communities of high nature conservation value, in particular those related to the MG8 *Cynosurus cristatus–Caltha palustris* grassland, M22 *Juncus subnodulosus–Cirsium palustre* fen-meadow and M24 *Molinia caerulea–Cirsium dissectum* fen-meadow communities of the British National Vegetation Classification (NVC) (Rodwell 1991, 1992). These correspond approximately with the *Cynosurion* pasture (C38.12), eutrophic humid grassland (C37.218 *Juncus subnodulosus* meadow) and oligotrophic humid grassland (C37.312 Acid *Molinia* grassland) CORINE biotopes of the EC habitat classification (Devillers *et al.* 1991). The surviving areas of species-rich swards are fragmented with only a few concentrations, especially on the wetter peat moors. These were estimated to cover around 1750 ha in 1987, based on aerial photographic interpretation and partial ground surveys (Glaves 1996a), representing around 35% of the estimated total English resource of unimproved lowland wet grassland (The UK Steering Group 1995). The area is considered of outstanding significance for such grassland in England (Jefferson 1996). Whilst there have been indications of a gradual lowering of water levels in the area (Green and Robins 1993), some concern has also been expressed about the possible detrimental effects of raising water levels on the botanical composition of some of these high-value communities (Evans *et al.* 1995; Leach and Cox 1995).

Although it is the grassland communities themselves that are of greatest nature conservation value, a number of nationally scarce plant species also occur, mostly associated with ditches or the remnant raised bogs (some of which are excluded from the ESA), including *Althaea officinalis, Lathyrus palustris, Peucedanum palustre, Sium latifolium* and *Thelypteris palustris*. The grasslands are, however, noteworthy for holding large populations of several nationally restricted or declining species including *Cirsium dissectum, Juncus subnodulosus, Oenanthe pimpinelloides, Silaum silaus* and *Thalictrum flavum*. Notable species associated with mud on ditch banks, droves and occasionally in fields (especially grassland subject to flooding) include *Baldellia ranunculoides, Myosurus minimus, Rumex maritimus* and *R. palustris*. A small relict population of *Eurodryas aurinia* (marsh fritillary), a threatened European butterfly that is listed in Annex II of the EC Habitats and Species Directive, is associated with an area of fen-meadow and mire communities.

Whilst much of the nature conservation interest of grassland has been lost or fragmented through agricultural improvement, the ditches are still important refugia for many species, including such nationally scarce or declining plants as *Hydrocharis morsus-ranae, Myriophyllum verticillatum, Potamogeton coloratus, Sium latifolium* and *Wolffia arrhiza*. Surveys indicate that ditches of high botanical interest occur widely across the ESA (e.g. Wolseley *et al.* 1984), although recent partial resurveys indicate a slight deterioration in quality, perhaps resulting from eutrophication (Cadbury 1995; Hughes 1995). Also associated with the ditches are a wide range of aquatic invertebrates, including 17 Red Data Book species (Shirt 1987), most notably a number of aquatic Coleoptera (water beetles) and large populations of the nationally scarce Odonata, *Coenagrion pulchellum* (variable damselfly), *Brachytron pratense* (hairy dragonfly) and *Sympetrum sanguineum* (ruddy darter). The ditch system and other watercourses also support a small but well-established population of *Lutra lutra* (otter).

THE THREAT FROM AGRICULTURAL INTENSIFICATION

The complex arterial land-drainage system, including embanked rivers and drains, tidal sluices, pump stations, water-penning structures and an intricate network of field ditches and rhynes, enables control of water levels and flooding, allowing the ESA to be farmed. The constraints of this system, along with poor access, fragmented ownership and traditional attitudes in the farming community, controlled the pace of agricultural change and led to the development and continuation of extensive livestock farming systems. Permanent grassland predominated, traditionally used mainly for summer cattle-grazing and hay-cutting. However, in the post-war period, and particularly during the 1970s and early 1980s, traditionally managed grassland came under increasing threat from agricultural intensification, particularly from the installation of pumping stations and field under-drainage. Large areas were cultivated and resown, some land was brought into arable production and much of the remaining grassland was improved by the use of fertilizers and herbicides (Purseglove 1988). Under the MAFF Agricultural Land Classification, which uses a 1–5 scale, the peat moors are Grade 2, capable of very high agricultural potential after drainage (which is, however, expensive), and much of the remainder is Grade 3, capable of sustaining good grass and cereal crops with adequate drainage. Thus there seemed every prospect that this improvement would continue to the detriment of the special environmental character of the grazing marsh landscape.

The threat was countered by the designation of 10 SSSIs between 1983 and 1985 and by the establishment in 1987 of the Somerset Levels and Moors ESA. The ESA scheme was reviewed by MAFF after five years of operation. Evidence of some decline in environmental quality, particularly reductions in breeding wading bird populations, perhaps associated with gradual lowering of water levels (Robins and Green 1988; Ministry of Agriculture, Fisheries and Food 1991; Robins et al. 1991) led to revisions to the scheme. It was relaunched in 1992, offering new support for raising water levels and marginally extending the designated area by 530 ha. The monitoring results described in this chapter informed the second review of the scheme, which resulted in a number of mostly minor changes (Ministry of Agriculture, Fisheries and Food 1997) and extended the ESA by a further 1590 ha in 1997.

ENVIRONMENTAL OBJECTIVES

Following the review of the scheme in 1991, MAFF specified an overall environmental aim common to all ESAs: 'to maintain and enhance the landscape, wildlife and historic value of the area by encouraging beneficial agricultural practices'. This was accompanied by ESA-specific environmental objectives and associated performance indicators, which focused on the priorities for achieving the environmental aim (Ministry of Agriculture, Fisheries and Food 1994a). The environmental objectives for the Somerset Levels and Moors ESA were:

1. To maintain the wildlife conservation value and landscape quality of grassland;
2. To enhance the wildlife conservation value of wet grassland without detriment to the landscape by maintaining higher water levels in ditches and rhynes;
3. To maintain and enhance landscape quality through management of characteristic landscape elements;
4. To maintain and enhance archaeological and historic features.

A full list of the associated performance indicators is given in Ministry of Agriculture, Fisheries and Food (1994a).

ESA MANAGEMENT PRESCRIPTIONS

The Somerset Levels and Moors ESA is a 'part farm' scheme with land managers able to enter the whole or any part of their land. The initial scheme offered two tiers of agreement. Tier 1 agreements provided for a relatively extensive grassland management regime, with restrictions on cultivation, under-drainage and the use of inorganic fertilizers, and Tier 2 imposed additional restrictions on stocking rates, winter sheep-grazing, cultivation, mowing dates and fertilizer use. Included in the prescriptions was provision for the management of ditches and gutters and the maintenance of trees and pollarded willows. The scheme placed an obligation on agreement holders not to damage or destroy any features of historical interest.

The revised scheme from 1992 introduced some extra prescriptions for grassland management and for the maintenance of a range of landscape features. Restrictions were placed on the planting or natural establishment of trees or shrubs. A requirement to maintain existing traditional gates and a restriction on the use of permanent fencing were added. In both Tier 1 and Tier 2, restrictions were placed on the use of organic manures, and basic requirements to maintain ditch water levels were introduced (higher in Tier 2). Tier 2 also imposed a requirement to mow at least one-third of the land. An additional tier (Tier 3) was introduced to enhance further the ecological interest of the grassland by the creation of wet winter and spring conditions. This tier provided for the raising of water levels and low-intensity management of the wet grassland, particularly for the benefit of birds, including no use of inorganic fertilizer, no grazing before 20 May and a later cutting date. In addition, supplements were introduced to Tiers 1 and 2 to raise water levels to those specified in Tier 3. Coherent blocks of land under Tier 3, and under Tiers 1 and 2 with raised water-level supplements, formed Raised Water-Level Areas (RWLAs).

Capital works that could be included in the conservation plan introduced in 1992 included tree planting, hedge planting, laying and coppicing, creation or reinstatement of ditches, construction of water-penning structures and the reintroduction of willow pollarding. An Access Tier was introduced to offer new opportunities for public use of agreement land for walking and other quiet recreation.

A list of management prescriptions in place between 1992 and 1997 was given in Ministry of Agriculture, Fisheries and Food (1994b) together with a list of capital works that can be included in a conservation plan. The payments to participants are calculated on the basis of profit forgone (change in output and costs) with an incentive element. They are reviewed on a biannual basis and in 1996 were £130 ha^{-1} for Tier 1, £215 ha^{-1} for Tier 2, £415 ha^{-1} for Tier 3 and £80 ha^{-1} for the raised water-level supplements to Tiers 1 and 2.

SCHEME UPTAKE

Several of the performance indicators set targets for the uptake of land into the various tiers of agreement for the five-year period from 1992. It is not possible to distinguish

Table 6.2 Uptake of eligible land to the tiers of Environmentally Sensitive Area agreement (after ADAS 1996a)

Tier of agreement	Area eligible to enter tier (ha)	'Target' uptake[1]	Area under agreement[2] (ha)	% eligible area under agreement
1	23 840	40%	10 949	46
2	23 840	15%	2 413	10
3, 1S, 2S	–	2 500 ha	992[3]	–

[1]Percentage of eligible area (Tiers 1 and 2) or hectarage (Tiers 3, 1S and 2S).
[2]Includes all applications and signed agreements as at end of December 1995.
[3]The area for land under Tier 1S and Tier 2S is also included in the total area of Tier 1 and Tier 2 respectively. Uptake of Tier 3 alone was 880 ha.

between the grassland that might be eligible for each of the three tiers, as eligibility is dependent on the ability to achieve the specified water levels and an assessment of the ecological interest or potential of the land (which is not comprehensively known). Therefore, the uptake targets for Tier 1 and Tier 2 were expressed as a proportion of the total grassland resource and a hectarage target has been set for Tier 3 and the raised water-level supplementary payments under the first two tiers (Tier 1S and Tier 2S) (Table 6.2).

By the end of 1995 the uptake of grassland under Tier 1 had already exceeded the target uptake, but the area of grassland under Tier 2 was short of the target by 1163 ha. The total amount of grassland under these two tiers was 56% of the eligible area, which is just above the combined uptake target of 55%. Grassland within RWLAs was only at 40% of the target. In total, 60% of the eligible area was under agreement at the end of 1995. In addition it was estimated that 69% of 'traditional grassland' and 48% of the land in Areas of High Archaeological Potential identified by Somerset County Council were under agreement, both slightly below the uptake targets. Of the 923 signed agreements at the end of December 1995, 13% had conservation plans, well below the 50% target.

ENVIRONMENTAL MONITORING

LANDSCAPE

Landscape assessment

A landscape assessment was carried out at the commencement of the ESA scheme in 1987 (Ministry of Agriculture, Fisheries and Food 1995). It provides a broad overview of the landscape character of the whole area, together with a description and map of the constituent 'landscape types'. The 'key characteristics' of each landscape type are identified and described in terms of the landscape elements that are present and their spatial distribution. For example the key characteristics of the 'open moor' landscape type include 'an open and expansive, homogeneous, natural character created by extensive areas of low intensity grassland and species-rich pasture, with a lack of scrub, woodland or fencing, with trees restricted to occasional lines of pollarded willows and isolated shelters'. Landscape types vary in strength of character. As a general rule, where the character is strong the landscape value is high and, conversely, where it is weaker the value is lower. It follows, therefore, that changes which strengthen landscape character are usually beneficial, whereas those which weaken the character are detrimental. Thus, the landscape assessment provides a benchmark for evaluating the impact of change occurring to the landscape elements, enabling judgements to be made about the performance of the scheme in maintaining and enhancing landscape value. A detailed review of changes in landscape elements by landscape type is given in Bolton and Shaw (1996).

Land cover

Land cover was mapped using aerial photographic interpretation of 1:10 000 false colour infra-red photographs taken in May 1987. Changes in land cover were identified in a

Table 6.3 Changes in the area of the main land cover classes from 1987 to 1994 in a 31% sample of the Environmentally Sensitive Area (based on Bolton and Shaw 1996)

Land cover class	Total area (ha)		Change in total area 1987/94	
	1987	1994	ha	%
Arable	623	707	+84	+14
Grassland	7563	7473	−92	−1
Orchard	36	31	−5	−14
Peat workings/water	30	28	−2	−6
Withy (*Salix viminalis*) bed	37	25	−12	−32
Woodland/scrub	19	36	+17	+89

random sample covering 31% of the ESA area, by comparison of aerial photographs taken in 1990 and 1995 with the baseline map. In addition changes in the patterns and structure of grassland in RWLAs were monitored using fixed-point ground photography of 15 fields on a clay moor and 11 fields on a peat moor in January, May and September of 1993, 1994 and 1995.

At designation in 1987 94% of the ESA was agricultural land, of which grassland accounted for 23 695 ha (92%). The remainder was mainly under arable cultivation. By the end of 1995 74% of the grassland area was under agreement, the majority under Tier 1. This level of uptake of the scheme represented a substantial area of grassland that was successfully protected from conversion to arable land and other inappropriate land management. From 1987 to 1994 more grassland was converted to arable (362 ha) than arable changed to grassland (261 ha) within the sample area. This resulted in a 14% increase in the arable area and a 1% decrease in the area of grassland (Table 6.3). This had the effect of weakening the landscape character in places by fragmenting the areas of open grassland. All of these changes occurred on non-agreement land, but over 40% of the new grassland subsequently came under ESA agreement (indicating reversion rather than arable/ley grassland rotation).

Photographic monitoring of the appearance of grassland in two RWLAs showed that a substantial number of fields (35%) had become 'rougher' in appearance between 1993 and 1995. This was due to the encroachment of scattered clumps of *Juncus effusus* and the spread of tussocky grasses, especially *Deschampsia cespitosa*. This was attributed to both the enhanced wetness and the low-intensity management of the grassland encouraged by the scheme. The two RWLAs lie within the open-moor landscape type and these changes broadly indicated a strengthening of the underlying wet, natural character of the grazing marsh landscape.

Linear and point features

A sample-based ground survey of linear and point features was undertaken in 30 random 25-ha squares (covering 3% of the ESA area) in 1989 and 1995. This identified a substantial overall increase (29%) in the length of ditches classed as 'wet' and 'clear', although a few others (2%) (on both agreement and non-agreement land) were infilled.

Changes to other landscape features were generally small and localized, and included a small increase in lines of *Salix* (willows), although the number of pollarded willows declined, and a small increase in post and wire fencing. The numbers of traditional gates increased (134%) throughout the ESA, many of which replaced modern ones.

GRASSLAND

Botanical monitoring of grassland has formed the main ecological appraisal of change in most ESAs in England (Critchley and Poulton 1994). Standard methods have been applied based on the establishment of permanent quadrats, or more recently 'stands', in a sample of fields (Critchley *et al.* 1996; Critchley and Poulton, in prep.).

In the Somerset Levels and Moors ESA general botanical monitoring was based on a stratified sample covering the range of grassland communities, soil type and the two tiers of ESA agreement under the original scheme (Glaves 1996a). From a sample of 100 fields under agreement in 1988, a total of 11 community 'endgroups' were described from an analysis of the species composition of the quadrats, using the TWINSPAN computer program (Hill 1979). Within fields, vegetation was sampled in five objectively placed 2 m × 2 m permanent quadrats. The vegetation communities represented by the 'end-groups' were described using the National Vegetation Classification (NVC) (Rodwell 1991, 1992). These varied from unimproved wet communities to wet improved or reverting grasslands (Table 6.4). In addition the botanical impact of raising water levels under the revised scheme was monitored as a separate exercise. A random sample of 25 fields were surveyed in 1993 with randomly located 8 m × 4 m stands in two RWLAs.

Most of the fields monitored were managed relatively extensively before entry into ESA agreement and the majority had never been ploughed. Many sites were subjected to high frequencies of flooding, which resulted in periodic 'damage' to the grassland (through aeration stress), in extreme cases killing swards. The RWLA agreements resulted in greater changes in management, particularly reductions in both organic and inorganic fertilizer inputs and a reduction in cutting for hay and silage.

Changes in vegetation from 1988 to 1995 (1993 to 1995 for the RWLAs) were assessed to determine how they reflected changes in species suited to prevailing high soil moisture content, low soil nutrient availability, grazing and absence of poaching (poaching is excessive trampling by livestock), and for RWLAs tussock-forming species were also monitored. All of these are to some extent influenced by ESA management prescriptions. The method, described in Critchley *et al.* (1996), in outline involved compiling a list of 'suited-species' for each of the criteria on the basis of published autecological information. The monitoring data were then analysed for changes in the proportion of suited-species to other species for each criterion. A scoring system was used to provide, for each criterion, a measure of the proportion of suited-species per quadrat or stand at each survey. The scores for the W (soil moisture content), Nu (nutrient availability) and G (grazing) criteria were calculated as the difference between the scores for the extremes of the condition. For example, the score for G was the difference between the score for species suited to grazed conditions minus the score for species suited to ungrazed conditions (dereliction). Scores for these criteria, therefore, could range from -1 to 1. The score for the P criterion (poaching) was simply the proportion of suited-species, giving a possible range of values from 0 to 1. The use of the proportion of suited-species for scoring a quadrat ensured that all species present in the quadrat, not just suited-

Table 6.4 Summary of changes in 'suited-species' scores for each ecological criterion (W = soil moisture content, Nu = nutrient availability, G = grazing and P = poaching) applied to endgroups, using scores based on presence/absence (P/A) and Domin data, 1988 (after Glaves 1996a)

Endgroup[1]	Vegetation type[2]	W		Nu		G		P	
		Domin	P/A	Domin	P/A	Domin	P/A	Domin	P/A
B	*Juncus* meadows/mires (M23a/b–MG10a)	NS	NS	NS	NS	NS	NS	NS	NS
C	*Juncus subnodulosus* fen meadows (M22b)	NS	NS	+	+	NS	NS	+	+
D	Unimproved wet-peat pastures/fen meadows (M22b–MG8)	+	+	+	NS	−	−	+	+
E	Unimproved wet-peat pastures/meadows (MG8)	+	+	+	NS	−	−	+	NS
F	Semi-improved wet pastures/meadows (MG8–MG7c)	+	+	+	+	−	−	NS	NS
G	Semi-improved damp meadows (MG8–MG5a)	+	+	+	+	−	−	NS	NS
H	Improved damp pastures/meadows (MG7c/d)	+	+	+	+	−	−	NS	NS
I	Semi-improved dry meadows (MG6a)	NS	NS	+	NS	NS	NS	NS	NS
J	*Deschampsia cespitosa* pastures (MG9a)	+	+	+	+	NS	NS	+	+
K	Semi-improved wet-clay inundation pastures (MG13)	+	+	+	+	−	−	+	+

+ = significant increase ($P < 0.05$); − = significant decrease ($P < 0.05$); NS = no significant change.
[1] Endgroup A, representing tall-herb fen vegetation, contained only three quadrats and was not included in the statistical analysis.
[2] Closest National Vegetation Classification subcommunity given in parentheses (Rodwell 1991, 1992).

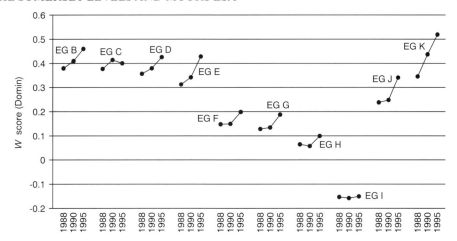

Figure 6.2 Comparison of Domin-weighted *W* 'suited-species' scores between endgroups for 1988, 1990 and 1995 (after Glaves 1996a). Data are means for each year and all changes were significant ($P<0.05$) except for endgroups B, C and I

species, contributed to the analysis. For the general survey, scores were calculated at two different scales; the first was based only on the presence of species in a quadrat and the second included weighting according to Domin cover abundance scores. This was because some changes may have been detectable at one of these scales but not at the other.

For the general sample of grassland, statistically significant changes were detected in the composition of most communities (using Generalized Linear Modelling and Analysis of Variance in combination with randomization testing), the main exceptions being a species-poor mire community and a dry 'semi-improved' hay meadow community (Table 6.4). None of the changes was sufficient to represent a change in the NVC community/subcommunity, but there was a general trend towards species associated with high soil moisture content and, to a lesser extent, towards those suited to higher soil nutrient status and poaching. There was a general decline in species associated with grazing. The changes relating to soil moisture (Figure 6.2) were the largest and most ecologically important.

The results obtained in the two RWLAs sampled were similar to those in the more general survey of grassland, although perhaps of greater magnitude. However, care is needed in making the comparison owing to differences in survey methodology, calculation of the scores and survey years. The main change in the vegetation was a significant increase ($P<0.0001$) in species suited to high soil moisture. Although this resulted in a shift towards inundation communities, in only seven (out of 25) stands was it considered sufficient to represent a change in NVC community/subcommunity. Overall, the change was greatest on the moor, which experienced more widespread and severe flooding in 1995, resulting in much bare ground. Although there was evidence that one flood-affected sward recovered to some extent in one growing season, these circumstances seem to have favoured tussock-forming wetland grasses, sedges and rushes. In this respect the changes were potentially beneficial in providing a diversity of structure and cover for nesting birds, although they may, if continued, pose a threat to floristically rich wet grassland related to the MG8 community type.

   Additional botanical monitoring and research work in other RWLAs in the ESA has identified similar trends, especially a slight decline in species richness, increases in a group of wetland species including *Carex nigra, Glyceria fluitans, Polygonum amphibium* and *Ranunculus repens* and declines in a group of common dry grassland species including *Cerastium fontanum, Holcus lanatus, Rumex acetosa, Taraxacum officinale* agg. and *Trifolium repens* (Evans *et al.* 1995; Leach and Cox 1995; Mountford and Treweek 1996). In addition, the results of hydrological modelling suggest that the RWLA regime may result in wetter conditions than those favoured by a number of characteristic MG8 species (Gowing 1996; Spoor *et al.* 1996). However, other communities, especially those with mire elements, may be better able to tolerate high water levels. The changes detected have occurred in most cases after only a few years of RWLA management and may represent only the initial transition to communities more in equilibrium with the new field water regimes.

   It seems likely that the changes in suited-species scores are linked and represent a response to the same factors, most likely water levels and flooding. These factors are, to some extent, influenced by the scheme prescriptions; however, weather is probably more important. Raising water levels in RWLAs is intended to create conditions of surface splashing, thus exacerbating any trend towards wetter weather. Inundation favours bulky wetland monocotyledons (e.g. *Carex riparia* and *Glyceria fluitans*) that are able to put on rapid growth early in the season when floodwater recedes, and before grazing management is introduced. It also results in nutrient enrichment. Thus, these conditions favour wetland and dereliction (as opposed to grazing) suited-species and species suited to moderate soil nutrient levels which, in combination, probably result in the observed changes in scores. Standing water in spring or early summer can lead to anaerobic conditions, killing swards; this gives a few wetland and poaching suited-species a competitive edge and sometimes eliminates species associated with drier conditions. The incidence of spring flooding was greater during the monitoring period (coinciding with the resurveys in 1990 and 1995) than in previous years (e.g. Tatem 1995). However, it was noted that the same general trends occurred in communities that were not subject to regular flooding and in a partial resurvey in 1989 during which it was particularly dry. This suggests that higher field water tables, perhaps in response to changes in ditch water-level regimes, or reduced field drainage and maintenance, may have influenced the changes in the vegetation. Both these factors are influenced to some extent by ESA prescriptions. It remains to be seen whether the changes, particularly in RWLAs, are permanent or temporary effects in what appear to be relatively dynamic communities.

   Thus, grassland communities in the ESA have shown major consistent change, representing a shift towards wetter conditions. These changes have occurred across most of the range of communities and ESA management agreement tiers. Such large changes are likely to have hidden any more subtle ones resulting from changes in grassland management.

BIRDS

Breeding wading birds

A total of 324 wader territories were located in 7444 ha (27% of the ESA) surveyed in 1995, representing an overall density of 4.4 km^{-2} (Glaves 1996b). This represented an

Table 6.5 Breeding wading bird territories (after Glaves and Trump 1996)

Species	No. of territories in 1995	Territory density in 1995 (km^{-2})	% change from 1987 to 1995
Vanellus vanellus	205	2.7	+48%
Gallinago gallinago	71	1.0	+13%
Tringa totanus	43	0.6	+19%
Numenius arquata	31	0.4	+5%

overall increase of 29% from 1987 to 1995, resulting from increases in all species, but particularly *Vanellus vanellus* (Table 6.5). However, this trend was not consistent across moors and increases were only recorded at four of the 11 moors surveyed. Additionally the increase in 1995 followed a decline of 5% from 1987 to 1992 and a longer-term decline since the first comprehensive survey in 1977 (Figure 6.3).

From 1987 to 1995, overall wader territories increased by 30% on six peat moors and by 27% on five clay moors. The overall increase on clay moors was due mainly to increases in *Vanellus vanellus* (60%) and *Tringa totanus* (53%), mostly at West Moor. However, *Gallinago gallinago* declined almost to extinction (from 12 in 1987 to one in 1992 and two in 1995) on clay moors, whereas on peat moors the overall numbers in 1995 increased to more than the 1987 level (despite having decreased from 1987 to 1992). However, excluding West Sedgemoor, only 19 *Gallinago gallinago* territories were located on peat moors (down by 41% from 1987). Indeed away from West Sedgemoor, overall wader territories on peat moors declined by 27%, including declines in *Vanellus vanellus* (39%) and *Tringa totanus* (45%).

The relationship between breeding wader populations and agreement status was investigated for 12 sites surveyed in 1992 and 1995, six of which included RWLA blocks by 1995. In both years most birds were on ESA agreement land. Tier 2 had more birds than Tier 1 and non-agreement land, but the relatively small area of RWLA land had the highest densities for all species. The effect was most marked for *Gallinago gallinago*, with an apparent shift from non-agreement and Tier 2 land into RWLAs.

The importance of suitable breeding habitat is indicated by a number of features of the population trends (Figure 6.3). The declines, although widespread, were greatest on the more improved moors, resulting in a retreat to a few (particularly wetter peat) moors. Even within these sites, there is evidence of redistribution to the most suitable fields (Hancock and McGeoch 1993; Evans *et al.* 1995). Similarly, the first indications of increases have been very localized, particularly at West Sedgemoor, where raised water levels have been in place longest (since the late 1980s). Even here the increase has been slow, perhaps resulting from the fact that, as the regional populations of waders else-where are low, the potential for recruitment is low. Population growth is thus most likely to be determined by breeding success. This is known to be affected by intensity of agricultural management, which in spring is also linked to water levels.

Grazing livestock can destroy wader nests and young birds by trampling. Data from the 1992 and 1995 surveys suggest a slight reduction in the area of grassland grazed in spring and early summer. The associated habitat data also indicate a slight increase in *Juncus* and *Carex* species, which may be of some importance in providing the well-structured swards favoured by some breeding waders and wintering birds. It may also indicate wetter conditions, which could also favour breeding waders.

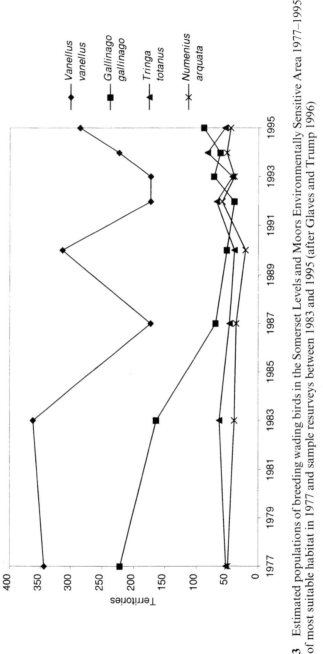

Figure 6.3 Estimated populations of breeding wading birds in the Somerset Levels and Moors Environmentally Sensitive Area 1977–1995, based on a survey of most suitable habitat in 1977 and sample resurveys between 1983 and 1995 (after Glaves and Trump 1996)

Wintering birds

Reduction in the extent and duration of winter flooding prior to the designation of the ESA in 1987 led to some reduction in wintering wildfowl numbers and their length of stay. For example, Owen *et al.* (1986) estimated an 80% reduction in usage over the preceding 25 years. More recent surveys indicate a continuing decline up to the late 1980s, with some recovery subsequently (Quinn and Bell 1994). There was no significant change in wintering bird populations between the two winters 1992/93 and 1994/95 (coinciding with the introduction of RWLAs), although total populations increased on three of the four sites surveyed in both winters (Glaves and Trump 1996). A longer-term comparison of counts at West Sedgemoor indicated that wintering birds have responded rapidly to water-level management (Evans *et al.* 1995), although the increase also coincided with a run of wetter winters. However, the recent increases detected on West Sedgemoor, and slight increases elsewhere, suggest that the existing RWLAs may already be having a beneficial effect for wintering waterfowl.

HISTORICAL

The land cover changes recorded as part of the landscape monitoring programme, from 1987 to 1994, were reviewed to assess which areas affected known historical sites (McCrone 1996). An assessment was made of whether the change was potentially beneficial or detrimental to the condition and long-term survival of the features affected. The majority of sites were unaffected by land cover changes and a large number (43%) were protected under the scheme from the threat of damage by cultivation. The maintenance, and particularly the raising, of water levels was considered to offer protection to buried features. Potentially detrimental changes occurred to a number of sites (4%) that were not under agreement, largely as a result of the continued attrition of earthworks and buried features by cultivation on arable land.

WATER LEVELS

Monitoring of field water levels in RWLAs involved the measurement of ditch water levels, field water tables and soil penetrability (Woode and Armstrong 1996). This was done weekly on two clay moors over a period of six years from 1989 to 1994, and a water balance model described by Armstrong *et al.* (1996) was used to predict the effect of raising ditch water levels on the water table within fields. The model predicted a consistently higher field water table under the RWLA regime, especially in spring and early summer (Figure 6.4), coinciding with the critical period for breeding wading birds. Similar models applied to peat soils in the ESA predict an even stronger relationship due to the greater hydraulic conductivity of these soils (e.g. Youngs *et al.* 1989; Armstrong *et al.* 1996; Gowing 1996). However, examination of actual spring and early summer water levels on one clay moor subject to the RWLA regime for two years suggested that the effects of drainage and recharge only affected the water table close to the ditches. Thus the effect of the RWLA regime on spring and early summer water levels may (at least on clay soils) be due in part to fields starting the season in a saturated condition, rather than

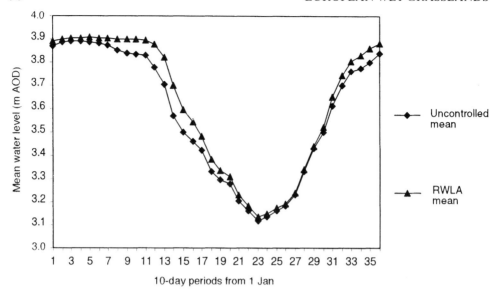

Figure 6.4 Modelled mid-field mean water levels (m above Ordnance datum, AOD) for uncontrolled and Raised Water-Level Area (RWLA) management regimes on Southlake Moor, by time of year (after Woode and Armstrong 1996)

to drainage and recharge via the ditches. The models also highlight the importance of ditch spacing, soil characteristics (especially hydraulic conductivity), microtopography and weather, as well as ditch water levels, in determining field water table. This suggests that blanket water-level prescriptions are unlikely to have a uniform effect (which could be beneficial) and that if fine control over water levels is required, then account needs to be taken of soil type and ditch (or other water-filled drain or gutter) spacing.

Soil strength was measured with a penetrometer incorporating a probe that mimicked the bill of a *Gallinago gallinago* (snipe). For both sites monitored there was a significant correlation ($P = 0.001$) between soil strength and field water table. Thus, despite the effect of other factors such as recent weather, vegetation cover, stock trampling and compaction due to agricultural machinery, the ground water table was shown to have the overriding controlling influence on the average soil penetrability on the two clay moors monitored.

CONCLUSION

The monitoring programme suggests that three of the four objectives and most of the associated performance indicators for the ESA have been, or are likely to be, met over the first five years of the revised scheme. The exception is Objective 3, relating to landscape elements, with in particular relatively small localized losses of ditches and pollarded willows recorded in the sample squares. Together with some losses of orchards and withy *(Salix viminalis)* beds and an increase in arable land, this has resulted in localized weakening of the landscape character. However, overall the scheme has been largely successful in maintaining the landscape, wildlife and historical value and there has been some limited evidence of enhancement. In particular there are indications that the RWLAs could have a beneficial effect in restoring the grazing marsh, although it is too early fully to assess their impact. It is likely that the relatively high uptake of the scheme since 1987 has made a substantial contribution to reducing the threat of habitat loss through agricultural improvement that led to the designation of the area as an ESA.

The more than 900 agreements achieved up to the end of 1995 clearly represent an important step in safeguarding the high environmental interest of the largest remaining grazing marsh in Britain. The second review of the scheme in 1996 offered the opportunity to take account of the results of monitoring, together with experience gained in managing the agreements and the views of other statutory and voluntary bodies.

ACKNOWLEDGEMENTS

The work described in this chapter was funded by the Ministry of Agriculture, Fisheries and Food (MAFF). The ESA monitoring programme was developed and implemented by ADAS in consultation with MAFF and the statutory conservation agencies. Advice and information received from a number of organizations and individuals in connection with the work are gratefully acknowledged, particularly that from the Countryside Commission, English Nature, the National Rivers Authority (now Environment Agency), Royal Society for the Protection of Birds and Somerset County Council.

The national monitoring programme was co-ordinated by A.J. Hooper, with the assistance of K.M. Joyce and J.A. Slater. The monitoring programme for the Somerset Levels and Moors ESA was initially co-ordinated by M.R. Watson and subsequently by R.G. Symes. Thanks go to the many ADAS staff involved in the monitoring programme, particularly the following, who compiled and edited the activity reports on which this paper is based: A.C. Armstrong, C.C. Bolton, A.J. Hooper, K.M. Joyce, P. McCrone, R.G. Symes, G. Shaw, D. Smallshire, D.P.C. Trump and P.R. Woode. Helpful comments on an earlier draft of this paper were provided by C.N.R. Critchley and M.D.K. Harrison.

Finally, but importantly, thanks are due to the farmers, landowners and land managers who allowed access to their land. This work would not have been possible without their co-operation.

SUMMARY

The impact of lowland wet grassland management agreements in the Somerset Levels and Moors Environmentally Sensitive Area (ESA), England, was studied in an environmental monitoring programme carried out between 1987 and 1995. The ESA is the largest remaining grazing marsh in Britain and is of outstanding environmental interest, including being of international importance for birds and supporting over a third of the total English resource of floristically rich wet grassland. The results of the monitoring were assessed in relation to the published environmental objectives and overall environmental aim of the scheme. It was concluded that the scheme has been largely successful in maintaining the landscape, wildlife and historical interest and that there has been some limited evidence of enhancement. In particular there are indications that Raised Water-Level Areas could have a beneficial effect in restoring the grazing marsh. It is likely that the relatively high uptake of the scheme since 1987 has made a substantial contribution to reducing the threat of

habitat loss, through agricultural improvement, that led to the designation of the area as an ESA.
Running title: Monitoring management in the Somerset Levels and Moors ESA

Keywords: Agriculture, Conservation, Environmentally Sensitive Area, Grazing marsh, Management agreement, Monitoring, Water levels, Wet grassland.

REFERENCES

ADAS (1995) *ADAS national strategy for ESA monitoring*. ADAS, Oxford.

ADAS (1996a) *Environmental monitoring in the Somerset Levels and Moors ESA 1987–1995*. Ministry of Agriculture, Fisheries and Food, London.

ADAS (1996b) *Environmental monitoring in the Broads ESA 1987–1995*. Ministry of Agriculture, Fisheries and Food, London.

Armstrong, A.C., Rose, S.C. and Miles, D.R. (1996) *Effects of managing water levels to maintain or enhance ecological diversity within discrete catchments*. Unpublished report to Ministry of Agriculture, Fisheries and Food. ADAS, Cambridge.

Bolton, C.C. and Shaw, G. (1996) *Landscape monitoring in the Somerset Levels and Moors ESA 1987–1995*. Unpublished report to Ministry of Agriculture, Fisheries and Food. ADAS, Oxford.

Cadbury, C.J. (1995) *The ditch flora of West Sedgemoor (Somerset) 1984 and 1994*. Unpublished report. Royal Society for the Protection of Birds, Sandy.

Critchley, C.N.R. and Poulton, S.M.C. (1994) Biological monitoring of grasslands in Environmentally Sensitive Areas in England and Wales. In: Hagger, R.J. and Peel, S. (eds), *Grassland management and nature conservation*, pp. 254–255. Occasional Symposium No. 28. British Grassland Society, Reading.

Critchley, C.N.R. and Poulton, S.M.C. (in prep.) *A method to optimise precision and scale in vegetation monitoring and survey*.

Critchley, C.N.R., Smart, S., Poulton, S.M.C. and Myers, G.M. (1996) Monitoring the consequences of vegetation management in Environmentally Sensitive Areas. *Aspects of Applied Biology*, **44**, 193–201.

Dargie, T.C. (1993) *The distribution of lowland wet grassland in England*. English Nature Research Report No. 49. English Nature, Peterborough.

Devillers, P., Devillers-Terschuren, J. and Ledant, J.P. (1991) *Habitats of the European Community. CORINE biotopes manual. Data specifications*, Part 2. Office for Official Publications of the European Community, Luxembourg.

Evans, C., Street, L., Benstead, P., Cadbury, J., Hirons, G., Self, M. and Wallace, H. (1995) Water and sward management for nature conservation: a case study of the RSPB's West Sedgemoor reserve. *RSPB Conservation Review*, **9**, 60–72.

Ferns, P.N., Green, G.H. and Round, P.D. (1979) The significance of the Somerset and Gwent Levels in Britain as feeding areas for migrant whimbrel. *Biological Conservation*, **7**, 22.

Gibbons, D.W., Reid, J.B. and Chapman, R.A. (1993) *The new atlas of breeding birds in Britain and Ireland: 1988–1991*. Poyser, London.

Glaves, D.J. (1996a) *Botanical monitoring of grassland in the Somerset Levels and Moors ESA, 1987–1995*. Unpublished report to Ministry of Agriculture, Fisheries and Food. ADAS, Oxford.

Glaves, D.J. (1996b) Survey of breeding waders in the Somerset Levels and Moors ESA, 1995. In: Gibbs, B.D. (ed.), *Somerset Birds 1995*, pp. 113–123.

Glaves, D.J. and Trump, D.P.C. (1996) *Monitoring of breeding and wintering birds in the Somerset Levels and Moors ESA 1987–1995*. Unpublished report to Ministry of Agriculture, Fisheries and Food. ADAS, Oxford.

Gowing, D.J.G. (1996) *Examination of the potential impacts of alternative management regimes in the Somerset Levels and Moors ESA*. Unpublished report to Ministry of Agriculture, Fisheries and Food. Silsoe College, Cranfield University, Silsoe.

Green, R.E. and Cadbury, J. (1987) Breeding waders of lowland wet grasslands. *RSPB Conserva-*

tion Review, **1**, 10–13.

Green, R.E. and Robins, M. (1993) The decline in the ornithological importance of the Somerset Levels and Moors, England and changes in the management of water levels. *Biological Conservation*, **66**, 95–106.

Grimmett, R.F.A. and Jones, T.A. (1989) *Important Bird Areas in Europe*. International Council for Bird Preservation, Cambridge.

Hancock, C.G. and McGeoch, J.A. (1993) The decline in breeding wader populations on Tealham and Tadham Moors on the Somerset Levels, 1984–90. *Somerset Archaeology and Natural History: Ecology in Somerset*, **137**, 207–215.

Harrison, M.D.K. (1997) Country case studies: English ESAs. *Proceedings of the OECD Seminar on Environmental Benefits from Agriculture, Helsinki, September 1996. OECD/GD (97) 110, pp. 133–142.*

Hill, M.O. (1979) *TWINSPAN – a FORTRAN program for arranging multivariate data in an ordered two-way table by classification of individuals and attributes.* Cornell University, New York.

Hughes, M.R. (1995) *A botanical survey of ditches, North Moor and Southlake Moor SSSIs, Somerset, 1994.* Unpublished report. English Nature, Taunton.

Leach, S.J. and Cox, J.H.S. (1995) *Effects of raised water-levels on grasslands at Southlake Moor SSSI.* Unpublished report. English Nature, Taunton.

Jefferson, R.G. (1996) *Lowland grassland in natural areas: national assessment of significance.* English Nature, Peterborough.

McCrone, P. (1996) *Historical monitoring in the Somerset Levels and Moors ESA 1987–1995.* Unpublished report to Ministry of Agriculture, Fisheries and Food. ADAS, Oxford.

Ministry of Agriculture, Fisheries and Food (1989) *Environmentally Sensitive Areas. First report.* HMSO, London.

Ministry of Agriculture, Fisheries and Food (1991) *Somerset Levels and Moors Environmentally Sensitive Area report of monitoring 1991.* Ministry of Agriculture, Fisheries and Food, London.

Ministry of Agriculture, Fisheries and Food (1993a) *Suffolk River Valleys Environmentally Sensitive Area report of monitoring 1992.* Ministry of Agriculture, Fisheries and Food, London.

Ministry of Agriculture, Fisheries and Food (1993b) *The Test Valley Environmentally Sensitive Area report of monitoring 1992.* Ministry of Agriculture, Fisheries and Food, London.

Ministry of Agriculture, Fisheries and Food (1994a) *Environmental objectives and performance indicators for ESAs in England.* Ministry of Agriculture, Fisheries and Food, London.

Ministry of Agriculture, Fisheries and Food (1994b) *The Somerset Levels and Moors ESA: guidelines for farmers*, revised edn. Ministry of Agriculture, Fisheries and Food, London.

Ministry of Agriculture, Fisheries and Food (1995) *Somerset Levels and Moors Environmentally Sensitive Area: landscape assessment*, revised edn. Ministry of Agriculture, Fisheries and Food/ADAS, Oxford.

Ministry of Agriculture, Fisheries and Food (1997) *Somerset Levels and Moors ESA: guidelines for farmers.* Revised edition. Ministry of Agriculture, Fisheries and Food, London.

Mountford, O.J. and Treweek, J.R. (eds) (1996) *Assessment of the effects of managing water levels to enhance ecological diversity.* Unpublished report to Ministry of Agriculture, Fisheries and Food, Institute of Terrestrial Ecology, Abbots Ripton.

Owen, M., Atkinson-Willes, G.L. and Salmon, D.G. (1986) *Wildfowl in Great Britain*, 2nd edn. Cambridge University Press, Cambridge.

Pritchard, D.E., Housden, S.D., Mudge, G.P., Galbraith, C.A. and Pienkowski, M.W. (eds) (1992) *Important Bird Areas in the UK including the Channel Islands and the Isle of Man.* Royal Society for the Protection of Birds, Sandy.

Purseglove, J. (1988) *Taming the flood.* Oxford University Press, Oxford.

Quinn, J.L. and Bell, M.C. (1994) *Waterfowl usage of the Somerset Levels and Moors IBA: assessing trends and the efficacy of current counting procedures.* Unpublished report to Royal Society for the Protection of Birds. Wildfowl and Wetlands Trust, Slimbridge.

Robins, M. and Green, R.E. (1988) *Changes in the management of water levels on the Somerset Moors.* Unpublished report. Royal Society for the Protection of Birds, Sandy.

Robins, M., Davies, S.G.F. and Buisson, R.S.K. (1991) *An internationally important wetland in crisis: the Somerset Levels and Moors.* Royal Society for the Protection of Birds, Sandy.

Rodwell, J.S. (ed.) (1991) *British plant communities,* Vol. 2, *Mires and heaths* Cambridge University Press, Cambridge.

Rodwell, J.S. (ed.) (1992) *British plant communities,* Vol. 3, *Grasslands and montane communities.* Cambridge University Press, Cambridge.

Shirt, D.B. (1987) *British Red Data Books: 2. Insects.* Nature Conservancy Council, Peterborough.

Spoor, G., Youngs, E.G., Gowing, D.J.G. and Gilbert, J.C. (1996) *Water regime requirements of the native flora – with particular reference to ESAs.* Unpublished report to Ministry of Agriculture, Fisheries and Food. Silsoe College, Cranfield University, Silsoe.

Tatem, K.W. (1995) *Somerset Moors and Levels studies.* Unpublished report. National Rivers Authority, Bridgewater.

UK Steering Group, The (1995) *Biodiversity: the UK Steering Group report.* HMSO, London.

Wolseley, P.A., Palmer, M.A. and Williams, R. (1984) *The aquatic flora of the Somerset Levels and Moors.* Unpublished report. Nature Conservancy Council, Taunton.

Woode, P.R. and Armstrong, A.C. (1996) *Monitoring of raised water levels in the Somerset Levels and Moors ESA 1989–1995.* Unpublished report to Ministry of Agriculture, Fisheries and Food. ADAS, Oxford.

Youngs, E.G., Leeds-Harrison, P.B. and Chapman, J.M. (1989) Modelling water-table movement in flat low-lying lands. *Hydrological Processes,* **3,** 301–315.

Part Two

BIODIVERSITY

7 The Formation, Vegetation and Management of Sea-Shore Grasslands in West Estonia

ELLE PUURMANN and URVE RATAS
Institute of Ecology, Tallinn, Estonia

INTRODUCTION

The coast of Estonia, a small country on the eastern coast of the Baltic Sea, is rich in banks, shoals, bays and peninsulas (Figure 7.1). About 1500 islands are found in Estonian coastal waters. The length of the indented Estonian coastline, including these islands, is 3790 km. The structure, dynamics and development of this sea coast are closely connected with the geological and geomorphological history of the area. Coastal development has been influenced by the uplift of the earth's crust, which in north-west Estonia at present is up to 3 mm per year (Vallner *et al.* 1988). The differentiation of the coastline has been strongly affected by the lithology of the sediments and contemporary shores are very variable with considerable regional differences. Orviku (1993) distinguishes six main shore types in Estonia (Figure 7.1).

Silty and clayey shores are extensive, especially in west Estonia. On these flat and gently sloping areas, where wave action is negligible even during strong surges, and where fine-grained sediments slowly accumulate, sea-shore grasslands develop with a vegetation of grasses, sedges and herbs. Extending up to 0.8 m above mean sea level, the width of the grasslands can be from several metres to some hundreds of metres, depending chiefly on the gradient of the shore. Sea-shore grasslands are more widely distributed in west Estonia, while in north Estonia usually only fragments of a few communities remain (Rebassoo 1975). On the smaller islands, sea-shore grasslands cover between 20 and 100% of the area. According to an agricultural land inventory of Estonia in the 1980s, sea-shore grasslands covered 5180 ha and paludified coastal grasslands (where the water table is above ground for most of the year) 4333 ha, accounting for approximately 3% of the total area of natural grasslands in the country (Aug and Kokk 1983).

Sea-shore grasslands in Estonia are valuable wildlife habitats in terms of botanical and ornithological diversity. Estonian sea-shore grasslands belong to the northern European group of maritime saltmarshes (Chapman 1977) and the constituent plant communities have been well studied by Lippmaa (1935), Laasimer (1965) and Krall *et al.* (1980). The most detailed classification is given in Rebassoo (1975, 1987), where the vegetation of halophilous sea-shore meadows is classified into 15 associations with subassociations and varieties. The most common and characteristic communities of Estonian sea-shore

---

See Glossary, p. 305, for explanation of technical terms. Scientific names of vascular plants follow Tutin, T. G. *et al.* (1964–80) *Flora Europaea* Volumes 1–5. Cambridge University Press. See p. 319.

European Wet Grasslands: Biodiversity, Management and Restoration. Edited by Chris B. Joyce and P. Max Wade.
© 1998 John Wiley & Sons Ltd.

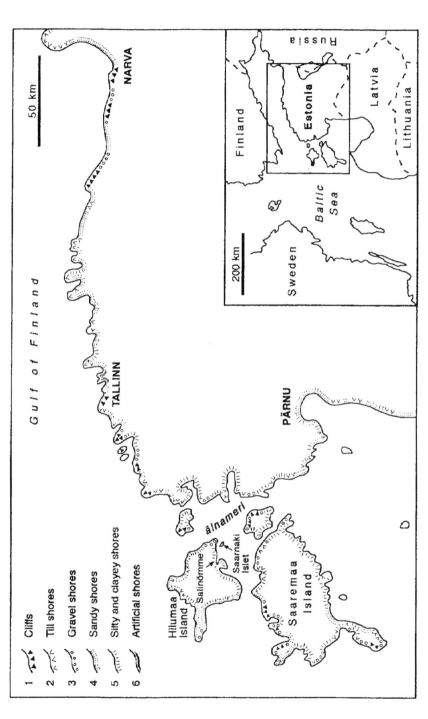

Figure 7.1 Location of the study area and distribution of shore types along the Estonian coast (From Orviku 1993)

grasslands include the following types with respect to the CORINE classification: Atlantic *Juncus gerardii* salt meadows (15.331), *Festuca rubra–Agrostis stolonifera* swards (15.333), *Carex nigra* salt meadows (15.33G) and *Salicornia europaea* swards (15.111). Also, on flat shores, communities of *Molinia caerulea* and *Sesleria caerulea* are widespread in addition to *Carex panicea* and *C. disticha*.

The phytogeography of these communities is interesting in relation to the geographical location of Estonia and there are a number of species that are on the edge of their international distribution. For example, *Salicornia europaea* and *Halimione pedunculata* are at their northeastern limits, *Sagina maritima* is at its eastern limit and *Carex distans* and *C. extensa* are at their northwestern limits (Rebassoo 1975).

Sea-shore grasslands are best developed and most widespread along Väinameri, a shallow sea between the mainland and the major islands of west Estonia (Figure 7.1). Studies were undertaken between 1985 and 1991 of the temporal and spatial changes in these grasslands on the sandy lowshores of Salinõmme (the southeastern coast of the Hiiumaa Island), on Saarnaki Islet (in the south-east of Hiiumaa), and on some other islets in Väinameri, with special reference to plant communities, subsurface water and soil (Puurmann and Ratas 1990). The landscape profile (i.e. transect) method and large-scale mapping (1:500) were used to describe the ecological and geochemical characteristics of these sea-shore grasslands. Sample points were located along the landscape profile on the basis of environmental and vegetation characteristics. These points were studied during the growing season in more detail, including:

- sampling of different soil horizons
- analysis of sea water and subsurface water
- collection of geobotanical data from 1-m^2 plots, with plant communities described on the basis of dominant species.

The littoral terminology of Du Rietz (1950) was used.

Figure 7.2 is an example of one such landscape profile, compiled for the Saarnaki Islet low shore. For further information on the methods see Puurmann and Ratas (1990). The information presented in this chapter is based on the above studies unless otherwise stated.

HYDROLOGICAL FACTORS

Important formation characteristics of Estonian sea-shore grasslands are their relatively young age, flat topography and fine-grained texture of sand and silt. However, the most important factor is the influence of sea water, which favours the growth of halophilous vegetation. This influence can occur in the shore through subsurface water, floods, waves and aerosols of salt spray (Ratas *et al.* 1988). Periodic inundation of sea water plays a decisive role in the formation of sea-shore grasslands, determining their specific moisture regime, as well as the content and distribution of chemical elements.

HYDROLOGY

There is no regular tide in the Baltic Sea but major fluctuations in sea level are caused by the seasonally changing meteorological conditions, including occasional fluctuations

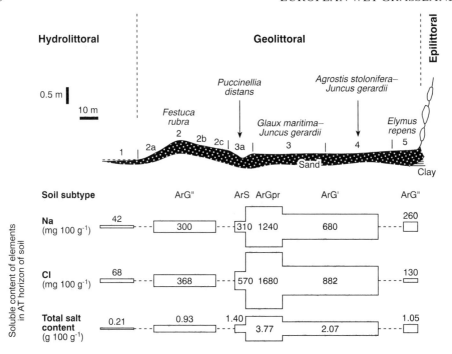

Figure 7.2 The landscape profile of the Saarnaki sea-shore grassland. 1, 2, 3, 3a, 4 and 5 are sample points. Soil subtypes: ArG″, rarely flooded saline littoral soil; ArS, salty littoral soil; ArGpr, primitive saline littoral soil; ArG′, frequently flooded saline littoral soil

induced by persistent strong winds and changes in air pressure. The annual precipitation in west Estonia is 500–550 mm, with most rainfall (70–80 mm) in August. The evaporation rate for this region is more than 400 mm yr^{-1}. The sea level is at its lowest in spring and early summer when easterly winds prevail. The salts in the soils of sea-shore grasslands are leached out by melting snow and rain. Figure 7.3 represents a typical pattern of sea-level changes over a five-month period during the growing season. Towards the end of June there is a rise in sea level and in summer short-term floods occur, extending over the lower and middle part of the shore and adding salts to the soils. Air temperature is relatively high and evaporation intense. In autumn there are characteristically frequent and prolonged sea-water floods along the whole shore due to the prevailing strong winds from the west and south-west. Variation in this seasonal pattern may occur, for example extreme low or high sea level in the summer period. Such extremes account for the long-term range of sea-level fluctuations on the coast of Väinameri of 2.5 m (Mardiste 1971).

SALINITY OF SUBSURFACE WATER

The high salinity levels of the subsurface water are one of the main properties characterizing sea-shore grasslands. The salinity of Estonian coastal waters can reach 7 g l^{-1} (Astok and Mardiste 1995). The water chemistry of the subsurface water is more variable than that of the sea water and maximum concentrations of marine elements are higher (Figure 7.4). This is likely to be due to the inflow of saline water and the relatively high

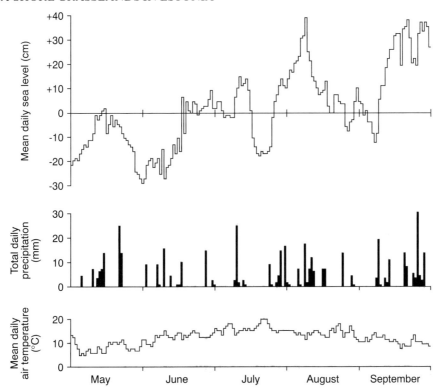

Figure 7.3   Climatic and hydrological conditions in the Väinameri coastal area during the growing season of 1987 (data of the Heltermaa Hydrometeorological Station in the south-east of Hiiumaa Island)

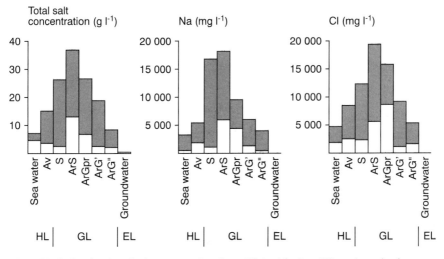

Figure 7.4   Variation in chemical content of sodium (Na), chlorine (Cl) and total salt concentration in sea water, groundwater and subsurface water on sandy lowshores of Väinameri. Shaded area indicates range of measured values. Soil subtypes: Av, underwater saline soil; S, salty patches; ArS, salty littoral soil; ArGpr, primitive saline littoral soil; ArG', frequently flooded saline littoral soil; ArG'', rarely flooded saline littoral soil. Littoral zones: HL, hydrolittoral; GL, geolittoral; EL, epilittoral

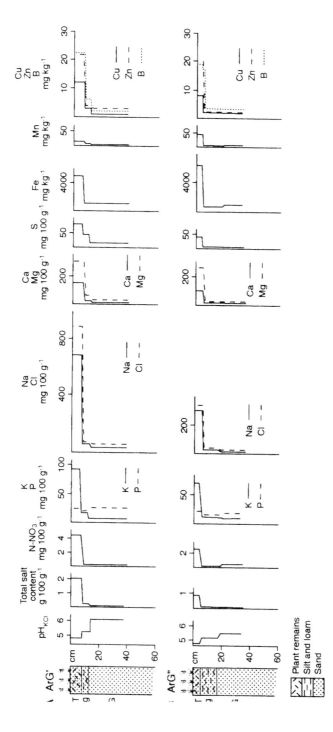

Figure 7.5 Characteristic soil profiles of saline littoral fluvisols and content of chemical elements in different soil horizons. (A) Saarnaki Island, *Agrostis stolonifera–Juncus gerardii* meadow; (B) Saarnaki Island, *Festuca rubra* meadow. Soil subtypes: ArG′, frequently flooded saline littoral soil; ArG″, rarely flooded saline littoral soil. Soil horizons: AT, humic horizon with partially decomposed plant material; Ag, humic horizon with all organic material fully decomposed; G, gley horizon

evaporation rate. During periods when the shore is inundated by sea water, the salinity of subsurface water becomes homogeneous throughout the whole shore.

SOIL PROPERTIES

The soils of west Estonian sea-shore grasslands are saline littoral soils that are young and developing. The typical soil profile consists of clearly distinguishable genetic horizons (AT, Ag, G) (Figure 7.5). The thickness of the humic horizons (AT and Ag) varies but can be up to 25 cm. They are often interrupted by thin layers of sand that have been deposited by sea and wind action. The blackish-brown AT horizon is characterized by the accumulation of partially decomposed plant material in saturated conditions, where organic matter is not tightly bound to the mineral fraction of the soil. Below the AT horizon is the Ag horizon, which is blackish-grey and of sticky texture and all organic material is fully decomposed and firmly bound to the mineral fraction. Below the AT horizon, the G horizon often occurs as two layers, i.e. clay covered by fine-grained sand.

Three different saline littoral soil subtypes can be distinguished, varying in their stages of development: primitive (ArGpr), frequently flooded (ArG') and rarely flooded (ArG"). They are characterized by high salinity and great spatial (Figure 7.2) and temporal variability of chemical content (Puurmann and Ratas 1990). The variation in the content of elements is most noticeably expressed in humic horizons due to their greater capacity to absorb organic matter. While sea water has a direct impact on the formation of soils, the saline littoral soils are characterized primarily by high content of sodium and chloride, although in every part of the geolittoral zone the variation in soil salinity is greater (Figure 7.6). However, the highest salinity is found in primitive and frequently flooded saline littoral soils.

VEGETATION

The spatial variation of the influence of sea water produces distinct hydrological and chemical gradients, which are responsible for the differentiation and usually very definite zonal arrangement of the littoral plant communities. The importance of the effects of hydrology and salinity on sea-shore vegetation variation is also stressed by Tyler (1971) and Vestergaard (1982).

The hydrolittoral zone of the west Estonian shore is usually occupied by reedbeds of predominantly *Phragmites australis*, which are comparatively species-poor. However, where the coast is more open to wave action this typical hydrolittoral vegetation is absent. There is a distinct boundary between the hydrolittoral and geolittoral vegetation zones, the geolittoral being composed of sea-shore grasslands.

In the geolittoral zone there is an intimate relationship between the frequency and duration of inundation and the spatial differentiation of vegetation. Each community occupies a definite vertical position in relation to mean sea level and the range of the main plant communities is very restricted.

Communities dominated by *Glaux maritima–Juncus gerardii* or *Eleocharis uniglumis* form a belt in the lower geolittoral zone, up to approximately 0.2 m above mean sea level.

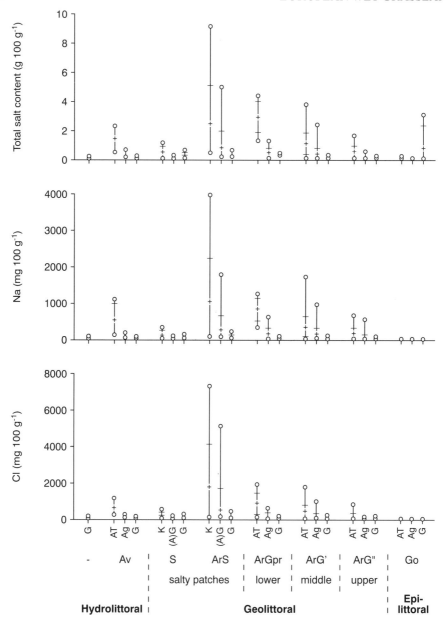

Figure 7.6 Variation in the soluble content of total salt, sodium (Na) and chlorine (Cl) in different soil subtypes on sandy lowshores of Väinameri. +, calculated mean; o, minimum–maximum; −, standard deviation. Soil subtypes: Av, underwater saline soil; S, salty patches; ArS, salty littoral soil; ArGpr, primitive saline littoral soil; ArG′, frequently flooded saline littoral soil; ArG″, rarely flooded saline littoral soil; Go, gley soil. Soil horizons: G, gley horizon; AT, humic horizon with partially decomposed plant material; Ag, humic horizon with all organic material fully decomposed; K, crust

This zone represents the first stage of vegetation development for most sandy sea-shore grasslands. These shore areas are waterlogged for long periods and the salinity of the soil is relatively high.

The most frequent community of the Estonian sea-shore grasslands is the *Juncus gerardii* community, which forms the middle geolittoral zone, 0.2–0.4 m above mean sea level. Sometimes *Agrostis stolonifera* is abundant. The zone is frequently flooded by sea water in summer and autumn and the salinity of soils is high.

Festuca rubra forms the main plant community in the upper geolittoral zone (0.4–0.8 m above mean sea level), situated further from the shoreline. This part of the sea-shore is rarely flooded and the composition of subsurface water is markedly different from sea water. The content of marine chemical elements in the soil is lower than that of the middle geolittoral zone.

Above mean high-water level, in the epilittoral zone, the *Festuca rubra* community is replaced by wet grasslands on gleyic or gleysols. The particular substrate type and soil moisture content of the area determine the kind of plant community that may develop.

The natural microtopography of the lowshores varies considerably and is characterized by low (up to 50 cm high) beach ridges and by shallow depressions (up to 10 cm in depth). This creates variation in the hydrological conditions (moisture regime) and chemical properties (salinity) within the lowshore and changes the zonal pattern of vegetation distribution into a mosaic. The depth of clay layers also has an important effect on the moisture regime, as it determines the thickness of the water-permeable layer.

In the shallow depressions, soil and vegetation develop under certain extreme influences, e.g. impeded drainage, long-term standing stagnant water, and salt accumulation. Indeed, for the greater part of the growing season these shallow depressions are usually filled with water. During dry periods, a thin (0.5 cm) reddish-brown crust of mainly organic matter and precipitated salts is formed on the sand surface, inhibiting plant growth. The middle parts of the shallow depressions – so-called salty patches – are plantless or have only sparse plant cover, usually a *Salicornia europaea* community. Around these, growing in concentric circles, are mostly annuals. At the edges of the depressions there is usually an open and species-poor *Puccinellia distans* community in which *Agrostis stolonifera* may be locally abundant. According to Tyler (1969) the vegetation of shallow depressions along the Baltic coast of Sweden is secondary, whereas in the Estonian coastal area it is probably primary vegetation.

In places along the Estonian shore where accumulation of organic debris (mainly drift litter) has been extensive, nitrophilous plant communities are likely to develop, for example those characterized by *Atriplex littoralis* and *Elymus repens*.

In some places away from the shoreline where the coast is very flat and the shore is extensive, *Carex*-rich paludified coastal grasslands on littoral gleysols have developed. In addition to *Carex* spp., *Molinia caerulea* and *Sesleria caerulea* are also widespread communities of these coastal grasslands. Halophytes occur only rarely. These grasslands have a water table above ground level for most of the year but they are not inundated by sea water and their plant communities are transitional units between the sea-shore grasslands and inland vegetation.

The distribution of vegetation on the shore may be regarded as a spatial expression of sea-shore grassland succession. As a result of land uplift, the soils and vegetation of Estonian sea-shore grasslands undergo a series of stages in their development, from hydrolittoral to epilittoral, in response to changes in the hydrological and chemical

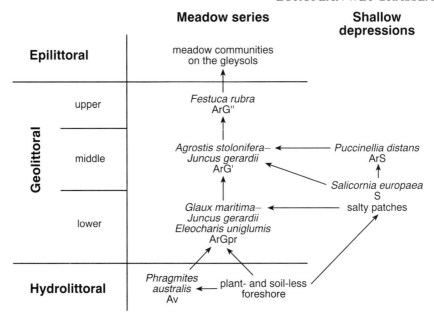

Figure 7.7 Generalized scheme of Estonian sea-shore grasslands on sandy lowshores. Arrows show the general direction of community development over time. Soil subtypes: Av, underwater saline soil; ArGpr, primitive saline littoral soil; S, salty patches, ArS, salty littoral soil; ArG', frequently flooded saline littoral soil; ArG", rarely flooded saline littoral soil

properties of the substrate over time. As these conditions change in an area, one plant community will be replaced by another. A generalized scheme of sea-shore grassland succession on sandy low shores is summarized in Figure 7.7. Hydrolittoral reedbeds or open salt pioneer communities are succeeded by dense salt meadow vegetation and the transition into terrestrial communities can follow. These changes may take place over hundreds of years.

Sea-shore grassland plant communities are intrinsically species-poor. High salinity and submerged conditions exert a powerful selective influence that relatively few species are able to survive. The diversity that does exist within this vegetation (Table 7.1) is due to the diversity of ecological conditions within the shore environment. However, Estonian sea-shore grasslands as a whole support considerable botanical diversity as many different plant communities in different developmental stages are found there.

BIRDS

Estonian sea-shore grasslands are important nesting sites for numerous birds including *Sterna paradisaea* (arctic tern), *S. hirundo* (common tern), *S. albifrons* (little tern), *Tringa totanus* (redshank), *Calidris alpina* (dunlin) and *Limosa limosa* (black-tailed godwit). Estonia is located in the east Atlantic flyway of migratory birds, and in spring and autumn large flocks of waterfowl, especially *Branta leucopsis* (barnacle goose) and *Anser anser* (greylag goose), stop to feed on the sea-shore grasslands during their migration. The activity of birds is a notable factor in the formation of the flora and plant cover of

Table 7.1 Plant cover (%) of the Saarnaki sea-shore grassland

Soil subtype	ArS		ArGpr		ArG'						ArG''			
Sampling point	2c	3a	3		2b		4		2a		2		5	
Total cover (%)	5	30	80	85	98	95	80	80	98	85	85	90	75	60
Number of species m^{-2}	6	6	6	6	7	7	6	6	11	11	8	8	15	15
Aster tripolium	0													
Salicornia europaea		0												
Spergularia salina	0	0							+	1				
Puccinellia sp.	0	0	+	+	3	2								
Glaux maritima	0		8	5	4	5	2	2					−	+
Juncus gerardii	0	0	70	65	15	25	30	35	2	5			−	2
Triglochin maritimum	0	0	+	+	+	−	+	+	+	+			−	1
Agrostis stolonifera		0	5	15	55	60	50	45	+	+	5	8	10	5
Atriplex calotheca			−	−	+	+			1	+	+	−		
Plantago maritima					25	5	−	1	1	1				
Atriplex hastata							−	−						
Trifolium fragiferum									+	+				
Poa pratensis									+	1	3	3		
Festuca rubra									20	15	2	3	20	5
Potentilla anserina									75	60	70	75	1	1
Leontodon autumnalis									1	2			+	+
Deschampsia cespitosa											+	−		
Carex nigra											5	3	1	5
Danthonia decumbens											+	+		
Plantago major													1	−
Elymus repens													40	40
Ophioglossum vulgatum													+	−
Filipendula vulgaris													+	−
Taraxacum sp.													+	−
Cirsium sp.													+	−
Equisetum arvense													−	−

The vegetation cover (%) is estimated from 1-m² quadrats.
Values for the dominant species of communities are surrounded by a continuous frame; values for important accessory species by a broken frame.
+ indicates the occurrence of a single individual of a species; − indicates species recorded in the zone, but not in the quadrat.
0 indicates area inundated by sea water and therefore impossible to determine percentage cover of individual species.
Soil subtypes: ArS, salty littoral soil; ArGpr, primitive saline littoral soil; ArG', frequently flooded saline littoral soil; ArG'', rarely flooded saline littoral soil.
See Figure 7.2 for locations of sampling points.

sea-shore grasslands. The thousands of waterfowl can, in places, destroy the vegetation. They can also enrich soils with phosphorus, nitrogen and zinc compounds (Mägi *et al.* 1995).

MANAGEMENT AS A CONSERVATION FACTOR

Animal husbandry has long been the most important agricultural sector in west Estonia. The extensive sea-shore grasslands found along the flat coasts have a long history of use for grazing by cattle, sheep and horses and, to a lesser extent, for hay-making. The sea-shore grasslands are widely known as well established but not particularly productive pastures, yielding between 0.4 and 1.5 t ha^{-1} (Aug and Kokk 1983).

Past human influence has brought about extensive alterations in the species composition of sea-shore grasslands. In the 1940s traditional farming was replaced by collectivization in Estonia. In the early years of collectivization (in the 1950s), the sea-shore grasslands in many places were overexploited, being used as grazing land for large herds of young cattle, especially on the islets. The mechanical effect of excessive trampling broke through the turf and caused the extinction of many sensitive species (Rebassoo 1975). Sea-shore grasslands are particularly sensitive to overgrazing, as soil formation and plant cover are at an early stage of development. Since 1957, however, many west Estonian islets and parts of the mainland coast have been saved from overgrazing through designation as nature protection areas. Earlier legislation gave partial protection but was not proactive in terms of conservation management.

A moderate grazing intensity, however, is of great importance to the maintenance of the characteristic vegetation of sea-shore grasslands. Not only is the floristic composition of the plant communities influenced by grazing, but some trampling compacts the humus layer and encourages the formation of a thick turf. Grazing can also ensure that the grasslands are manured at an appropriate level (Krall *et al.* 1980).

The problems of sea-shore grassland conservation have increased in recent years. Due to a decline in the market and changes in land ownership in Estonia there are too few stock animals to graze the grasslands effectively. Since 1991, decollectivization of agriculture and the privatization of land with its changes in land ownership have brought problems regarding land use and protection. A change back to family farming has now taken place, but only some of the new owners are interested in continuing traditional land-use practices. Without grazing or hay-making management, changes in sea-shore vegetation occur. The limits between the zonally arranged vegetation units become indistinct, but the most conspicuous feature of these ungrazed grasslands is that *Phragmites australis* extends inland into the upper vegetation zones. Thin reedbeds develop in the geolittoral and become so dense that the colonization of sea-shore grassland species is prevented. The expansion of reedbeds is also encouraged by the eutrophication of the Baltic Sea, which has been occurring over the last 30 years, largely due to an increase in pollution. This has resulted in the formation of extensive areas of reedbeds along Estonian shores, and some of the islets are completely covered with *P. australis*.

Thus, a key factor for preservation of biodiversity of sea-shore grasslands is the continuation of traditional land-use practices, especially regular grazing at a moderate intensity. There are still sea-shore grasslands in Estonia that are grazed. Management such as regular grazing will result in the elimination of the reedbeds, with a distinct limit being established at the transition between hydrolittoral and geolittoral zones. Grazing in early spring is of particular importance because it weakens or even kills patches of *P. australis* in the upper parts of the shore.

Estonia has recently produced new legislation for nature protection in general and for conservation in particular. This proposes regulation and management based on the principles of sustainable development and maintaining biodiversity. Estonia has joined with several international projects and agreements on conservation (e.g. CORINE, Ramsar and HELCOM (Helsinki Commission – the Baltic Marine Environment Protection Commission)) and it is proposed to designate valuable sea-shore grassland areas as

CORINE biotopes.

A management plan has been prepared and management for conservation objectives is being carried out on the sea-shore grasslands of Matsalu Nature Reserve (Lotman 1994). However, the most effective mechanism through which to preserve representative sea-shore grassland types in Estonia is to establish special management areas within which farmers are paid to manage the land in a manner sympathetic to nature. In the long term, management of sea-shore grasslands needs to be considered at the ecosystem level and appropriate management should be considered inseparable from the sustainable development of the whole region.

SUMMARY

Sea-shore grasslands are extensive in west Estonia. Studies were undertaken of the temporal and spatial changes in sea-shore grasslands on the sandy low shores on Salinõmme and Saarnaki Islet, with special reference to plant communities, subsurface water and soil. The important formation characteristics of Estonian sea-shore grasslands are their relatively young age, flat topography and the fine-grained texture of sand and silt. The spatial variation of sea water produces distinct hydrological and chemical gradients responsible for the differentiation and the usually very definite zonal arrangement of the plant communities in the littoral zone. The sea-shore grassland plant communities are relatively species-poor. The diversity that does exist within this vegetation is caused by the diversity of ecological conditions within the shore environment. A key factor for preservation of biodiversity of sea-shore grasslands is the continuation of traditional land-use practices, especially regular grazing.

Keywords: Coastal management, Estonia, Littoral soils, Sea-shore, Wet grasslands.

REFERENCES

Astok, V. and Mardiste, H. (1995) Nüüdismeri. In: Raukas, A. (ed.), *Eesti loodus*, pp. 228–237. Valgus, Tallinn.

Aug, H. and Kokk, R. (1983) *Eesti NSV looduslike rohumaade levik ja saagikus.* Eesti NSV ATK IJV, Tallinn.

Chapman, V.J. (ed.) (1977) *The wet coastal ecosystems.* Elsevier Scientific, Amsterdam.

Du Rietz, G.E. (1950) *Phytogeographical excursion to the maritime birch forest zone and the maritime forest limit in the outermost archipelago of Stockholm.* Excursion guide B1: pp. 125–172. Proceedings of the 7th International Botanical Congress, Stockholm.

Krall, H., Pork, K., Aug, H., Püss, Õ., Rooma, I. and Teras, T. (1980) *Eesti NSV looduslike rohumaade tüübid ja tähtsamad taimekooslused.* Eesti NSV PM IJV, Tallinn.

Laasimer, T. (1965) *Eesti NSV taimkate.* Valgus, Tallinn.

Lippmaa, T. (1935) Vegetatsiooni geneesist maapinna tõusu tõttu merest kerkivatel saartel Saaremaa lo01derannikul. *Acta Instituti et Horti Botanica Universitatis Tartuensis*, **4**, 2–38.

Lotman, A. (1994) Management plan for Matsalu Wetland. *WWF Baltic Bulletin,* **1**, 11–12.

Mägi, E., Ratas, U. and Puurmann, E. (1995) Landscape changes in the cormorant nesting ground of the small islet of Tondirahu. In: Mägi, E. and Kaljuste, T. (eds), *Loodusvaaatlusi 1994. Matsalu State Nature Reserve*, pp. 41–52. Tallinn. (In Estonian, summary in English.)

Mardiste, H. (1971) The Väinameri – inland sea, Estonia. Acta et Commentationes, University of Tartuensis, No. 282. *Publications of Geography*, **8**, 47–66.

Orviku, K. (1993) Contemporary coasts. In: Lutt, J. and Raukas, A. (eds), *Geology of the Estonian shelf*, pp. 29–39. Estonian Geological Society, Tallinn. (In Estonian, summary in English.)

Puurmann, E. and Ratas, U. (1990) Saline maritime soils of Estonia. *Soviet Soil Science*, **22**, 25–31.

Ratas, U., Puurmann, E. and Kokovkin, T. (1988) *Genesis of islet geocomplexes in the Väinameri (the west Estonian inland sea)*. Academy of Sciences of Estonia SSR, Tallinn.

Rebassoo, H.E. (1975) *Sea-shore plant communities of the Estonian islands, I and II.* Academy of Sciences of the Estonia SSR, Tartu.

Rebassoo, H.E. (1987) *Biocoenoses of eastern Baltic islets, their composition, classification and protection, I and II.* Valgus, Tallinn. (In Russian, summary in English.)

Tyler, G. (1969) Flora and vegetation. Studies in the ecology of Baltic sea-shore meadows II. *Opera Botanica*, **25**, 1–101.

Tyler, G. (1971) Hydrology and salinity of Baltic sea-shore meadows. Studies in the ecology of Baltic sea-shore meadows III. *Oikos*, **22**, 1–20.

Vallner, L., Sildvee, H. and Torim, A. (1988) Recent crustal movements in Estonia. *Journal of Geodynamics*, **9**, 215–223.

Vestergaard, P. (1982) Horizontal variability of some soil properties within homogeneous stands of coastal salt meadow vegetation. *Nordic Journal of Botany*, **2**, 343–351.

8   Inundation Grasslands of the Morava River, Slovakia: Plant Communities and Factors Affecting Biodiversity

VIERA BANÁSOVÁ, IVAN JAROLÍMEK, HELENA OTAHELOVÁ and MÁRIA ZALIBEROVÁ
Slovak Academy of Sciences, Bratislava, Slovakia

INTRODUCTION

The Morava River is an important tributary of the Danube, with the confluence near Bratislava (Figure 8.1). This location marks the beginning of the Devínska Brána Gate – a break valley of the Carpathians joining the Dunajská Kotlina basin (the Carpathian basin) with the Vienna basin and creating a commercially and strategically significant route between western and northern Europe and southern Europe. Human settlement dates from 5000 BC and the area has been home to Celtic and German tribes, occupied by Roman legions and, since the sixth century AD, settled by the Slavs. After the establishment of the Hungarian Empire (in the eleventh century AD), the southern section of the Morava River formed the border between the Empires of Hungary and Austria. In 1918 the river became the state border between Czechoslovakia and Austria, and since the Second World War became part of the Iron Curtain and remained so up until 1989. In this 40-year period, when floodplains throughout Europe were subjected to intensive anthropogenic transformations, the floodplains of the Morava River, lying within the Iron Curtain border zone, were almost completely closed to civil activities. This helped to protect the valuable and biodiverse semi-natural ecosystem. However, now that this border zone has been opened to the public, the area has been subjected to strong human pressure and there is an acute need to implement appropriate management in order to conserve this important area.

There are almost no botanical data available for the period during which the Morava River corridor was part of the Iron Curtain border, due to its inaccessibility. An exception is the information about the floodplain meadows, including that area on the river side of the flood embankment, i.e. the inundated area (Balátová-Tuláčková 1968, 1976). Attention was drawn to the high species diversity of this area immediately after the opening up of the border zone (Otahelová and Husák 1992; Otahelová et al. 1992; Šomšák 1992; Otahelová and Zlinská 1993; Ružičková et al. 1993; Valachovič 1993; Zaliberová et al. 1993; Feráková 1994). Such studies emphasized its conservation import-

See Glossary, p. 305, for explanation of technical terms. Scientific names of vascular plants follow Tutin, T. G. *et al.* (1964–80) *Flora Europaea* Volumes 1–5. Cambridge University Press. See p. 319.

European Wet Grasslands: Biodiversity, Management and Restoration. Edited by Chris B. Joyce and P. Max Wade.
© 1998 John Wiley & Sons Ltd.

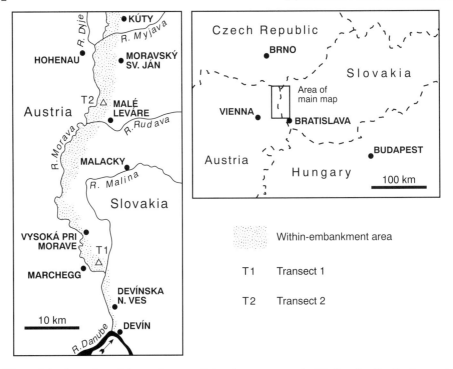

Figure 8.1 Location of the study area. Triangles correspond with the sites for the transects

ance and value. This work was followed by more detailed inventories and descriptions of the ecological characteristics of the plant communities: aquatic and wetland vegetation (Otahelová *et al.* 1994a), meadows (Ružičková 1994), psammophilous vegetation (Stanová and Šeffer 1994), shrub vegetation (Zaliberová 1994), forest communities (Jarolímek 1994) and a vegetation map of selected botanically valuable areas (Otahelová *et al.* 1995). The relationships between vegetation structure and ecological factors are described by Banásová *et al.* (1994a,b) and Otahelová *et al.* (1994b) and the ecology and conservation of the Morava River floodplain in Austria by Lazowski (1985) and Farasin and Lazowski (1990).

The aim of the present chapter is to present the results of three years of investigation into the ecological conditions of the main plant communities of those wet grasslands distributed in the Slovak floodplain of the Morava River. Attention is drawn to the biodiversity and richness of endangered plant species in this area and to the increasing human interference linked to the spread of neophytes within these communities.

CHARACTERISTICS OF THE STUDY AREA

NATURAL CONDITIONS

The study area is situated on the Slovak (eastern) side of the Morava River floodplain between the confluence of the Morava and Danube Rivers in the south and the confluence of the Dyje and Morava Rivers in the north. The river corridor is 70.8 km long and is bounded by the Morava River channel and the eastern flood embankment (Figure 8.1). This area is a part of a morphological subunit, the Dolnomoravská Niva floodplain, which is included in the Záhorská nížina lowland and Vienna basin. Its altitude varies from 135 to 152 m and the river gradient is low (0.02–0.03%). Numerous depressions remain in the floodplain as a result of former oxbows, providing evidence of the activity and migration of the river in the past.

The floodplain and its ecosystems are influenced mainly by the Morava River, the hydrological regime of which is typical of the highland rivers of central Europe. The annual average discharge at the village of Moravský Svätý Ján (Figure 8.1) is 95 $m^3 s^{-1}$. Peak runoff occurs in March. From approximately 19 km north of its confluence with the Danube, the hydrology is increasingly influenced by the Danube, with a hydrological regime more typical of alpine rivers. The annual mean discharge of the Danube at Bratislava, a few kilometres downriver of its confluence with the Morava, is 1993 $m^3 s^{-1}$, with the peak discharge usually in late June or in July.

The climate of the study area is relatively warm and dry. The highest daily average temperatures in July are above 20°C; the lowest fall to below −2°C in January. The average precipitation is 530–650 mm yr^{-1}.

Landscape diversity along the floodplain is due, in part, to the distribution of different soil types. Fluvisols (alluvial soils) are found close to the river; phaeozems (meadow soils) occur away from the river; sandier soils are typical of elevated areas; waterlogged soils with more clay (gley soils) occur in depressions in the alluvium (Račko and Bedrna 1994).

Phytogeographically, this area belongs to the Pannonian xerothermic flora. The original vegetation was composed of elm floodplain forests (*Ulmenion*), willow–poplar floodplain forests (*Salicion albae, Salicion triandrae*) and humid grasslands (*Molinion coeruleae*) (Michalko *et al.* 1987).

HUMAN IMPACT

The extent of settlement in the floodplain of the Morava River during the Sub-boreal and earlier Sub-Atlantic periods (1250–500 BC) corresponds with the present distribution of alluvium. Humans exploited this area for the grazing of cattle, hunting, fishing, timber harvesting and for seasonal agriculture (Katkinová 1994). The erosive and accumulative activity of the Morava River was limited by the construction of flood embankments from the nineteenth to the mid-twentieth century. Consequently, the width of the floodplain varies considerably, from some metres at the confluence of the Morava River with the Danube up to approximately 3 km. During the periods of high water level almost the whole within-embankment zone is flooded, although the total extent of flooding has been reduced by about 80% due to the construction of the embankments. The area outside of the flood embankments that is protected from flooding has been transformed into arable land.

A Czechoslovak–Austrian project undertaken between 1935 and 1964 to regulate the Morava River had a profound effect on the landscape. During this time 17 meanders were cut through, the river was straightened and the tributaries were also canalized, resulting generally in a gradual lowering of the Morava river bed.

Many other water courses in Europe, especially lowland rivers, were receiving similar treatment at this time. Typically, the floodplains were deforested, enabling their exploitation for agriculture. However, due to its inclusion within the Czechoslovak border zone after 1945, the eastern part of the Morava floodplain was inaccessible to the public and the only activities were a limited amount of timber harvesting and mowing of the meadows. Given that this area is near to large cities such as Bratislava and Vienna, it would otherwise have experienced intensive development from agriculture and tourism as has occurred on the Austrian side (Figure 8.1), where arable land predominates and forests and meadows are fragmented or have disappeared (Farasin and Lazowski 1990). However, due to its isolation, a valuable semi-natural ecosystem with a high biodiversity has been preserved in the within-embankment area, the most frequent habitats being meadows and wetlands.

The situation altered considerably after the political changes of 1989. The border zone on the Slovak side was re-opened, and many new environmental problems arose as a result. For example, an interest has been shown in mining gravel, and there have been efforts to make the Morava River navigable and to develop water-based leisure activities. An important issue is the restitution of land ownership and the associated problem of the new owners' desire to increase the agricultural production of their property, particularly the meadows.

METHODOLOGY

Field investigations were carried out in the within-flood embankment zone of the Morava River (Figure 8.1) during 1992–94. Two localities were chosen in this area in order to study the full range of inundated grassland communities. At each locality a transect was established across the different vegetation types ranging from the wettest to the driest habitats. The zone between two different community types is described as the transitional zone (ecotone).

The first transect (transect 1) was located south-south-west of Vysoká pri Morave village, approximately 15 km before the confluence with the Danube (16° 56′ 30″ E, 48° 17′ 00″ N) (Figure 8.1). It was situated in a riverine plain in a hydrological–tectonic depression at an altitude of 138 m, from which the remains of the Würmian terrace stand out (altitude 140 m). The area is interwoven by a network of silted river beds. The within-flood embankment zone is approximately 3 km wide. The 425-m-long transect was situated approximately 2.25 km from the Morava river channel.

The second transect (Transect 2) was located north-west of the Malé Leváre village, 58 km from the Danube (16° 57′ 10″ E, 48° 31′ 50″ N). It was situated in a riverine plain and at an altitude of 148 m. The within-flood embankment zone is relatively narrow at this point (about 1 km). This 375-m-long transect was situated 600 m from the river, lying almost parallel to the present river channel.

The main grassland types along the transects were distinguished on the basis of the dominant species. In each grassland type a borehole was dug, 1.5–2.5 m deep, to measure the changes in groundwater levels. Also at each sampling point, a 75-cm-high wooden post with a 10-cm calibration was placed. The above or below ground water level was measured manually every two weeks. In the event of high flooding, the water level was calculated from the dated records of a gauging station. Elevation was measured every 5 m along the transect by means of a theodolite.

In 1993, soil samples were taken from the rhizosphere in the different vegetation types for a depth of 0.5–10 cm (soil cores 1–6 in transect 1 and 1–11 in transect 2). The pH of the soil was measured potentiometrically, the anion (Cl^- and SO_4^{2-}) and cation (Na^+ and K^+) content was measured according to Hraško *et al.* (1962) and soil texture was established by the method of Hraško *et al.* (1962). The following soil fractions were distinguished: clay (<0.01 mm), silt (0.01–0.05 mm), fine sand (0.05–0.25 mm) and sand (0.25–2 mm).

Each transect was divided up on the basis of dominant plant species into its constituent community types (e.g. Figure 8.2). For each community type a representative section of the transect, 15 m in length, was then used to describe the characteristics of a given community. This section included, as well as the typical community, its transitional zones with neighbouring communities. The vegetation in each representative section was analysed using adjacent 1 × 1 m quadrats. In each quadrat the full list of taxa and their cover values on a 9-point scale were recorded (Barkman *et al.* 1964). Characteristics of the community noted included: a profile of the percentage cover of the main species, α diversity, (i.e. within community diversity) and equitability measured every metre along the section; a profile of microrelief; the pattern of fluctuation in water level (groundwater or floodwater) in 1992–94; and the structure of the soil based on a sampling point in or adjacent to that section of the transect. The index of species diversity was calculated from transformed percentage cover data in the 1 × 1 m quadrats using Shannon's formula with binary logarithms (Whittaker 1972). The cover data were transformed into mean percentage values. The equitability index was calculated after Pielou (1966).

The methods of the Zürich–Montpellier school (Braun-Blanquet 1964) were used for the community classification. Endangered and rare taxa of higher plants accord to Maglocký and Feráková's (1993) Red List for Slovakia.

CHARACTERISTICS OF VEGETATION TRANSECTS

TRANSECT 1

Transect 1 crossed the moderately undulating relief of the Morava River floodplain including as wide a range of vegetation types as possible. It began immediately beside the flood embankment within a woodland strongly influenced by humans and dominated by *Populus nigra* with *Phragmites australis* in the undergrowth. The community occupying the greatest part of the transect was *Alopecurus pratensis–Cnidium dubium* meadow, which alternated on the more elevated parts of the floodplain with a *Festuca nigrescens* community. In contrast, large sedges (*Carex acuta* community) occurred in the depressions, *Glyceria maxima* dominated in the deeper places, while *Scirpus lacustris* was found in the deepest ones (Figure 8.2).

The hydrological regime in this area is influenced by both the Morava and Danube Rivers. At high water levels the whole area is inundated (Plates 8.1 and 8.2). The floodplain was inundated for a short time in the spring of both 1992 and 1994 (Figure 8.3).

The timing and length of flooding and fluctuations in groundwater level have a considerable influence on pedogenesis. The soils in this area are heavy with a predominant clay fraction (gley soils) (Figure 8.4), while in the more elevated places the soil is a light, permeable, sandy loam with a lower horizon of gravelly sand (regosol) (Račko and

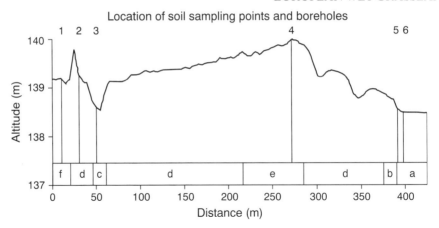

Figure 8.2   Vegetation transect 1. a, *Scirpus lacustris* community; b, *Glyceria maxima* community; c, *Carex acuta* community; d, *Alopecurus pratensis Cnidium dubium* community; e, *Festuca nigrescens* community; f, *Populus nigra* forest

Plate 8.1   Transect 1 during late summer (6.8.91)

Bedrna 1994). In the depressions, the clay horizon is situated high up in the soil, hindering the percolation of water. The highest concentrations of salts (Cl^-, SO_4^{2-}, Na^+ and K^+) are found in these depressions (Figure 8.5). The soils are generally moderately acid (pH 5.5–6.2) although in the deepest depressions they are more acidic (pH 4.8) (Figure 8.4).

   The meadows in transect 1 are mown once between the end of May and the beginning of June and sometimes there is a second cut in August (Plate 8.3). These meadows are not grazed by cattle and the reedbeds are not mown.

Plate 8.2   Transect 1 after flooding (28.8.91)

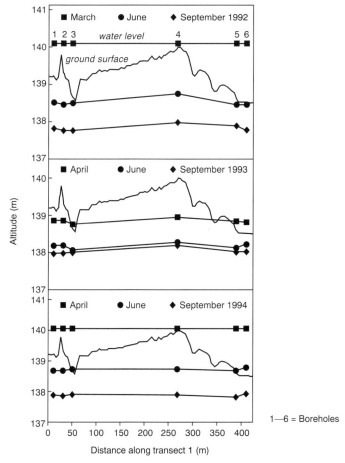

Figure 8.3   Water level at the beginning, middle and end of the vegetation season in 1992–94 along transect 1

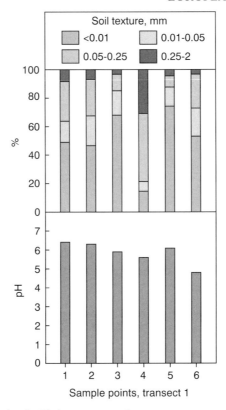

Figure 8.4   Soil texture and soil pH along transect 1

TRANSECT 2

Transect 2 begins at one end of an oxbow, passes through an aggradation wall of the former river bed and through the undulating surface of the alluvial plain, in the depressions of which are remainders of the ancient migrating river bed. The transect ends at the other end of the oxbow (Figure 8.6).

   At the beginning of the transect the oxbow supports irregular patches of emergent vegetation and either an extensive cover of *Lemna minor* and *Spirodela polyrhiza* or amphibious plant communities including *Rorippa amphibia*. The erosion margin bordering the oxbow is dominated by a *Populus tremula* woodland and the most elevated sections of the transect are occupied by meadows of *Alopecurus pratensis* and *Cnidium dubium*. The lower parts support the hygrophilous species *Carex acuta, C. riparia* and *Glyceria maxima*, with *Rorippa amphibia* occupying those deep depressions which retain water for relatively long periods. In the middle of the transect, microdepressions with no outlet contain meadow with *Potentilla anserina.* At the other end of the oxbow, a *Phragmites australis* community occurs along the accumulation littoral with a *Rorippa amphibia* community growing on the emerging bed of the oxbow. The water surface is covered by aquatic vegetation, being dominated by *Trapa natans.*

   The majority of the transect is regularly flooded in spring by sustained inundations (Figure 8.7; Plate 8.4). The soils are heavy and clayey, with an elevated horizon of grey

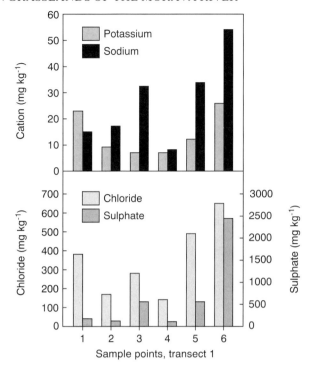

Figure 8.5 Soil cation (K^+, Na^+) and anion (Cl^-, SO_4^{2-}) content along transect 1

Plate 8.3 The meadows in transect 1 are regularly mown once or twice a year (14.6.96)

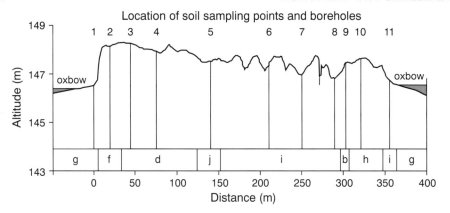

Figure 8.6 Vegetation transect 2. b, *Glyceria maxima* community; d, *Alopecurus pratensis–Cnidium dubium* community; f, *Populus tremula* forest; g, aquatic vegetation; h, *Phragmites australis* community; i, *Rorippa amphibia* community; j, *Potentilla anserina* community

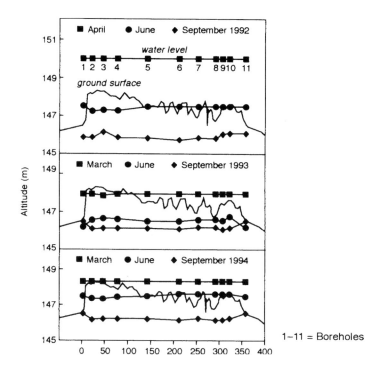

Figure 8.7 Water level at the beginning, middle and end of the vegetation season in 1992–94 along transect 2

clay (Figure 8.8), although on the aggradation wall there are also loamy clay soils. Waterlogged habitats exhibited increased levels of salts (Cl^-, SO_4^{2-}, Na^+ and K^+) (Figure 8.9). The soils have a moderately acid pH (5.6–5.8), being more acid in the depressions (pH 3.9–4.9) (Figure 8.8).

The meadows with communities of *Alopecurus pratensis–Cnidium dubium* and *Poten-*

Plate 8.4 Late spring flooding in the wet meadows of transect 2 (15.6.95)

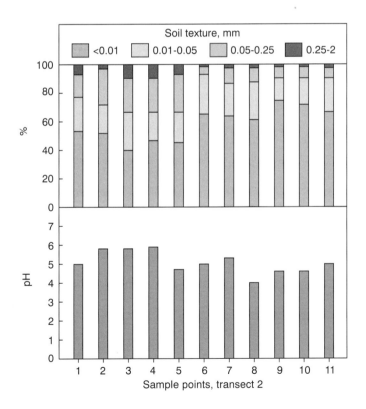

Figure 8.8 Soil texture and soil pH along transect 2

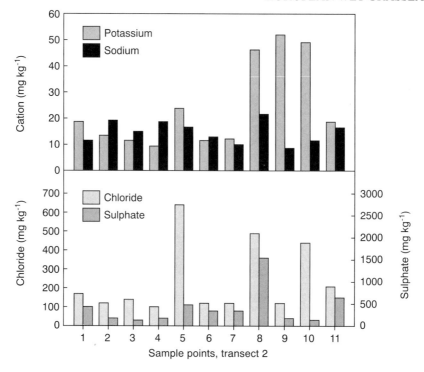

Figure 8.9   Soil cation (K^+, Na^+) and anion (Cl^-, SO_4^{2-}) content along transect 2

tilla anserina are mown in May/June and occasionally in August/September. Except for the stands of *Phragmites australis*, the other wetland vegetation is mown once a year, in August or September, depending upon the duration of flooding.

Together the two transects include the range of wet grasslands found in the lower Morava River floodplain. Whilst transect 1 included the driest type of meadows (*Festuca nigrescens* community), the diversity of wetland communities was higher in transect 2. This reflects the difference in frequency and duration of flooding and in the nature of the soils. For example, the lighter sandy loam soils which experienced less flooding were only found in transect 1.

ECOLOGY OF GRASSLANDS COMMUNITIES

SCIRPUS LACUSTRIS COMMUNITY

This community grows sparsely and in small patches in oxbows and depressions that are inundated from winter to June. In these areas the summer groundwater level fluctuates around −0.5 m so that the deep network of *Scirpus lacustris* rhizomes is well supplied with water during the whole growing period. *S. lacustris* represents the pioneer stage in the silting process of standing waters. The soils are clay (Figure 8.10) and retain the spring floodwater and rainwater for a long time.

The stands are monodominant with few other species. The α diversity and equitability

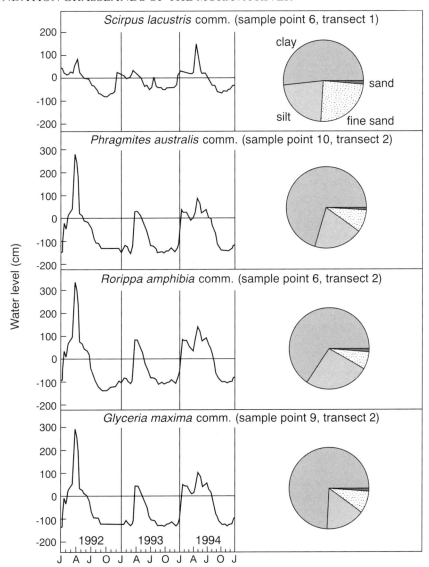

Figure 8.10 Groundwater level in 1992–94 and soil texture in communities of *Scirpus lacustris*, *Phragmites australis*, *Rorippa amphibia* and *Glyceria maxima*

are very low (Figure 8.11). Except for *Scirpus lacustris,* only *Glyceria maxima* occurs constantly with a low cover and usually forms the contact (adjacent) community (Figure 8.11). The cover of species such as *Rorippa amphibia, Polygonum amphibium* and *Galium palustre* is low. Other marsh species (e.g. *Oenanthe aquatica, Iris pseudacorus, Sium latifolium, Butomus umbellatus* and *Carex acuta*) only occur in this community occasionally, due to the competition from dominant species. Synusia of the lemnids (*Lemna minor* and *Spirodela polyrhiza*) appear for a short period.

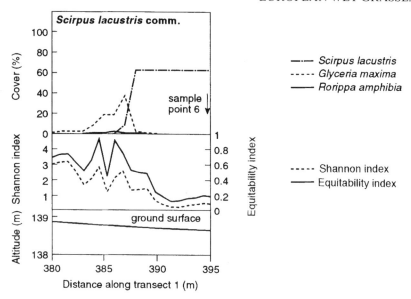

Figure 8.11 Characteristics of the *Scirpus lacustris* community and its transitional zone in transect 1 (sample point 6): percentage cover of main species; diversity and equitability indices; and microrelief

This community belongs to the association *Scirpetum lacustris* Chouard 1924 (CORINE biotope type 53.12 *Scirpus lacustris* beds).

PHRAGMITES AUSTRALIS COMMUNITY

The *Phragmites australis* community is the most frequent of the wetland communities in the study area. It grows in large beds in the accumulation zone of rivers or cut-off meanders and oxbows with open water, for example in the silted river beds of oxbows and in depressions (Figure 8.6). Such habitats are regularly inundated in spring and have a variable hydrological regime; periods when the soil is waterlogged alternate with those when the soil is dried out. The community tolerates both winter and summer drying and the groundwater levels at sampling point 10 (transect 2) fell to about −1.5 m (Figure 8.10). The soils are clayey (Figure 8.10) with moderately high salinity.

These stands are generally species-poor, being dominated by *Phragmites australis*, which reproduces mostly vegetatively by offshoots from the rhizome. Species diversity can be higher due to the frequent occurrence of *Calystegia sepium*, *Urtica dioica* and *Rorippa amphibia*, along with the accessory species *Carex riparia*, *Phalaris arundinacea*, *Symphytum officinale*, *Lysimachia vulgaris*, *Solanum dulcamara*, *Galium palustre*, *Myosotis scorpioides*, *Euphorbia palustris* and *Lythrum salicaria*. Due to the occurrence of these species the α diversity is higher than that of the *Scirpus lacustris* community (Figure 8.12).

The community belongs to the phytocoenological unit *Phragmitetum australis* von Soó 1927 (CORINE biotope type 53.11 *Phragmites australis* beds).

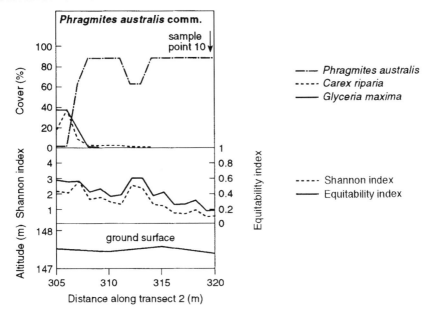

Figure 8.12   Characteristics of the *Phragmites australis* community and its transitional zone in transect 2 (sample point 10): percentage cover of main species; diversity and equitability indices; and microrelief

RORIPPA AMPHIBIA COMMUNITY

This is a very frequent community in the inundation area of the Morava River, which occurs as mosaic or linear stands on the denuded bottoms of oxbows and the shorelines of cut-off meanders. The development and structure of the community exhibit a close relationship with the hydrological regime. The stand in which water-level dynamics were studied was inundated predominantly in spring and summer. In the early summer the water level was +0.8 m falling rapidly to −0.8 m (Figure 8.10). Soil texture is mainly clay with a notable amount of silt (Figure 8.10) and there is a stratum of grey clay in the upper soil horizon. This community is a result of an unstable hydrological regime with a variable microrelief (Figure 8.6). *Rorippa amphibia* occurs constantly and dominates in the bottom of microdepressions. Accessory species are represented predominantly by the amphiphytes *Oenanthe aquatica* and *Polygonum amphibium*. There are also striking patches of tall plants (*Sparganium erectum*, *Phalaris arundinacea* and *Butomus umbellatus*) occurring in relatively elevated locations. The range of the accessory species is relatively wide: *Bidens frondosa*, *B. tripartita*, *Plantago major*, *Alisma plantago-aquatica* and *Rumex maritimus*. The contact zone is usually characterized by a *Glyceria maxima* community (Figure 8.6). The α diversity and equitability in well-developed stands are among the lowest in the communities studied (Figure 8.13).

   This phytocoenological unit includes the communities of the alliance *Oenanthion aquaticae* Hejný ex Neuhäusl 1959 (CORINE biotope type 53.146 *Oenanthe aquatica–Rorippa* communities).

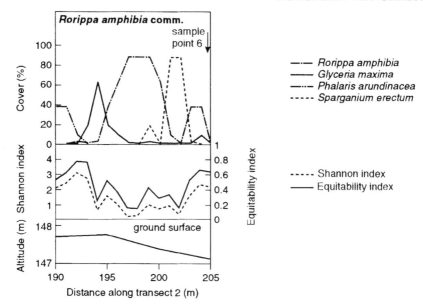

Figure 8.13 Characteristics of the *Rorippa amphibia* community and its transitional zone in transect 2 (sample point 6): percentage cover of main species; diversity and equitability indices; and microrelief

GLYCERIA MAXIMA COMMUNITY

This community occurs abundantly in the study area. It colonizes large areas in depressions or as a part of the hydroseries of marsh communities in the littoral zone, habitats characterized by an unstable water regime. Such habitats are regularly inundated in spring but in May or June the water level drops quickly to about −1.4 m (Figure 8.10). Balátová-Tuláčková *et al.* (1993) observed water-level fluctuations in this community from +0.7 m to −1.85 m in south Moravia. The soils are heavy with a large clay fraction (Figure 8.10) and deposits of organic detritus are found on the soil surface.

The nominate species, *Glyceria maxima,* is dominant in these homogeneous stands and the number of accessory species is low (Figure 8.14). *Polygonum amphibium* occurs regularly, but with low cover. Other associated species are *Oenanthe aquatica, Rorippa amphibia, Galium palustre, Leucojum aestivum, Sparganium erectum* and *Butomus umbellatus*. Patches of *Ranunculus aquatilis* occur regularly in spring. Therophytes such as *Bidens frondosa, B. tripartita, Chenopodium rubrum* and *C. polyspermum* occur frequently, especially in the transitional zones. In this community the diversity and equitability indices are very low (Figure 8.14). Towards moister parts, the community borders with the *Rorippa amphibia* community. Stands of *Glyceria maxima* are usually mown in August.

This community belongs to the association *Glycerietum aquaticae* Hueck 1931 (CO-RINE biotope type 53.15 *Glyceria maxima* beds).

CAREX ACUTA COMMUNITY

This community is very frequent in the bottom of shallow depressions or along the margins of oxbows, habitats inundated from winter until April/May (Figure 8.15).

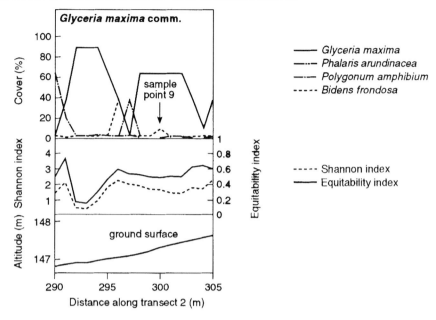

Figure 8.14 Characteristics of the *Glyceria maxima* community and its transitional zone in transect 2 (sample point 9): percentage cover of main species; diversity and equitability indices; and microrelief

Fluctuations in water level are less than in the preceding community and the groundwater level did not drop below −1 m during this study (Figure 8.15). The soil is clayey (Figure 8.15) and a 3-cm-thick layer of organic detritus covered the soil surface.

Diversity is low (Figure 8.16). The dominant species, *Carex acuta,* is accompanied in the wetter sites by *Glyceria maxima* and *Iris pseudacorus.* Accessory species include *Ranunculus repens, Galium palustre, Rorippa amphibia, Polygonum amphibium, Symphytum officinale, Carex vesicaria* and *Phalaris arundinacea.* In the drier parts of the transect, this community alternates with the more diverse *Alopecurus pratensis–Cnidium dubium* community (Figure 8.16).

This community belongs to the association *Caricetum gracilis* (Almquist 1929) R. Tx. 1937 (CORINE biotope type 53.2121 *Carex acuta* beds).

POTENTILLA ANSERINA COMMUNITY

This mesophilous mown meadow type is not widely distributed in the study area. It occurs in the form of small patches on the inundated bottoms of microdepressions and is characterized by trailing hemicryptophytes. In transect 2, the community covered the bottom of a shallow basin without outflow (Figure 8.6). Inundation occurred annually in spring, the site remaining flooded at least until the end of May, but at the beginning of June the groundwater level dropped markedly to −1.5 m (Figure 8.15). The upper soil horizon (5 cm) is loamy clay under which is a clay horizon (Figure 8.15). The soil texture of this community is similar to that of the *Alopecurus pratensis–Cnidium dubium* community. However, the habitats of the *Potentilla anserina* community experience longer

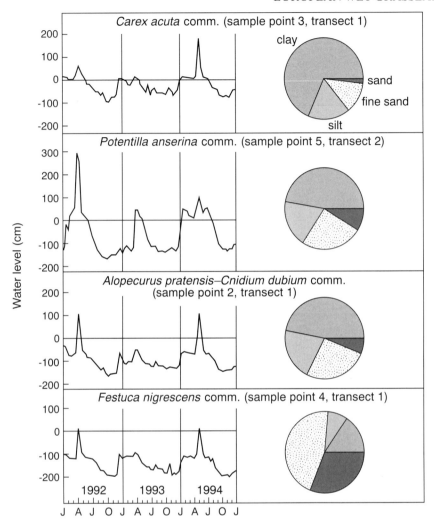

Figure 8.15 Groundwater level in 1992–94 and soil texture in communities of *Carex acuta*, *Potentilla anserina*, *Alopecurus pratensis–Cnidium dubium* and *Festuca nigrescens*

periods of inundation and higher concentrations of salts in the soils, manifested in summer by a white cover of salt deposits on the soil surface.

The community does not form closed stands and is floristically relatively rich. It includes species such as *Potentilla anserina*, *Ranunculus repens* and *Agrostis stolonifera*. *Veronica scutellata*, *Plantago major* and other species achieve only a low percentage cover, reflected by high diversity and equitability indices (Figure 8.17). Facultative halophytes are well represented and, in addition to those species already mentioned, *Eleocharis palustris*, *Mentha pulegium* and *Inula britannica* occur in these stands. The helophytes *Glyceria maxima*, *Alisma lanceolatum* and *A. plantago-aquatica* also occur frequently.

This community borders with that of the *Alopecurus pratensis–Cnidium dubium* com-

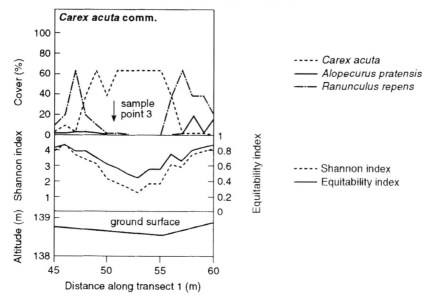

Figure 8.16 Characteristics of the *Carex acuta* community and its transitional zone in transect 1 (sample point 3): percentage cover of main species; diversity and equitability indices; and micro-relief

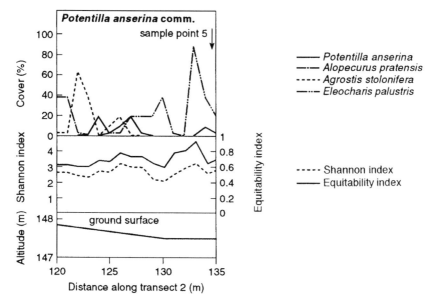

Figure 8.17 Characteristics of the *Potentilla anserina* community and its transitional zone in transect 2 (sample point 5): percentage cover of main species; diversity and equitability indices; and microrelief

munity, which grows in the more elevated parts of the transect (Figure 8.6). *Potentilla anserina* and *Eleocharis palustris* pass through the contact zone together, forming a physiognomically striking stand.

The *Potentilla anserina* community belongs to the alliance *Potentillion anserinae* R. Tx. 1947 (CORINE biotope type 37.242 *Agrostis stolonifera* and *Festuca arundinacea* swards).

ALOPECURUS PRATENSIS–CNIDIUM DUBIUM COMMUNITY

This community forms the most extensive meadows in the study area. The habitat is characterized by inundations (or at least by high groundwater levels) at the beginning of the growing season followed by a dry period. During this study the soils were water-logged until the end of April. In summer the groundwater level dropped to −1.5 m, and after a time the soil cracked. The soil is a loamy clay, with a relatively high proportion of fine sand, and the upper horizon has an abundance of roots (Figure 8.15).

This community exhibits the highest diversity of all the floodplain communities studied (Figure 8.18). A number of species exhibit a similar cover value, which results in a high equitability index. The grasses *Alopecurus pratensis, Poa angustifolia, Agrostis stolonifera* and *Elymus repens* dominate. *Cnidium dubium, Gratiola officinalis, Carex praecox, Galium boreale, Veronica longifolia, Serratula tinctoria, Inula salicina* and *Clematis integrifolia* are accessory species. Geophytes are represented by *Allium angulosum* and *Iris sibirica. Agrostis stolonifera, Gratiola officinalis* and *Symphytum officinale* occur in the depressions. These meadows provide a high-quality animal feed resource and are mown once, twice or occasionally three times a year.

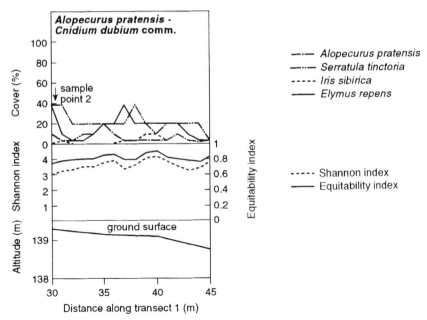

Figure 8.18 Characteristics of the *Alopecurus pratensis–Cnidium dubium* community and its transitional zone in transect 1 (sample point 2): percentage cover of main species; diversity and equitability indices; and microrelief

The grasslands belong to the communities of the alliance *Cnidion venosi* Bal.-Tul. 1966 (CORINE biotope type 37.23 Subcontinental *Cnidium* meadows).

FESTUCA NIGRESCENS COMMUNITY

This was the most xerophilous type of meadow community in the study area. It occurred on the most elevated parts of the fluvial relief, such as on river terraces and aggradation walls. Inundations here are very short and irregular (Figure 8.15). The moist soil horizon dries quickly and the groundwater level falls to just below −2 m in the autumn (Figure 8.15). In summer, soil moisture is dependent upon precipitation. The soils are light, sandy loams with a relatively low proportion of clay but with a high proportion of sand and fine sand (Figure 8.15).

The open horizontal structure of these stands allows the existence of a number of other species with some cover value even under the predominance of *Festuca nigrescens*. This is reflected in the intermediate values of diversity and equitability indices (Figure 8.19).

Serratula tinctoria, Galium verum, Colchicum autumnale, Sanguisorba officinalis and *Ranunculus auricomus* occur regularly in the stands. In spring, the community is indicated physiognomically by the white flowers of *Ornithogallum orthophyllum.* The characteristic species of drier meadows (*Arrhenatheretum elatioris* community), such as *Leucanthemum vulgare, Achillea millefolium, Centaurea jacea* and *Trifolium dubium,* are present. Hemicryptophytes predominate and geophytes are represented sporadically. The meadows are mown once or twice a year, and are not grazed.

This community belongs to the association *Serratulo–Festucetum commutatae* Bal.-

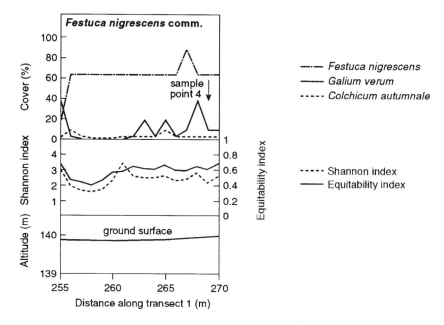

Figure 8.19 Characteristics of the *Festuca nigrescens* community and its transitional zone in transect 1 (sample point 4): percentage cover of main species; diversity and equitability indices; and microrelief

Tul. 1966 (CORINE biotope type 37.31 *Molinia caerulea* meadows and related communities).

ENDANGERED AND RARE PLANT SPECIES

A total of 540 species of higher plants has been recorded in the Morava River floodplain. Species of different phytogeographical regions occur: boreal (continental) species such as *Inula salicina*, *Veronica longifolia* and *Barbarea stricta* grow together with submediterranean (subcontinental) species such as *Plantago altissima*, *Viola pumila* and *Carex melanostachya,* and continental (euroasiatic) species such as *Cnidium dubium* and *Erysimum diffusum* are frequent. The warm climate enables the Pannonian endemic, *Dianthus pontederae*, to grow in the drier habitats such as flood embankments or terraces.

Biodiverse habitats of the lowland alluvia that have become rare in other parts of Europe have survived in this area. Consequently, a relatively large number of endangered and rare species are found here, representing more than 12% of the total number of species recorded. The relationships of the endangered and rare plant species to various habitats are presented in Figure 8.20. This shows that grasslands support an important proportion of such species. These include *Allium angulosum, Cerastium dubium, Clematis integrifolia, Gentiana pneumonanthe, Gratiola officinalis, Lathyrus palustris, L. pannonicus* subsp. *pannonicus, Leucojum aestivum, Iris sibirica, Molinia caerulea, Scirpus holoschoenus, Silaum silaus, Thalictrum flavum, T. lucidum, Veronica catenata, V. scutellata* and *Viola pumila.*

The grassland ecosystem, besides its value for agricultural production, has great significance in the protection of the gene pool of plants. The increasing efforts to improve the agricultural productivity of the meadows by fertilization represent a threat through the gradual increase in dominance of highly productive grasses like *Alopecurus pratensis* and a decrease in other species, including many rare ones. Ružičková (1994) found that *Clematis integrifolia* tolerates moderate fertilization, but *Cnidium dubium* is less tolerant and species like *Allium angulosum, Iris sibirica, Gratiola officinalis, Lathyrus palustris* and *Thalictrum flavum* decline after fertilization. These important observations need to be considered when developing management methods.

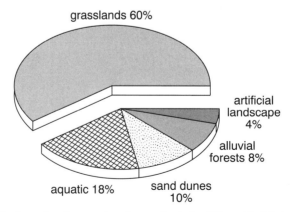

Figure 8.20 Distribution of endangered and rare plants in different habitats in the within-flood embankment area of the Morava River

HUMAN INFLUENCE ON VEGETATION

River floodplains have a long history of intensive use by humans. The Morava River floodplain had already been settled in the sub-boreal period (Katkinová 1994). Human activities resulted in a gradual synanthropization of the biota, including the vegetation. In terms of the physiognomy of the vegetation there has been a gradual change in the proportions of forest and non-forested areas in favour of the latter. At present, the majority of the study area is covered by herbaceous vegetation.

Human activities have also caused a significant shift in the species pool. A total of 143 alien plant taxa (80 archaeophytes and 63 neophytes) has been recorded in the narrowest part of the floodplain between the villages Devín and Devínská Nová Ves (Figure 8.1), which is the stretch most exposed to human activity (Feráková 1994). These adventive species contribute to the γ diversity (i.e. the total species diversity) of the region but wtih regard to the preservation of the original vegetation they are undoubtedly a negative influence. This is especially true of the neoindigenophytes which, in contrast to the other anthropophytes, are distributed mainly in semi-natural or natural plant communities and can limit the occurrence of truly indigenous species.

These invasive plant species, which due to their high competitive ability displace competitively weaker indigenous species, are especially damaging. The majority of neoindigenophytes in the Morava floodplain are invasive. The most widely distributed are *Acer negundo*, *Aster novi-belgii*, *Bidens frondosa*, *Echinocystis lobata*, *Helianthus decapetalus*, *Impatiens parviflora* and *Xanthium strumarium*. Less frequent are *Ailanthus altissima*, *Erigeron annus*, *Impatiens glandulifera*, *Solidago canadensis* and *S. gigantea*. There are also some indigenous taxa such as *Cirsium arvense* and *Urtica dioica* that have become invasive as a consequence of anthropogenically altered conditions.

The majority of the invasive species spread mainly in the forest communities and along river banks. In the Morava floodplain the river bank represents the main corridor for the spread of the propagules of many invaders. *Cirsium arvense*, *Urtica dioica* and *Aster novi-belgii* are also relatively frequent in the meadow communities.

THE PROTECTION OF BIODIVERSITY

The within-flood embankment area of the Morava River has an exceptional status among the floodplains of European lowland rivers. In spite of the fact that it has been exploited and transformed by human activities for a considerable time, its high value in terms of species diversity, range of habitats and landscape integrity have been preserved. In addition to aquatic habitats and the remnants of floodplain forests, the grasslands are maintained as important semi-natural habitat.

The semi-natural grasslands in the within-flood embankment area of the Morava River are considered extremely important habitats. The critical ecological factor for their existence is regular flooding which represents a resource of moisture and nutrients. Together with other ecological factors (e.g. relief, soils and management) this creates conditions for the occurrence of a wide range of various meadow communities. These meadows occupy almost 20 km^2 and only a small area has been converted into arable land. Among the total of 540 plant species recorded in the study area, the highest number of the rare or endangered species has been found in the grasslands themselves. These semi-natural grassland communities represent a reserve of biological information, stored in the genotypes of the different plants, animals and soil micro-organisms. The meadow stands in particular function as a refugium of biological diversity within this landscape (Rychnovská et al. 1985).

Various animal species are closely bound to the unique grassland communities. The grasslands are an important nesting place for many birds and provide a food resource for many other animals, including the thousands of invertebrate species and many amphibian species which occur there. Due to the general gradual decrease in area of these habitats, numbers of these animals are declining and many are endangered species (Kalivodová et al. 1991).

The Morava floodplain meadows in the Austrian territory have mostly been converted into arable land or reforested, and meadows have only been preserved in small areas (Farasin and Lazowski 1990). The significance of the Slovak side of the Morava River therefore increases due to the fact that it represents a rich food resource for the birds nesting on the Austrian part of the floodplain and also for the migratory species using this area (Kalivodová et al. 1994).

Following the opening of the Morava area for civil activities after 1989, pressure for its intensive exploitation is increasing. This is represented above all by recreation and commercial interests, and an increase in traffic. This could result in a considerable reduction of the species diversity in this area, the extinction of endangered species and communities, and the spread of neophytes and synanthropic vegetation.

At present, the territory is included in the Protected Landscape Area, Záhorie. At the international scale, protection has been guaranteed since 1993 by including the area in the Ramsar list. In addition, there are efforts to create a trilateral (i.e. Slovakia, Austria, Hungary) 'national' park, Podunajsko, which would include the floodplain on both sides of the Morava River. However, it is recognized that preservation of the grassland diversity in the within-embankment area of the Morava River is possible only under the following conditions:

- maintenance of the inundation regime
- prevention of a lowering of the groundwater level
- regular mowing of meadows (once or twice a year), but respecting the bird breeding season
- limitation of fertilization
- regulation of recreation activities (e.g. tourism, fishing, boating and construction of recreation facilities)
- regulation of commercial activities (e.g. mining of gravel, construction of roads, bridges and moorings)

ACKNOWLEDGEMENTS

This research was supported by the Slovak Academy of Sciences, Grant No. 2/1177/95. The authors are greatly indebted to Dr Šustek for English improvement and critical comments and Mrs B. Wolfová for technical help.

SUMMARY

The Slovak floodplain of the Morava River was closed to civil activities for political reasons for about 40 years up to 1989. This helped to protect the valuable semi-natural ecosystem with a high biodiversity that forms the regularly inundated (within-embankment) part of the area. Between 1992 and 1994, the main inundation grassland communities, *Scirpus lacustris, Phragmites australis, Rorippa amphibia, Glyceria maxima, Carex acuta, Potentilla anserina, Alopecurus pratensis–Cnidium dubium* and *Festuca nigrescens*, were investigated at two localities in the within-embankment area. At each locality a transect was established crossing the different vegetation types ranging from the wettest to the driest habitats. This enabled the communities to be characterized by floristic composition, percentage cover of main species, diversity and equitability indices, and by ecological conditions such as groundwater level, dynamics, elevation, soil properties, and management. Attention is drawn to the biodiversity and richness of endangered plant species in the Slovak Morava area and to the increasing human interference linked to the spread of neophytes. Measures are proposed for the preservation of biodiversity in the within-embankment area of the Morava River.

Keywords: Biodiversity, Ecological conditions, Groundwater level, Human influence, Inundation grasslands, Morava River, Plant communities, Soil properties.

REFERENCES

Balátová-Tuláčková, E. (1968) Grundwasserganglinien und Wiesengesellschaften. *Acta Scientiarum Naturalium, Brno*, **2**, 1–37.

Balátová-Tuláčková, E. (1976) *Rieder- und Sumpfwiesen der Ordnung Magnocaricetalia in der Zahorie-Tiefebene und dem nordlich angrenzenden Gebiete.* Vegetácia ČSSR B3. Verlag d. SAW, Bratislava.

Balátová-Tuláčková, E., Mucina, L., Ellmauer, T. and Wallnîer, S. (1993) *Phragmiti-Magnocaricetea.* In: Grabher, G. and Mucina, L. (eds), *Die Pflanzengesellschaften österreichs, Teil II*, pp. 79–130. Fischer Verlag, Jena.

Banásová, V., Otahelová, H., Jarolímek, I., Zaliberová, M. and Husák, Š. (1994a) Morava river floodplain vegetation in relation to limiting ecological factors. *Ekológia, Bratislava*, **3**, 247–262.

Banásová, V., Otahelová, H., Jarolímek, I., Zaliberová, M., Janauer, G.A and Husák, Š. (1994b) The influence of important environmental factors on the vegetation structure in the alluvial plain of the Morava River. *Ekológia, Bratislava*, Suppl. 1, 115–133.

Barkman, J.J., Doing, H. and Segal, S. (1964) Kritische Bemerkungen und Vorschläge zur quantitativen Vegetationsanalyse. *Acta Botanica Neerlandica*, **13**, 394–419.

Braun-Blanquet, J. (1964) *Pflanzensoziologie, Gründzuge der Vegetationskunde*, 3rd edn. Springer Verlag, Wien.

Farasin, K. and Lazowski, W. (1990) *Verteilung und Veränderung der Vegetation und Biotopstruktur in den unteren Marchauen.* Ramsar, Bericht 1, pp. 161–198. Umweltbundesamt, Wien.

Feráková, V. (1994) Floristic remarks to the lowest part of Morava River floodplain area with special attention to naturalization of neophytes. *Ekológia, Bratislava*, Suppl. 1, 29–36.

Hraško, J., Červenka, C., Facek, Z., Komár, J., Němeček, J., Pospíšil, F. and Sirový, V. (1962) *Soil analyses.* (In Slovak.) Slovenské vydavatelstro pôdohospodárskej literatúry, Bratislava.

Jarolímek, I. (1994) Contribution to knowledge of forest communities along Morava River. *Ekológia, Bratislava*, Suppl. 1, 115–124.

Kalivodová, E., Ružičková, H. and Kozová, M. (1991) Does the nature of the River Morava get a chance? (In Slovak.) *Životné prostredie, Bratislava*, **25**, 191–194.

Kalivodová, E., Feriancová-Masárová, Z. and Darolová, E. (1994) Birds of the floodplains of the River Morava (March). *Ekológia, Bratislava*, Suppl. 1, 189–201.

Katkinová, J. (1994) Settlement of alluvium of the River Morava in the Urnfield and Hallstatt periods in dependence of natural conditions. *Ekológia, Bratislava*, Suppl. 1, 15–20.

Lazowski, W. (1985) Altwässer in den Augebieten von March und Thaya mit einer Gegenüberstellung der Donau-Altwässer, In: Gepp, J. (ed.), *Auengewässer als Ökozellen*, Grüne Reihe, Vol. 4, pp. 159–222. Bundesministerium für Gesundheit und Umweltschutz, Wien.

Maglocký, S. and Feráková, V. (1993) Red List of ferns and flowering plants (*Pteridophyta* and *Spermatophyta*) of the flora of Slovakia (the second draft). *Biológia, Bratislava*, **4**, 361–385.

Michalko, J., Magic, D., Berta, J., Rybníček, K. and Rybníčková, E. (1987) *Geobotanical map of ČSSR*. Veda, Bratislava.

Otahelová, H. and Husák, Š. (1992) A contribution to knowledge of the flora of Morava River. (In Slovak.) *Bulletin Slovenskej botanickey spoločnosti, Bratislava*, **14**, 36–42.

Otahelová, H. and Zlinská, J. (1993) *Lindernia procumbens* (Krock.) Philcox im Marchüber-schwemmungsgebiet in der Slowakei. *Biológia, Bratislava*, **1**, 61–5.

Otahelová, H., Banásová , V., Jarolímek, I., Husák, Š., Zaliberová, M. and Zlinská, J. (1992) On the occurrence of endangered taxa of Slovak flora on the flood-plain of lower flow of Morava River. (In Slovak.) *Bulletin Slovenskej botanickey spoločnosti, Bratislava*, **14**, 34–35.

Otahelová, H., Janauer, G.A. and Husák, Š. (1994a) Beitrag zur Wasser- und Sumpfvegetation Marchinundationsgebiet (Slowakei). *Ekológia, Bratislava*, Suppl. 1, 43–54.

Otahelová, H., Banásová, V., Jarolímek, I., Zaliberová, M., Otahel, J., Feranec, J. and Husák, Š. (1994b) Vegetation and ecological conditions in the floodplain of the Morava River (Slovakia). In: Aubrecht, G., Dick, G. and Prentice, C. (eds), *Monitoring of ecological change in wetlands of middle Europe*. Proceedings of an International Workshop, Linz, Austria, 1993. *Stapfia*, **31**, 121–127.

Otahelová, H., Banásová, V., Jarolímek, I., Zaliberová, M., Janauer, G.A., Otahel, J. and Feranec, J. (1995) Vegetation units of the Morava River floodplain ecotones area. *Biológia, Bratislava*, **50**, 367–375.

Pielou, E.C. (1966) The measurement of diversity in different types of biological collections. *Journal of Theoretical Biology*, **13**, 131–144.

Račko, J. and Bedrna, Z. (1994) Soils in the floodplain of lower Morava. *Ekológia, Bratislava*, Suppl. 1, 5–13.

Ružičková, H. (1994) Wiesenvegetation des Inundationsgebietes des Unterlaufes des March-Flusses südlich von Vysoká pri Morave. *Ekológia, Bratislava*, Suppl. 1, 89–98.

Ružičková, H., Kalivodová, E. and Otahelová, H. (1993) Flora- und Faunaforschung der Mar-chau mit Hinsicht auf den Umweltschutz. In: *Sammlung des Symposium 'Wasser im Pannonis-chen Raum', Sopron, Hungary, 18 May 1993*, pp. 214–217. Universität für Forstwirtschaft und Holzindustrie Sporon.

Rychnovská, M., Balátová-Tuláčková, E., Ulehlová, B. and Pelikán, J. (1985) *Ecology of grass-lands*. (In Czech.) Academia, Praha, 292 pp.

Šomšák, L. (1992) *Leucojum aestivum* L. in the interior part of the Záhorská nížina lowland. (In Slovak.) *Biológia, Bratislava*, **47**, 591–592.

Stanová, V. and Šeffer, J. (1994) Acidophilous sand vegetation of dune system on locality Borová – description and indirect gradient analysis. *Ekológia, Bratislava*, Suppl. 1, 99–106.

Valachovič, M. (1993) The community with *Berula angustifolia* in Záhorská nížina lowland. *Bulletin Slovenskej botanickey spoločnosti, Bratislava*, **15**, 41–43.

Whittaker, R.H. (1972) Evolution and measurements of species diversity. *Taxon, Utrecht*, **21**, 213–251.

Zaliberová, M. (1994) Die Strauchweidengesellschaften im March Alluvium. *Ekológia, Bratislava*, Suppl. 1, 107–114.

Zaliberová, M., Otahelová, H. and Banásová, V. (1993) The interesting locality of psammophytes in Morava River alluvium. (In Slovak.) *Bulletin Slovenskej botanickey spoločnosti, Bratislava*, **15**, 61–63.

9 The Important Habitats and Characteristic Rare Invertebrates of Lowland Wet Grassland in England

MARTIN DRAKE

English Nature, Peterborough, UK

INTRODUCTION

Little has been published on the conservation interest of the invertebrate fauna of British lowland wet grasslands considering the extent and variety of the habitat (Jefferson and Grice 1998). This is in strong contrast to the considerable attention given to calcareous grasslands, which have long been recognized to support important invertebrate communities. In his resource review of lowland wet grassland, Dargie (1995) recorded the existence of data on nationally uncommon invertebrates for approximately a quarter of the blocks of lowland wet grassland greater than 10 ha in Essex and Norfolk, and indicated that data existed for many other blocks within England. The significance of these data has yet to be summarized. Luff *et al.* (1989) investigated Araneae (spiders) and Carabidae (ground beetles) of many grasslands and showed that soil moisture, pH and vegetation structure are the three variables that best explain differences between assemblages. Lowland wet grassland would therefore be expected to have a distinct community, and indeed Eyre and Luff (1990) and Luff *et al.* (1992) showed this to be true for widespread carabids in a number of lowland wet grassland sites in both Britain and northern Europe. There is also a growing body of data for Carabidae and Araneae of lowland grasslands in Belgium (papers summarized in Maelfait *et al.* 1988). Some studies, for example that of Blake and Foster (1998), consider the conservation significance of carabids and other invertebrates not in terms of the invertebrate fauna but as food items for bird species. Apart from these conservation-related studies, much of the remaining literature on lowland grasslands concentrates on intensively farmed and not necessarily wet pasture (e.g. Curry 1994).

The ditch systems of lowland wet grasslands are recognized as being an important element for invertebrates. Driscoll (in press) provides a bibliography of the many surveys of ditch systems but these are only one of the habitats found on lowland wet grasslands. Results have been published for individual grazing marshes, e.g. Clemons (1982), Clare and Edwards (1983), Charman *et al.* (1985) and Drake (in press), and in Davidson *et al.*

See Glossary, p. 305, for explanation of technical terms. Scientific names of vascular plants follow Tutin, T. G. *et al.* (1964–80) *Flora Europaea* Volumes 1–5. Cambridge University Press. See p. 319.

European Wet Grasslands: Biodiversity, Management and Restoration. Edited by Chris B. Joyce and P. Max Wade.
© 1998 John Wiley & Sons Ltd.

(1991). In The Netherlands, too, ditches are also a species-rich feature of the landscape (e.g. Verdonschott 1990).

Interest in the conservation of lowland wet grasslands has been stimulated by their ornithological and botanical value, which has consequently dictated the management. Foremost among the requirements for breeding and overwintering birds is a large area of grassland. However, it is becoming clear that the considerable invertebrate interest of lowland wet grasslands depends on other aspects of the sites, notably ditches but also other relatively small features. While the advice for managing birds may also be adequate for the invertebrate fauna of the grass sward, additional thought must be given to patchiness and small-scale features needed by invertebrates. Kirby (1992) gives the best advice to date, the basis for this being an amalgam of unattributed natural history experience and literature.

The main aim of this chapter is to show the relative importance of the habitat features most frequently found in lowland wet grasslands. This clarifies where conservation action is most needed. A second aim is to identify some of the most frequently occurring of the nationally rare species found on wet grasslands which, by inference, are characteristic of lowland wet grasslands.

SOURCES OF INFORMATION

One main source was used to derive information on the fauna of wet grasslands: the Invertebrate Site Register (ISR) maintained by the Joint Nature Conservation Committee (JNCC). This is a database of sites where nationally rare or scarce terrestrial and freshwater invertebrates have been recorded (Ball 1994). In simplified terms, nationally rare species (hereafter referred to as rare) are those with Red Data Book status. These occur, or are considered to occur, in 15 or fewer 10-km squares of the British Ordnance Survey National Grid (with a few exceptions that are undergoing rapid decline towards this state). Nationally scarce species (referred to as scarce) are those that occur, or are thought to occur, in 16–100 10-km squares. Full criteria and definitions are given in Ball (1994).

Much of the information on the status, biology and distribution of rare and scarce invertebrates has been published by the JNCC and the Nature Conservancy Council in the form of reviews covering many major groups (see Ball (1994) for a list of these). Using these reviews, Kirby (1994) allocated codes indicating the habitats where each species had been recorded. The allocation was based on a fairly literal interpretation of the ecological accounts and did not attempt to indicate the preferred habitat of each species. Groups reviewed by Parsons (1993) and Hyman and Parsons (1994) were not covered by Kirby and have not been included in this paper. The ISR gives a short account of the biology and distribution of each species, based on literature and the experience of entomologists contributing to the database, and this has been used to help interpretation of the results presented here.

Buisson and Williams (1991) and Jefferson and Grice (1998) provide an initial list of nationally important lowland wet grasslands and this was supplemented by sites with a high number of rare and scarce invertebrates listed in the ISR. It is not possible to assess from which habitat or even exactly where on a site invertebrate records were made because many of the records in the ISR pertain to whole sites and originate from a wide variety of sources including casual collecting, surveys, museum collections and publications. Because of this limitation, sites with extensive areas of habitat other than wet grassland were excluded. Also, to make the results manageable, the largest unit was used. For example, the Somerset Levels National Nature Reserve (NNR) includes five distinct Sites of Special Scientific Interest, whereas West Sedgemoor, also on the Somerset Levels but outside the NNR, was treated as a separate site. Within these limitations, a sample was produced of 24 lowland wet grassland sites of variable extent.

Using the species accounts in the ISR and Kirby (1994), rare and scarce species from the 24 sites were allocated to the single habitat that they were most likely to occupy at such sites. While it is recognized that this is an oversimplification of the ecologies of these species, the purpose was to describe the habitats most frequently used on wet grasslands. Thus, a species whose habitats include fen, carr and wet heath was classed as a fenland species. Several categories were amalgamated because they are imprecise, for example fen and marsh, or because the habitats would not be expected on wet sites, so no information was lost. For instance, the categories heathland, sandy places, shingle, dunes and other dry coastal habitats were classified as 'dry places'. This simplified classification allowed the proportion of the fauna using different habitats to be calculated easily by avoiding the problem of some species using several habitats. Red Data Book species were not distributed among the habitats any differently from scarce species, so species of both statuses were combined in the results.

TYPES OF LOWLAND WET GRASSLAND

Lowland wet grasslands in England can be divided into coastal grazing marshes, flood-plain meadows, and grasslands derived from fens. None consists solely of grassland. Coastal grasslands characteristically consist of large tracts of superficially uniform terrain on clay soils with extensive drainage-ditch systems but few hedges and trees. They tend to flood infrequently because few sites have large rivers passing through them so the grassland is usually rather dry. Wet habitat is mostly confined to the ditches and their immediate banks. Compared to coastal marshes, grasslands on old fens usually have a diversity of features such as patches of fen, bushes and isolated trees, and are more prone to flooding because they are land-locked. Poorly drained examples have genuinely wet grassland that remains saturated throughout the year. River floodplain grasslands can be even more varied, and often include areas of carr woodland, derelict water meadows and fen, and some may flood annually. Lowland wet grassland sites are therefore not solely grassland; they contain a variety of other habitats and features.

HABITATS OF VALUE ON WET GRASSLAND SITES

In the sample of 24 sites, 242 invertebrate species recorded were rare and 700 species were scarce. Most of the sites are clearly of national importance for their invertebrates (Table 9.1).

The habitats used by most species of these sites are the relic fen, water margins and standing water bodies, which together comprise the habitats for between one- and

Table 9.1 Percentage of rare and scarce invertebrates in lowland wet grasslands, categorized into habitat types

Site type	County	Site	S	Total no. of spp.	Grass, wet	Grass	Grass, dry	Dry	Fen	Mar	Reed	Pond	Strea	Salt	Scrub	Wood	Gen	Unkn
Coast	Essex	Benfleet Marshes	51	146	0	9	7	2	10	14	2	14	0	16	12	8	5	2
	Essex	Blackwater Estuary	26	47	0	0	2	17	9	21	0	26	0	23	2	0	0	0
	Essex	Colne Estuary	39	166	1	7	5	34	10	19	4	30	0	24	11	6	5	3
	Essex	Hamford Water	19	46	0	9	2	15	7	9	7	24	0	15	9	0	2	2
	Essex	Inner Thames Marshes	18	69	0	0	10	10	10	20	4	33	0	6	1	1	1	1
	Kent	South Thames Marshes	9	123	1	5	6	5	8	34	3	18	0	15	2	1	2	0
	Kent	Swale	21	206	0	7	6	14	13	14	1	13	0	20	2	4	4	1
	Suffolk	Minsmere RSPB Reserve	13	53	2	0	2	21	23	6	6	0	0	8	9	23	2	0
Fen	Avon	Gordano	15	25	0	8	0	4	16	28	0	28	0	0	0	8	4	4
	Gloucs	Martin Mere	6	3	3	3	0	33	0	0	0	0	0	0	0	0	33	0
	Gloucs	Walmore Common	1	9	0	11	0	0	22	33	0	33	0	0	0	0	0	0
	Kent	Romney Marsh	10	51	2	0	0	6	4	35	0	49	0	0	2	0	2	0
	Kent	Stodmarsh	8	67	0	1	3	7	42	9	4	7	1	1	9	9	3	1
	Oxon	North Meadow	20	4	25	0	0	0	25	25	0	0	25	0	0	0	0	0
	Somerset	West Sedgemoor	4	36	0	6	0	3	8	22	0	56	0	0	0	0	0	6
	Somerset	Somerset Levels NNR	17	56	2	5	0	0	16	23	0	41	0	0	4	4	4	2
	Sussex	Amberley Wild Brooks	12	24	0	0	0	8	13	29	0	25	8	4	0	4	4	4
	Sussex	Pevensey Levels	33	103	1	1	0	0	12	29	7	36	1	3	5	3	1	2
River	Cambs	Castor Flood Meadow	2	11	0	9	0	0	0	64	0	18	0	0	9	9	0	0
	Cambs	Nene Washes	5	19	5	0	5	0	0	37	0	47	0	0	0	0	0	5
	Cambs	Ouse Washes	7	36	3	0	0	3	19	22	0	25	6	0	11	6	6	0
	Hants	Itchen Valley	19	59	0	3	7	3	36	10	0	3	5	0	14	17	2	0
	Hants	Test Valley	30	180	1	8	18	10	8	6	2	2	3	0	13	22	2	6
	Yorks	Derwent Ings	49	91	5	3	2	7	29	30	0	4	2	1	2	9	4	1

Grass, wet; *Grass*, unspecified; *Grass, dry*; *Dry* places (heathland, dune, shingle etc.); *Fen*; *Margin* of water; *Reedbeds*; *Pond* and standing water including ditches; *Stream* and flowing water; *Salt/marsh*; *Scrub* and trees; *Woodland*; *Generalists*; *Unknown*. S, number of sources for the information in the Invertebrate Site Register. RSPB, Royal Society for the Protection of Birds; NNR, National Nature Reserve.

two-thirds of the rare and scarce species on coastal marshes and usually more than two-thirds of these species on fen and floodplain sites.

Species of wet grassland itself make up a minor proportion of the fauna on any type of site. There are very few invertebrates classified as wet grassland species on any of the three types of wet grasslands. It is significant that most 'wet grassland' species are from river floodplains, including three sites that flood annually and often have standing water for several weeks or months in winter (Nene Washes (Jenman and Kitchin 1998), Ouse Washes and Derwent Ings). Even a species such as *Selatosomus nigricornis* (click beetle), which is probably one of the most characteristic of the nationally uncommon species of wet swards, has been recorded at few sites.

The remaining grassland species that were not allocated to the wet grassland category were classified as typical of either grassland of an unspecified nature or dry grassland (Table 9.1). The former may include invertebrates of moist pasture but there are unlikely to be misclassified wetland species here. Together, these two grassland types usually support about 10–15% of the rare and scarce species of coastal sites, less than 10% of species on river floodplains and virtually none on fenland grasslands (Table 9.1).

Features of grasslands sites that are often not included in botanical evaluations are isolated trees, scrub and hedgerows. Invertebrate species associated with these apparently minor features (grouped under scrub in Table 9.1) are an important component of a number of sites from all three types of lowland wet grasslands. Only exposed coastal sites such as those of the Thames Estuary support a very small number of these species. The low number recorded on the moors of Somerset and Avon is almost certainly due to the small amount of data entered to the ISR to date. An example of a group of species dependent on trees are those feeding on *Salix* (willows), a characteristic feature of fen and floodplain grasslands where these trees serve more than an aesthetic purpose. Some of these invertebrates are dead-wood beetles and moths, for example *Sesia bembeciformis* (lunar hornet moth), *Cossus cossus* (goat moth), *Synanthedon formiciformis* (red-tipped clearwing moth) and *Oberea oculata* (long-horn beetle), whereas *Cryptorhynchus lapathi* (weevil), *Lithophane socia* (pale pinion moth) and *Earias clorana* (cream-bordered green pea moth) feed on the leaves or in galls on willows.

The proportions of scarce and rare invertebrate species in some of the more important habitats differ between coastal, fenland or floodplain grassland types (Table 9.1). Some of this variation is an artefact of combining records for all habitats on a site. Thus, species of running water are a small but characteristic component of floodplain sites, and species of saltmarsh, intertidal and strongly brackish-water habitats are an important component on coastal marshes (usually about 15–25% of rare and scarce species). Other differences between the three grassland types are probably real, for instance species of dry habitats are frequent in coastal grassland sites (usually about 10–20% of species) and for only a few sites can these be attributed to large tracts of non-grassland habitat, such as at Minsmere in eastern England.

CHARACTERISTIC RARE INVERTEBRATES OF LOWLAND WET GRASSLANDS

Some invertebrate species have been recorded sufficiently often to suggest that they are characteristic of lowland wet grassland sites (Table 9.2). This list was not the result of

Table 9.2 Nationally scarce and rare species that occur widely on lowland wet grassland sites

ODONATA	*Brachytron pratense, Coenagrion pulchellum*

COLEOPTERA	
Carabidae	*Bembidion clarki, B. fumigatum, Pterostichus anthracinus, Stenolophus skrimshiranus, Chlaenius nigricornis, Odacantha melanura, Demetrias imperialis*
Haliplidae	*Peltodytes caesus, Haliplus apicalis*
Noteridae	*Noterus crassicornis*
Dytiscidae	*Agabus conspersus, Dytiscus circumflexus, Coelambus parallelogrammus, Rhantus frontalis, R. grapii*
Hydrophilidae	*Hydrochus elongatus, Helophorus alternans, H. nanus, Cercyon convexiusculus, C. sternalis, C. tristis, C. ustulatus, Limnoxenus niger, Anacaena bipustulata, Helochares lividus, Enochrus bicolor, E. melanocephalus, E. ochropterus, E. halophilus, Hydrophilus piceus, Berosus affinis, B. signaticollis*
Hydraenidae	*Ochthebius marinus, O. nanus, O. viridis*
Staphylinidae	*Paederus fuscipes, Philonthus punctus, Gabrius bishopi*
Cantharidae	*Silis ruficollis*
Curculionidae	*Bagous cylindrus, B. subcarinatus, B. temphestivus, Hydronomus alismatis, Notaris bimaculatus, Litodactylus leucogaster, Drupenatus nasturtii, Gymnetron villosulum*

LEPIDOPTERA:	
Cossidae	*Cossus cossus*
Sessiidae	*Sesia bembiciformis*
Cochyllidae	*Phalonidia alismana*
Pyralidae	*Calamotropha paludella, Schoenobius gigantella, Evergestis extimalis, Synaphe punctalis*
Vanessidae	*Eurodryas aurinia*
Geometridae	*Scopula emutaria, Eupithecia subumbrata*
Arctiidae	*Spilosoma urticae*
Noctuidae	*Mythimna obsoleta, Senta flammea, Simyra albovenosa, Apamea oblonga, Archanara geminipuncta, A. dissoluta, A. sparganii, Chilodes maritimus, Deltote uncula, Earias clorana, Macrochilo cribrumalis*

DIPTERA	
Tipulidae	*Limonia ventralis, Helius pallirostris, Pilaria scutellata, Erioptera bivittata*
Stratiomyidae	*Vanoyia tenuicornis, Odontomyia ornata, O. tigrina, Stratiomys singularior*
Syrphidae	*Neoascia geniculata, N. interrupta, Lejogaster splendida, Lejops vittata*
Lauxanidae	*Sapromyza opaca*
Sciomyzidae	*Colobaea punctata, Pherbellia brunnipes, P. grisescens, P. dorsata, Sciomyza simplex*
Anthomyzidae	*Anagnota collini*
Chloropidae	*Elachiptera pubescens*

rigorous analysis – the records were collected in too disparate a way for this – but it is offered as a first appraisal. There is considerable regional variation in the occurrence of some of these species, with a large proportion being apparently confined to the south-eastern coastal marshes. This may be the result of a genuine scarcity of brackish marshes elsewhere in Britain or may be due to the warmer summer climate favouring species that are rare because they are at the edge of their climatic range.

Members of the aquatic fauna that are characteristic of lowland wet grassland sites, and are often common in this habitat despite their restricted national distribution, are the Odonata (dragonflies and damselflies) *Brachytron pratense* (hairy dragonfly) and *Coenagrion pulchellum* (variable damselfly); the Stratiomyidae (soldier flies) *Odontomyia ornata*, *O. tigrina* and *Stratiomys singularior*; and many aquatic Coleoptera (water beetles), notably *Hydrophilus piceus*, *Limnoxenus niger*, *Peltodytes caesus*, *Rhantus frontalis* and *Dytiscus circumflexus*. Of these, *O. ornata* has not been recorded away from ditch systems on grazing marshes for many decades, and *H. piceus* and *L. niger* are infrequently recorded elsewhere. Frequently recorded aquatic weevils feeding on water plants or living at the water's edge are *Bagous tempestivus*, *B. subcarinatus*, *Hydronomus alismatis* feeding on *Alisma plantago-aquatica*, and *Litodactylus leucogaster* feeding on *Myriophyllum*, mainly in southeastern sites.

The fauna of water margins consists of several guilds with different requirements. An example of a group with aquatic or semi-aquatic larvae and whose adults usually spend their time in tall emergent vegetation are the Sciomyzidae (snail-killing flies), many of which feed on aquatic snails. Not only has a large proportion of the 66 British species been recorded on wet grassland sites, but about half of the 31 scarce and rare species have been recorded minimally on one of the sites listed in Table 9.1. At least five species, *Pherbellia dorsata*, *P. grisescens*, *P. brunnipes*, *Colobaea punctata* and *Sciomyza simplex*, are widely distributed on lowland wet grassland sites. Carabidae (ground beetles) exemplify a group whose habitat preference is determined less by the structure of the emergent vegetation than by soil saturation. Surprisingly little information for Carabidae is held on the ISR for these wet grasslands but two species frequently recorded are *Bembidion clarki* and *B. fumigatum*. These, and several mentioned below, may be among the few scarce species that can be found in the wet swards of grassland sites. A great many Diptera (true flies) are probably directly associated with the saturated soil found in and near water margins because this is where their larvae live. Some Tipulidae (craneflies) that may rely on wet grassland sites, mainly grazing marshes, are *Erioptera bivittata*, confined to brackish eastern marshes, *Limonia ventralis* and *Pilaria scutellata*. The rare *Limnophila pictipennis* may prefer washlands with regular winter flooding.

Emergent plants support a suite of species that either feed on them or make use of the litter they generate. *Phragmites australis* reedbeds have a distinct fauna and are taken as an example of the importance of a non-grassland plant in wet grassland sites. Their fauna includes species feeding directly on the plants, examples among the Lepidoptera being the pyralid moth *Schoenobius gigantella*, *Simyra albovenosa* (reed dagger) and several wainscots (*Mythimna obsoleta*, *Archanara geminipuncta*, *A. dissoluta* and *Chilodes maritimus*), which all occur on many wet grassland sites. Some non-herbivorous species are often found in association with reed, for example the predatory carabid beetle *Demetrias imperialis*, which is the only very widely recorded carabid at sites in Table 9.1. Other carabids often associated with reedbeds but only found at a few sites include *Chlaenius nigricornis*, *Odacantha melanura* and *Stenolophus skrimshiranus*. *Typha* (reedmace) is another plant that supports not only phytophagous species such as *Archanara sparganii* (Webb's wainscot) and *A. algae* (rush wainscot) (not commonly found on wet grassland) but also detritivorous species whose larvae are associated with the stem bases, for example, the cranefly, *Helius pallirostris* and the hoverfly, *Lejogaster splendida*.

INVERTEBRATES OF WET GRASS SWARDS

Kirby (1994) lists 40 nationally rare or scarce species that published sources suggest are found on wet grassland. Many of these are clearly not primarily grassland species, and in some cases the information is too scanty to assess whether wet grassland is their favoured habitat. After excluding these species and a few of upland wet grasslands, the list is halved (Table 9.3). The sites listed in the ISR for these 21 species were appraised for their importance as lowland wet grassland using the site description. Only four of the species were recorded from sites where wet grassland was obviously an extensive habitat: *Capsus wagneri, Apion difforme, Selatosomus angustulus* and *S. nigricornis*. Many others in the list are doubtfully truly characteristic of wet grassland alone but may also be found in fens, at water margins and perhaps in wet grassland on mosaic sites.

An attempt was made to predict phytophagous species that ought to be found on lowland wet grasslands by searching the ISR for those species whose foodplant is a common component of wet grassland communities. Plants with a frequency of more than 40% (Rodwell 1992) were identified from the lowland wet grassland plant communities listed by Jefferson and Grice (1998; Table 3.1), ignoring grass species. *Phragmites australis* was also included, which is not a frequent member of these communities. This produced a list of 29 plants. The ISR is not an ideal database for this purpose because the species accounts are brief and mention only commonly used foodplants. However, this search found 31 rare or scarce insects that fed on these plants and which were likely to be found on grassland sites (species found in, for example, wet woodland were not included). Another 16 species are known to feed on *P. australis* or are likely to do so. Of these species, 47 in total, 37 have been recorded at the sites listed in Table 9.1 (Table 9.4). Ignoring those feeding on *P. australis*, few species have been recorded from more than

Table 9.3 Provisional list of rare and scarce species that may occur on lowland wet grassland as their primary habitat (adapted from Kirby 1994)

ORTHOPTERA	
Gryllotalpidae	?*Gryllotalpa gryllotalpa*
HEMIPTERA	
Cicadellidae	*Agallia brachyptera,* ?*Aphrodes albiger*
Miridae	**Capsus wagneri*
COLEOPTERA	
Apionidae	**Apion difforme, Apion vicinum*
Chrysomelidae	*Chrysolina graminis*
Curculionidae	*Chaetocnema subcoerulea,* ?*Phytobius muricatus*
Cryptophagidae	?*Atomaria rubricollis*
Elateridae	??*Fleutiauxellus quadripustulatus,* **Selatosomus angustulus,* **S. nigricornis*
Scarabaeidae	?*Aphodius consputus*
Staphylinidae	?*Atheta scotica,* ?*Calodera protensa, Philonthus atratus,* ?*Stenus circularis,* ??*Zyras haworthi*
DIPTERA	
Dolichopodidae	*Dolichopus cilifemoratus*

?=the degree of uncertainty about a species' dependence on lowland wet grassland; an *=strong association.

Table 9.4   Insects herbivorous on frequently occurring plants of wet grassland communities which have been recorded from sites listed in Table 9.1

Foodplant	Order	Species
Phragmites australis	Hemiptera	*Paralimnus phragmitis?, Chloriona dorsata, C. vasconica*
	Lepidoptera	*Schoenobius gigantella, Mythimna obsoleta, Simyra albovenosa, Archanara geminipuncta, A. dissoluta*
	Diptera	*Anagnota collini, Cryptonevra nigritarsis?, Elachiptera pubescens?, E. rufifrons?, Oscinosoma gilvipes?*
Typha angustifolia	Diptera	*Anthomyza bifasciata*
Typha latifolia, Iris pseudacorus, Scirpus lacustris	Lepidoptera	*Archanara algae*
Typha spp., *Iris pseudacorus, Sparganium erectum, Scirpus lacustris*	Lepidoptera	*Archanara sparganii*
Typha spp., *Carex* spp., *Juncus* spp.	Coleoptera	*Notaris waltoni*
Carex spp., *Juncus* spp.	Coleoptera	*Notaris bimaculata*
Eupatorium cannabinum	Lepidoptera	*Diachysia chryson*
	Diptera	*Vidalia cornuta*
Alisma plantago-aquatica	Coleoptera	*Hydronomus alismatis*
	Lepidoptera	*Phalonidia alismana*
Equisetum fluviatile	Coleoptera	*Bagous lutulentus*
Equisetum spp.	Coleoptera	*Grypnus equiseti*
Polygonum spp.	Coleoptera	*Phytobius quadrinodosus?, Apium difforme*
Polygonum amphibium	Coleoptera	*Phytobius muricatus*
Succisa pratensis	Lepidoptera	*Eurodryas aurinia, Aethes piercei*
Thalictrum flavum	Lepidoptera	*Perizoma sagittata*
Valeriana officinalis	Lepidoptera	*Eupithecia valerianata*
Althaea officinalis	Lepidoptera	*Hydraecia osseola*
Mentha aquatica, Lycopus europaeus	Lepidoptera	*Phalonidia manniana*
Lysimachia vulgaris	Lepidoptera	*Spilosoma urticae*
Filipendula ulmaria	Symphyta	*Hartigia xanthostoma*
Sparganium spp.	Coleoptera	*Donacia bicolora*
Potentilla palustris	Coleoptera	*Phytobius comari*

The species of some genera of plant were not given in the source. A question mark indicates uncertain ecology.

one site. These include the weevils *Hydronomus alismatis* and *Notaris bimaculata*, the cochylid micromoth *Phalonidia manniana*, *Spilosoma urticae* (water ermine moth) and *Eurodryas aurinia* (marsh fritillary butterfly). Most of the 29 plants initially selected that do not appear in Table 9.4 are those most characteristic of grass swards such as *Crepis paludosa, Caltha palustris* and *Ranunculus acris*. This analysis may be criticized for using an incomplete database but it does suggest that common plants of wet grassland swards support relatively few species of rare and scarce herbivorous invertebrates compared to the numbers feeding on common wetland plants of ditches, water margins and fens.

DISCUSSION

Lowland wet grassland sites support nationally important numbers and species of invertebrates, but on most sites the grassland itself is of relatively low interest in terms of rare and scarce species restricted to it. The most valuable features are the true wetland habitats of fen, water margin, and standing water, yet these features occupy only a small proportion of the area of the most important lowland wet grasslands. Often all that remains of these habitats are ditches and their margins, and it is mainly along ditches that isolated trees, hedges and remnants of fen vegetation manage to survive grazing and mowing. Despite the data being not strictly comparable, having been collected by many entomologists with varying interests (indicated by the number of sources in Table 9.1), this analysis shows that the trend is consistent over many sites. Management needs to take these features into account if the range of uncommon invertebrates is to be conserved.

The importance of field boundaries to Carabidae (ground beetles) and Araneae (spiders) has been demonstrated in grazed *Lolium* pastures in Belgium (de Keer *et al*. 1986; Maelfait *et al*. 1988) where the numbers of species were higher in the ditches and in rough grass under fences compared with the species richness within the sward. These authors suggest that the boundary features provide greater shelter and litter than the sward, warm hibernation sites and well-aerated soils. Beyond these obvious aspects of natural history, a number of reasons may be advanced to explain why wet grassland appears to have a limited fauna. These include the effects of past history of pasture management and the inherent low structural complexity of plagioclimax neutral grassland swards.

Several studies show a negative relationship between stocking density of sheep and the abundance and biomass of some invertebrates in leys (e.g. Walsingham 1978; Hutchinson and King 1980). Walsingham (1978) identifies a similar set of factors to those listed by Maelfait *et al*. (1988) for the reduction in invertebrates in intensively grazed swards, including removal of surface cover and food supplies as herbage and litter, and an unfavourable change in the microclimate. Rushton (1987) found that the species richness of Araneae in unmanaged grasslands was higher than in grazed pasture, and that of Carabidae was highest in lightly grazed or unmanaged grasslands. Intensive pasture management, especially soil disturbance and pesticide use, has been shown to result in large changes in the composition of the Carabidae and Araneae assemblages and in overall loss of species (Rushton *et al*. 1989). Although some of these results were obtained at sites with high stocking densities, they do suggest that even semi-natural grasslands may be a difficult habitat for an invertebrate at times of intensive grazing and trampling. Another factor that may have reduced niche opportunities within the sward is reduced plant species diversity (and a correspondingly reduced structural diversity) as a result of fertilizer application (Smith 1993). The relevance of agricultural practice to lowland wet grassland is that many sites are working farms where sward improvements have taken place in the past and may be partly responsible for currently low invertebrate interest.

Structural complexity at different scales is likely to be the main reason for the greater usage of boundary features. At the scale of single plants, their architecture is positively correlated with insect species richness (Lawton 1978). Grazed grass is one of the least complex structures to be found on lowland wet grasslands, so is not likely to support a large fauna. As many invertebrates, particularly insects, use more than one part of a habitat during their life cycle, then, at a scale larger than individual plants habitat mosaics become important, and a patchy environment will support more species. Patchiness has also been identified as an essential requirement to maintain persistent host–parasite associations in the wild (May 1993). If this conclusion can be extrapolated to more complex associations of interacting species, then the patchiness, represented by non-grassland features on grassland sites, should result in populations of more species surviving for longer, so species richness will be greater than in the uniform sward where local extinction rates are postulated to be higher. The scale of mosaics that are likely to benefit invertebrates will be in the order of a few square metres in size, rather than of hectares which, in contrast, is the patch size required by several birds of importance on lowland wet grasslands.

An increase in species richness of Coleoptera (beetles) and Hemiptera (plant bugs) has been shown to correlate with an increase in structural diversity of the plants as succession proceeds from a ruderal field to secondary woodland (Southwood *et al.* 1979). It is possible, therefore, that plagioclimax grass swards give few opportunities for changes in the communities over time compared to, for example, ditches and long-ungrazed grass where succession involves a sequence of assemblages.

The analysis in this chapter has concentrated on rare species and, in comparing with other work, on species richness which is itself likely to be correlated with the number of rare species. Species richness of invertebrates may not be correlated with their abundance and biomass, yet the biomass of invertebrates is a key determinant of a wet grassland's value to birds (Blake and Foster 1998); hence much conservation effort is spent on raising the abundance of important food items. If the analysis of important habitat features presented here is correct, then superficially there is little conflict between managing grass swards to make them more appealing to birds, for instance by raising groundwater levels. Conflict may arise if management simplifies the habitat, for instance by felling isolated trees to remove corvid hunting perches, or removing bushes and tall reeds to improve long-range visibility for birds such as *Vanellus vanellus* (lapwing).

The conclusions of this chapter are based on many disparate sources of data that reveal a number of similar trends. It is felt, nevertheless, that field work should be undertaken to test comprehensively these conclusions.

SUMMARY

A selection of nationally important lowland wet grassland sites are shown to support large numbers of nationally rare species of invertebrates. These are associated mainly with habitat features that are usually confined to field boundaries, for example ditches, patches of fenland, hedges and trees. Few rare species appear to be dependent upon the grass sward itself. Wetland plants appear to support more rare invertebrate species than do grassland forbs in these habitats. Despite the small area occupied by these features, they are responsible for much of the conservation interest for invertebrates and need to be taken into account when managing wet grasslands. The probable reason for the features supporting more species than does the grassland is their greater structural complexity and relatively sympathetic management.

Keywords: Coastal grazing marshes, Fenland grasslands, Floodplain meadows, Habitat features, Invertebrates, Rarity.

REFERENCES

Ball, S.G. (1994) The invertebrate site register – objectives and achievements. *British Journal of Entomology and Natural History*, **7** (Suppl. 1), 2–14.

Blake, S. and Foster, G.N. (998) The influence of grassland management on body size in Carabidae (ground beetles) and its bearing on the conservation of wading birds. In: Joyce, C.B. and Wade, P.M. (eds), *European wet grasslands: biodiversity, management and restoration*, pp. 163–169. John Wiley, Chichester.

Buisson, R. and Williams, G. (1991) RSPB action for lowland wet grasslands. *RSPB Conservation Review*, **5**, 60–64.

Charman, K., Palmer, M. and Philp, E.G. (1985) Survey of the aquatic habitats in the North Kent marshes. *Transactions of the Kent Field Club*, **10**, 19–32.

Clare, P. and Edwards, R.W. (1983) The macroinvertebrate fauna of the drainage channels of the Gwent Levels, South Wales. *Freshwater Biology*, **13**, 205–225.

Clemons, L. (1982) A survey of the flora and fauna of Murston Marshes. *Transactions of the Kent Field Club*, **9**, 31–64.

Curry, J.P. (1994) *Grassland invertebrates. Ecology, influence on soil fertility and effects on plant growth*. Chapman & Hall, London.

Dargie, T.C. (1995) *Lowland wet grassland resource survey: module 2*. English Nature Research Report No. 149. English Nature, Peterborough.

Davidson, N.C., Laffoley, D. d'A., Doody, J.P., Way, L.S., Gordon, J., Key, R., Drake, C.M., Pienkowski, M.W., Mitchell, R. and Duff, K.L. (1991) *Nature conservation and estuaries in Great Britain*. Nature Conservancy Council, Peterborough.

de Keer, R., Desender, R., D'Hulster, M. and Maelfait, J.P. (1986) The importance of edges for the spider and beetle fauna of a pasture. *Annales de la Société royale zoologique de Belgique*, **116**, 92–93.

Drake, C.M. (in press) Factors influencing the species richness of aquatic invertebrates in grazing marsh ditches. In: Harpley, A.J. and Wade, P.M. (eds), *Nature conservation and the management of drainage system habitat*. Wiley, Chichester.

Driscoll, R.J. (in press) A bibliography of ditch surveys in England and Wales. In: Harpley, A.J. and Wade, P.M. (eds), *Nature conservation and the management of drainage system habitat*. Wiley, Chichester.

Eyre, M.D. and Luff, M.L. (1990) A preliminary classification of European grassland habitats using carabid beetles. In: Stork, N.E. (ed.), *The role of ground beetles in ecological and environmental studies*, pp. 227–236. Intercept, Andover.

Hutchinson, K.L. and King, K.L. (1980) The effect of sheep stocking level on invertebrate abundance, biomass and energy utilization in a temperate, sown grassland. *Journal of Applied Ecology*, **17**, 369–387.

Hyman, P.S. and Parsons, M.S. (1994) *A review of the scarce and threatened Coleoptera of Great Britain*, Part 2. UK Nature Conservation, No. 12. Joint Nature Conservation Committee, Peterborough.

Jefferson, R.G. and Grice, R.V. (1998) The conservation of lowland wet grassland in England. In: Joyce, C.B. and Wade, P.M. (eds), *European wet grasslands: biodiversity, management and restoration*, pp. 31–48. John Wiley, Chichester.

Jenman, W. and Kitchin, C. (1998) A comparison of the management and rehabilitation of two wet grassland nature reserves: the Nene Washes and Pevensey Levels, England. In: Joyce, C.B. and Wade, P.M. (eds), *European wet grasslands: biodiversity, management and restoration*, pp. 229–245. John Wiley, Chichester.

Kirby, P. (1992) *Habitat management for invertebrates: a practical handbook*. Royal Society for the Protection of Birds, Sandy.

Kirby, P. (1994) *Habitat fragmentation: species at risk, invertebrate group identification*. English Nature Research Report No. 89. English Nature, Peterborough.

Lawton, J.H. (1978) Host–plant influences on insect diversity: the effects of time and space. In: Mound, L.A. and Waloff, N. (eds), *Diversity of insect faunas. Symposium of the Royal Entomological Society of London*, **9**, 105–125.

Luff, M.L., Eyre, M.D. and Rushton, S.P. (1989) Classification and ordination of habitats of ground beetles (Coleoptera, Carabidae) in north-east England. *Journal of Biogeography*, **16**, 121–130.

Luff, M.L., Eyre, M.D. and Rushton, S.P. (1992) Classification and prediction of grassland habitats using ground beetles (Coleoptera, Carabidae). *Journal of Environmental Management*, **35**, 301–315.

Maelfait, J.P., Desender, H.K., de Keer, R. and Pollet, M. (1988) Investigations on the arthropod fauna of grasslands. In: Park, J.R. (ed.), *Environmental management in agriculture: European perspectives*, pp. 170–177. Belhaven Press, London.

May, R.M. (1993) The effects of spatial scale on ecological questions and answers. In: Edwards, P.J., May, R.M. and Webb, N.R. (eds), *Large-scale ecology and conservation biology*, pp. 1–19. Blackwell, London.

Parsons, M.S. (1993) *A review of the scarce and threatened pyralid moths of Great Britain*. UK Nature Conservation, No. 11. Joint Nature Conservation Committee, Peterborough.

Rodwell, J.S. (ed.) (1992) *British plant communities*, Vol. 3, *Grasslands and montane communities*. Cambridge University Press, Cambridge.

Rushton, S. (1987) Terrestrial site assessment using multivariate techniques. In: Luff, M.L. (ed.), *The use of invertebrates in site assessment for conservation*, pp. 62–75. Agricultural Environmental Research Group, University of Newcastle-upon-Tyne.

Rushton, S., Luff, M.L. and Eyre, M.D. (1989) Effects of pasture improvement and management on the ground beetle and spider communities of upland grasslands. *Journal of Applied Ecology*, **26**, 489–503.

Smith, R.S. (1993) Effects of fertilizers on plant species composition and conservation interest. In: Haggar, R.J. and Peel, S. (eds), *Grassland management and nature conservation.* Occasional Symposium No. 28. British Grassland Society, Reading.

Southwood, T.R.E., Brown, V.K. and Reader, P.M. (1979) The relationships of plant and insect diversities in succession. *Biological Journal of the Linnean Society*, **12**, 327–348.

Verdonschott, P.F.M. (1990) *Ecological characterization of surface waters in the province of Overijssel (The Netherlands)*. Unpublished PhD thesis, University of Wageningen.

Walsingham, J.M. (1978) Effects of sheep grazing on the invertebrate population of agricultural grassland. *Scientific Proceedings of the Royal Dublin Society, Series A*, **6**, 297–304.

10   The Role of Invertebrate Communities as Indicators of Environmental Characteristics of European River Margins

BRUNO MAIOLINI, ALESSANDRA FRANCESCHINI and ADRIANO BOSCAINI

Museo Tridentino di Scienze Naturali, Trento, Italy

INTRODUCTION

The animal community of a given area can be regarded both as the result and the synthesis of the biogeographical and the ecological characteristics of that area. This is particularly true of riparian ecotone communities, which are strongly influenced by the dynamics of the river channel and by the frequency of flood events (Décamps *et al.* 1988; Chemini and Pizzolotto 1990; Calow and Petts 1992). The invertebrate element of the animal community has been widely used as a fundamental component of standard methods in the biological monitoring of environmental quality, and communities of benthic invertebrates have been used in biological monitoring of water quality (Ghetti 1986; Rosenberg and Resh 1992; Maiolini *et al.* 1993). Terrestrial invertebrate communities can also provide important ecological information about an area and can be very sensitive indicators of environmental changes both at the regional and continental scale (Chemini and Pizzolotto 1990).

Existing literature on soil-dwelling invertebrates caught by pitfall traps has focused mainly on the Coleoptera (beetles) and, within this order, on the families Carabidae (ground beetles) and Staphylinidae (rove beetles) (see, in particular, Casale *et al.* 1982; Chemini and Perini 1982; Chemini and Zanetti 1982; Zanetti 1985; Greenwood *et al.* 1991). These insects are generally poor fliers and tend to be restricted to a given area. Thus, their communities are indicative of the ecological conditions of that area. Other strict soil surface dwellers that have proved to be important in the understanding of soil ecology are Araneae (spiders) (Locket *et al.* 1975), Opiliones (harvestmen) (Chemini 1979), Gastropoda (snails and slugs) (Germain 1930), Chilopoda (centipedes) and Diplopoda (millipedes) (Demange 1981).

Flying insects such as Diptera (true flies), winged Hymenoptera (bees, wasps and allied insects) and Hemiptera (e.g. bugs and leaf-hoppers) are found in pitfall traps but their presence is less significant in terms of representation of ecological conditions in the area. However, they have been considered in this study on account of their role in enhancing

See Glossary, p. 305, for explanation of technical terms. Scientific names of vascular plants follow Tutin, T. G. *et al.* (1964–80) *Flora Europaea* Volumes 1–5. Cambridge University Press. See p. 319.

biodiversity and in assessing the degree to which marginal zones act as refuge habitats and migration corridors for such species (Maiolini 1993).

A protocol was established for sampling the invertebrates within the European Union European River Margins Study (ERMAS) project. The aim was to assess the composition, structure and functional role of invertebrate communities from European river margin ecotones and to investigate local changes in community parameters along a spatial gradient from frequently flooded areas through to terrestrial ones that are never flooded.

MATERIALS AND METHODS

STUDY SITES

The overall aim of the ERMAS project is to compare the structure and functioning of river margins through a European south to north gradient. To achieve this, five middle-sized rivers were selected: the Garonne (France), Adige-Noce (Italy), Trent (England), Helgeån (south Sweden) and Vindel (north Sweden). The latitude of the river sites ranges from 44° (Garonne) to 64° (Vindel). The Vindel apart, all the rivers are more or less regulated and in all cases their floodplains experience some degree of annual flooding.

For each river system a total of 15 sites were located in three series: F1–F5, frequently flooded sites near the river and submerged for more than 10 weeks per year; R1–R5, rarely flooded, with flood events occurring for between one and 10 weeks per year; and T1–T5, strictly terrestrial or never-flooded sites. In the case of the Adige-Noce, five sites (R2, R4, R5, T4 and T5) were located on the Adige River, the remaining 10 sites being on its tributary, the Noce, 2 km upstream of the confluence of the two rivers. The Adige sites were 1 km downstream of the confluence. On the Helgeån River, six sites (F1, F2, R1, R2, T1 and T2) were located 15 km upstream of the remaining nine sites.

FIELD PROCEDURES

At each site, nine pitfall (Barber) traps were set up, consisting of a plastic cup with an opening of 8 cm and a depth 10 cm, partially filled with a 50% solution of ethylene glycol (commercial antifreeze) as a preservative and covered by a wooden raised tile (15 × 15 cm) kept in place by four long nails. No bait was used. Eight of the traps were arranged to form a circle of about 8 m in diameter and the ninth was placed in the middle of the circle. Altogether, 135 traps were placed in each river floodplain.

Sampling started in March 1993 in Italy, in April in England and France, and in May in Sweden, and ended in either October or November, depending on local climatic conditions. Every 14 days the contents of the pitfall traps were collected by site, i.e. the contents of the nine traps were combined. Thus, each sample covered an area of about 50 m². The unsorted samples were then sent to the Museo Tridentino di Scienze Naturali in Trento, Italy for identification and analysis.

LABORATORY PROCEDURES

In order to describe the whole community for each site, individuals were initially identified to order or family level. In a second phase, species identification will be carried out for certain selected taxonomic groups, starting with Carabidae (Coleoptera) and then Isopoda, Chilopoda and Diplopoda.

All the material collected during the study is being deposited in the collections of the Museo Tridentino di Scienze Naturali, preserved in vials containing 70% ethanol.

Data collated so far have been analysed using SYN-TAX IV (Podani 1990) in order to classify the sites into homogeneous groups.

RESULTS

DISTRIBUTION OF INVERTEBRATE TAXA

In all, more than 400 000 individual invertebrates have been collected, sorted and identified. The distribution of the total number of individuals collected from the five research river systems during the sampling period is summarized in Table 10.1. The highest activity in terms of total numbers of individuals was recorded for the Garonne system, followed by the Trent, the Helgeån, the Adige-Noce and the Vindel respectively. However, the period of sampling differed from one river to the next. Sampling on the Adige-Noce sites stopped at the end of September due to a very high and long-lasting flood that submerged all R and F sites for almost a month. This was due to heavy rain and the release of water from upstream reservoirs. A severe flood at the end of June affected all F and R sites of the Trent and on the same river four of the five T sites were vandalized in August. On the Vindel the F and part of the R sites flooded for most of the sampling period due to an exceptionally rainy year. The number of individuals was highest in R sites on the Adige-Noce and the Trent while it was highest in T sites on the Garonne, Helgeån and Vindel. Dominant taxa in the communities varied both at river and site level. Table 10.1 and Figure 10.1 provide summaries of the invertebrate data, comparing the different rivers; the former also compares T, R and F sites.

The Garonne system is characterized by larger numbers of individuals being found in the terrestrial (T) sites as compared with the rarely and frequently flooded areas (Table 10.1). Formicidae (ants) were numerically dominant in the T sites but were also well presented in the R and F ones. Although Formicidae represented 73% of individuals captured in the T sites, their presence was restricted to a few particular sampling areas, probably near to nests. Carabidae (Coleoptera) increased in number from inland sites towards the river channel. The same was true for the Araneae and Limacidae (slugs). The reverse trend was observed for other taxa, namely Silphidae (burying beetles), Collembola (springtails), Arionidae (slugs) and other Gastropoda, Opiliones, Chilopoda, Diplopoda (represented by the order Polydesmida) and Isopoda (woodlice).

On the River Adige-Noce system the numerically dominant taxa were Formicidae and Isopoda, followed by Collembola, Carabidae (Coleoptera), Araneae and Opiliones. Diplopoda, Carabidae and Silphidae were more active in the R sites, while Staphylinidae were uniformly distributed across all habitats (Table 10.1). Among Gastropoda, the family Limacidae was more abundant in the T sites whereas Arionidae and other Gastropoda such as Helicidae (snails) favoured the R ones. Polydesmida and Juliformia (Diplopoda) were both well represented in the R sites, less in the T sites and almost absent from the F ones. The Lithobiomorpha were the only well-dispersed order of Chilopoda, their numbers decreasing from the T sites to the F ones. Isopoda were dominated by the family Porcellionidae and all were particularly abundant in the R sites.

Table 10.1 Total numbers of individuals collected in five European river margin areas.

Taxa	Garonne			Adige-Noce			Trent			Helgeån			Vindel		
	T	R	F	T	R	F	T	R	F	T	R	F	T	R	F
INSECTA															
COLEOPTERA															
Carabidae	811	2518	5029	1451	2490	736	675	3192	4030	1688	1197	1224	391	636	13
Staphylinidae	1224	1206	1003	296	293	219	1893	2788	2958	1667	1503	1337	3514	2082	401
Pselaphidae				1	2	14				11	318	56			
Silphidae	319	236	130	14	103	36	2	744	13	70	236	23	16	2	
Scarabaeidae	21	26	27	56	47	16		32	1	1	1			2	
Curculionidae	10	12	10	26	21	18	90	62	11	52	43	14	10	18	
Elateridae	5	40	19	12	11	1	1	7	10	8	122	115	4	8	
Dytiscidae					1			2		1	2				
Chrysomelidae	17	23	22	22	54	51		8	6					15	4
Lucanidae	37	24	13	10	4		60	1						3	
Cerambycidae	1	1		4	1		1	16	1						
Other Coleoptera	231	350	212	33	43	24	288	371	367	241	141	152	105	63	58
DIPTERA	2674	3231	2445	935	1862	1280	5436	10984	6059	7706	3555	3544	969	955	
HEMIPTERA	400	359	456	122	69	228	213	405	343	151	305	187	107	41	1
HYMENOPTERA															
Formicidae	53376	5051	8642	3661	3409	6185	4	6	2	582	575	403	12811	10951	34
Other Hymenoptera	482	763	911	137	158	91	223	366	341	462	237	341	123	99	3
COLLEMBOLA	1402	993	290	3170	3837	565	3047	25719	19481	13758	9198	8596	1732	733	71
TRICHOPTERA	5	25	28	3	12		11	10	2	10	6	22	16	74	
MECOPTERA	100	98	11	8	12	1		12	13	1	1	5			
DERMAPTERA	4	1		5	3			3	3	273					
Other Insecta	31	35	34	3	10	2	17	29	2	18	23	33	12	20	40
Insecta larvae	2449	2403	1353	511	479	358	547	1080	928	875	589	469	383	513	
ARACHNIDA															
Araneae	1213	2195	2588	1844	1559	1847	725	1648	1723	1961	2402	2418	1481	2106	198
Opiliones	405	218	297	1043	1407	485	1113	689	313	7058	2609	1724	1117	683	122
Acarina	2552	2004	1138	33	88	183	2017	482	194	1897	1187	1383	214	247	12
Pseudoscorpiones	12	4	1	18	23	14	63	1	3	10	5	2			

MYRIAPODA

	1	2	3	4	5	6	7	8	9	10	11	12	13	14	15
DIPLOPODA															
Polydesmida	135	105	8	132	465	7	1273	308	146	172	174	155			1
Juliformia	39	39	10	263	424	2	5718	159	133	438	252	224			
Glomerida										149	110	32			
CHILOPODA															
Geophilomorpha	12	1		3	1	3	3	2	9	1	1	1			
Scolopendromorpha	10			5	2		1								
Lithobiomorpha	52	25	5	163	70	15	269	78	56	16	11	11			11
CRUSTACEA															
ISOPODA															
Armadillidae	1198	602	457	1	10	5	1788	3		1	3				
Oniscoideae	476	186	15	127	370	18		30	2	19	3	2			
Porcellionidae	906	727	747	2902	6714	1777	463	306	82		1		1		
Triconiscidae	12		1	8	61	13	6			287	271	555			
Other Isopoda	2	4	2	1	1		95	69	35	14	13	18		1	
GASTROPODA															
Limacidae	250	323	501	41	29	25	48	497	955	94	92	103	6	6	
Arionidae	432	106	47	130	332	68	74	86	85	49	28	39	50	9	
Other Gastropoda	1793	1215	390	215	471	136	501	620	140	36	70	70	87	17	
OLIGOCHAETA	107	164	104	19	15	12	60	75	281	16	17	19	14	24	16
Totals for site	73205	25313	26946	17428	24962	14436	26731	50889	38730	39795	25300	23276	23163	19308	975
TOTAL	125464			56826			116350			88371			43446		

T, terrestrial/never flooded; R, rarely flooded (1–10 weeks yr^{-1}); F, frequently flooded (10+ weeks yr^{-1}).

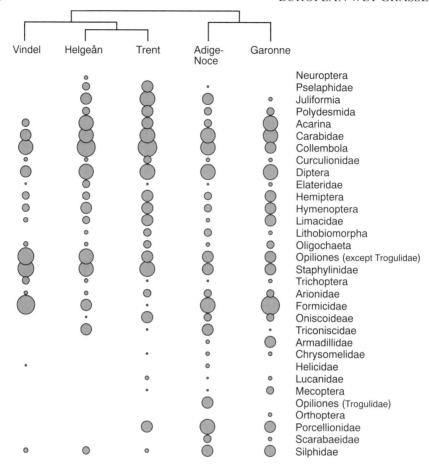

Figure 10.1 Classification analysis of pitfall trap invertebrate data. The hierarchical order of divisions separating samples from the five rivers is shown along with the relative abundances of selected taxa (those which occurred in at least 50% of samples in one or more rivers, and were important in distinguishing between rivers)

On the River Trent, Collembola and Diptera were the best represented taxa while Formicidae were practically absent (Table 10.1). Among the Coleoptera, Carabidae and Staphylinidae increased from terrestrial habitats towards the river's edge, whilst Pselaphidae were restricted to the R sites. However, within the Gastropoda, the Limacidae increased towards the wetter habitats whereas the Arionidae showed no preference and were evenly distributed in all sites. The shell-bearing Gastropoda, mainly Helicidae, were more frequent in the T and R sites. Within the Arachnida, the number of Araneae increased from the terrestrial sites to the wetter habitats of the river edge. Pseudoscorpiones, however, were restricted to the T sites. Diplopoda were present in considerable numbers on the Trent sites, mainly Polydesmida and Juliformia, both decreasing markedly from dry to wet habitats as did Opiliones, Acarina and Lithobiomorpha (Chilopoda). Two families of Isopoda were recorded in considerable numbers: Oniscoideae and Porcellionidae. The first was restricted to the T sites whereas the second was also present in the wetter habitats, though decreasing in numbers towards the river's edge.

Collembola was the numerically dominant taxon at all of the Helgeån sites, followed by Diptera, Opiliones, Araneae, and Staphylinidae and Carabidae (Coleoptera) respectively (Table 10.1). All Gastropoda were poorly represented and those found were evenly distributed in all sites. Araneae and Opiliones showed opposite trends: the first increased from T to F sites, while the second decreased. Diplopoda were well represented and besides Polydesmida and Juliformia, there were also Glomerida. This group, not recorded in any of the other river systems, decreased from T to F sites. Chilopoda were practically absent. The only family of Isopoda with a consistent presence was Triconiscidae, with a marked preference for the wetter habitats.

Individuals from sites on the Vindel River represent overall only 2% of the captures due to the F sites having been flooded for a very long period. Both T and R sites were dominated by Formicidae, followed by Staphylinidae (Coleoptera), Araneae, Opiliones and Diptera (Table 10.1). It is not possible to explain the very high number of Staphylinidae relative to the Carabidae and the complete absence of Diplopoda, Chilopoda and Isopoda.

CLASSIFICATION OF SITES

The classification of the sites using data for all invertebrate taxa for each of the five rivers revealed two important and interesting factors dictating the groupings of the species: location along the river corridor and location in relation to extent of flooding. The sites from the Adige-Noce, Trent and Helgeån Rivers exemplify these patterns (Figures 10.2, 10.3 and 10.4 respectively). For example, in the Adige-Noce and the Helgeån Rivers, in both of which the sites were located in different parts of the river corridor, the classification distinguished between upstream and downstream sites. In the Helgeån all six upstream sites were differentiated from the downstream sites (Figure 10.4); and four of the five Adige sites (R2, R4, R5 and T4) plus one Noce site (T2) were grouped together (Figure 10.2).

Within the sites from any one part of the river corridor, the F (frequently flooded), R (rarely flooded) and T (terrestrial/never flooded) sites were differentiated from each other. This is strikingly shown in the Trent sites (Figure 10.3) and in the upstream and downstream parts of the Helgeån, though in the latter, the F and R sites were not quite so distinctive (Figure 10.4). This pattern can also be seen in the Noce sites, apart from site F2 (Figure 10.2), but is not apparent in the Adige sites. An explanation for the absence of pattern in the latter could be that as the Adige River is embanked, flooding of the riparian areas is a relatively rare event. The presence of a frequently flooded site (e.g. F2) among a group comprising dry T sites could be due to the structure of the substrate, which in this particular site is composed of loose pebbles, with a very low moisture retention capacity.

DISCUSSION

River ecotones are an important part of the floodplain ecosystem. They are unique features linking most of the constituent habitats and thus functioning as corridors for animals, plants and energy fluxes. More information regarding invertebrate communities

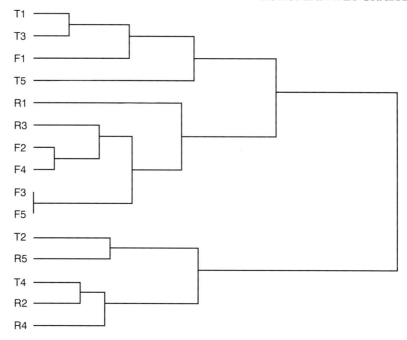

Figure 10.2 Classification of experimental sites on the River Adige-Noce system, Italy. T, terrestrial/never flooded; R, rarely flooded (1–10 weeks yr^{-1}); F, frequently flooded (10+ weeks yr^{-1})

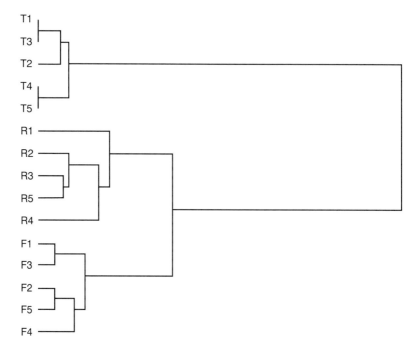

Figure 10.3 Classification of experimental sites on the River Trent, England. T, terrestrial/never flooded; R, rarely flooded (1–10 weeks yr^{-1}); F, frequently flooded (10+ weeks yr^{-1})

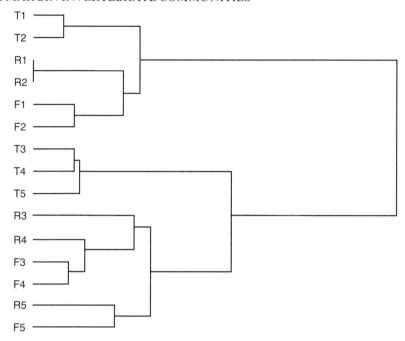

Figure 10.4 Classification of experimental sites on the River Helgeån, south Sweden. T, terrestrial/never flooded; R, rarely flooded (1–10 weeks yr^{-1}); F, frequently flooded (10+ weeks yr^{-1})

living in river corridor ecotones is needed for a better understanding of the ecology and associated human-induced or natural changes occurring in these delicate and sensitive ecosystems. The river ecotone, with its range of more or less flooded areas and associated vegetation, comprises different invertebrate communities in rather restricted areas, thus stressing the importance of river margins in enhancing biodiversity.

In this research the use of only one capture system (Barber traps) has been selective, sampling only endogenous and plant-dwelling invertebrates. This standard methodology across all rivers and all sites ensured that samples were comparable and repeatable. The study has demonstrated the value of using a whole-community approach to the differentiation of river margin habitats on the basis of the families or orders of invertebrates. Identification of the individuals to species has not been necessary in order to identify differences between rivers and between different types of sites within rivers. The Vindel River sites are relatively different from other sites, with the Trent and Helgeån Rivers being similar to each other, and likewise the Adige-Noce and Garonne (Figure 10.1). As would have been expected, the number of recorded taxa overall decreased from south to north. For example, Pseudoscorpiones, Diplopoda, Chilopoda and Isopoda are absent from the River Vindel study area. However, mean numbers of individuals alone for each of the five research areas show a significant correlation with the Continentality Index (CI: mean July air temperature minus January air temperature) (Petts and Décamps 1995). The highest abundance was found at the Trent sites (CI°C = 13) followed by the Garonne (CI°C = 16), the Helgeån (CI°C = 17), the Adige-Noce (CI°C = 22) and the Vindel (CI°C = 27).

Carabids were well distributed in all sites and, owing to their ecological role as predators and the fact that the autecology of the species is well known, this family represents a suitable group for more specific studies. The identification of samples from this study is being undertaken to species level and initial classification of the sites using carabids yields results very similar to those described above based on community data at the family or order level.

Classification of the sites based on community data identified the importance of location along the river corridor and of local environmental conditions, notably frequency of flooding. Further analyses of the data are needed to explore the importance of other environmental variables such as soil structure and moisture type, age and structure of vegetation cover, and the presence of shelters (e.g. stones and wooden debris), all of which are potentially influenced by distance of a site from the river and frequency of flooding.

CONCLUSIONS

In this research some aims have been achieved and new questions have been posed.

1.   The composition of riparian invertebrate communities in ecotones of some middle-sized rivers along a European latitudinal gradient has been assessed with regard to the soil surface-dwelling component.
2.   The role of spatial and temporal gradients in changing local community parameters has been tested but the importance of physical, chemical and biological variables needs further study before drawing conclusions. In particular, we expect more evidence from species analysis of the carabid communities in terms of evaluating the influence of flooding on terrestrial invertebrates. Most attention will be paid to hygrophilous species (*sensu lato*) i.e. all those species with habitats related to water and/or moisture.

It is important to investigate any changes that might be apparent at a species level as existing literature has shown that orders or families considered as a whole have a high ability to maintain stability and function (Vanhara, 1994).
3.   The data collected in this study may provide a useful comparison for future studies aimed at assessing variation of biota in riparian areas due to global change.

ACKNOWLEDGEMENTS

We wish to thank the Project Coordinator of ERMAS Henrì Décamps, and also Serenella Marchetti, Elisabetta Grigolli, Cinzia Roat, Barbara Del Prete and Ornella Casari for help in the field on Italian sites and for sorting the samples from all of the rivers.

Eric Chauvet, Gilles Pinay, Anne Marie Jean-Louis, Aimé Pech and Corinne Calvo undertook the field work on the Garonne; Malcolm Greenwood, John Taylor and Helen Ruffel from Loughborough University, on the River Trent; Lena Vought and Marie Svensson, on the Helgeån River; and Christer Nilsson, Mats Johansen, Shaojun Xiong, Ulf Larsson, Magnus Svedmark and Maria Danvind, on the Vindel River.

The ERMAS Project (Contract number EV 5 VCT 920100) was funded by the European Union.

SUMMARY

Riparian invertebrate communities were studied as part of the European River Margins Study (ERMAS). Sampling areas were chosen in the river margins of five floodplains along a latitudinal gradient across Europe (France, Italy, England, southern and northern Sweden). In each study area, five replicate sets of pitfall traps were set along a gradient from the river's edge to the inland unflooded habitats. Results are presented for whole communities with specimens identified to family or order level. Data have been used to classify sites according to their invertebrate taxa, with distinct communities inhabiting areas with different flooding regimes and similarly for different reaches of a given river system. The distribution of the main taxa at local and latitudinal level is discussed.

Keywords: Biodiversity, Entomology, Invertebrate communities, Pitfall traps, Riparian ecotones, River ecology, River margin, Similarity index.

REFERENCES

Calow, P. and Petts, G.E. (1992) *The rivers handbook*. Blackwell Scientific, London.

Casale, A., Sturani, M. and Vigna Taglianti, A. (1982) *Coleoptera Carabidae. Fauna d'Italia*, Vol. 18. Calderini, Bologna.

Chemini, C. (1979) Phalangids by pitfall trapping from Favogna, Province of Bolzano, Northern Italy (Arachnida, Opiliones). *Studi Trentini di Scienze Naturali*, **56**, 46–61.

Chemini, C. and Perini, G. (1982) Il popolamento di Carabidi in un bosco a Carpino Bianco presso Pergine (Trento) (Insecta: Coleoptera: Carabidae). *Studi Trentini di Scienze Naturali*, **59**, 66–195.

Chemini, C. and Pizzolotto, R. (1990) Carabid communities in woodland sites from the Lessini Mountains, Trentino, Italian Alps (Coleoptera: Carabidae). *Studi Trentini di Scienze Naturali*, **67**, 197–227.

Chemini, C. and Zanetti, A. (1982) Censimenti di Coleotteri Stafilinidi in tre ambienti forestali di Magré e Favogna (Provincia di Bolzano) (Insecta: Coleoptera: Staphylinidae). *Studi Trentini di Scienze Naturali*, **59**, 202–213.

Décamps, H., Fortune, M., Gazelle, F. and Pantou, G. (1988) Influence of man on the riparian dynamics of a fluvial landscape. *Landscape Ecology*, **1**, 163–173.

Demange, J. (1981) *Les milles pattes*. Boubée, Paris.

Germain, L. (1930) Mollusques terrestres et fluviatiles. *Faune de France*. Lechevalier, Paris.

Ghetti, P.F. (1986) *I macroinvertebrati nell'analisi di qualità dei corsi d'acqua*. Provincia Autonoma, Trento.

Greenwood, M.T., Bickerton, M.A., Castella, E., Large, A.R.G. and Petts, G.E. (1991) The use of Coleoptera (Arthropoda: Insecta) for floodplain characterization on the River Trent, UK. *Regulated Rivers: Research and Management*, **6**, 321–332.

Locket, G.H., Millidge, A.F. and Merret, P. (1975) *British spiders*. Ray Society, London.

Maiolini, B. (1993) *The role of terrestrial invertebrates in the riparian corridor in the Adige River*. Feasibility study for the cleaning plant Trento 3. Unpublished data.

Maiolini, B., Betti, L., Dorigoni, E., Franceschini, A. and Grigolli, E. (1993) *Le acque del Parco Adamello-Brenta*. Parco Documenti, No. 4.

Petts, G.E. and Décamps, H. (1995) *European river margins as indicators of global change*. Unpublished European Union Report.

Podani, J. (1990) SYN-TAX IV. *Computer programs for data analysis in ecology and systematics*. United Nations Industrial Development Organization. International Centre for Science and High Technology, Trieste.

Rosenberg, D.M. and Resh, V.H. (1992) *Freshwater biomonitoring and benthic macroinvertebrates*. Chapman & Hall, London.

Vanhara, J. (1994) Long-term ecological studies of terrestrial arthropods in the floodplain area

along the lower reaches of the Moravia and Dyje Rivers, with regard to floodplain forest Diptera. *Quaderni Stazione Ecologica civico Museo Storia Naturale Ferrara*, **6**, 185–204.
Zanetti, A. (1985) *Coleoptera Staphylinidae. Fauna d'Italia*, Vol. 25. Calderini, Bologna.

11 The Influence of Grassland Management on Body Size in Carabidae (Ground Beetles) and its Bearing on the Conservation of Wading Birds

SHONA BLAKE and GARTH N. FOSTER
The Scottish Agricultural College, Auchincruive, UK

INTRODUCTION

Assessment of conservation value and appropriate management strategy for any site should ideally take account of all components of the ecosystem, including vegetation and both vertebrate and invertebrate faunas. Nevertheless, it is perhaps unavoidable that the greatest efforts are often directed towards those groups which have the highest profile in the public perception. In lowland wet grasslands, the wading birds (waders) (Charadriiformes) are one such high-profile group, and the value of a wetland site is generally linked, in the public mind at least, with the numbers of breeding or overwintering birds it supports. José and Self (1994) list 14 breeding and another 12 wintering Red Data Book bird species at least partly dependent on British lowland wet grasslands. The British populations of 14 of these 26 species are internationally important. The health of a lowland wet grassland site may be judged, to some extent, by the numbers of wet grassland bird species it supports. Due largely to unsympathetic management, only 7% of the potential wet grassland in England and Wales has been found to support any breeding waders at all (Smith 1983).

If these habitats are to be managed for the benefit of their avian fauna, it is important to consider the effects of management practices on the invertebrates on which the birds depend for prey. Many species, such as *Vanellus vanellus* (lapwing), forage opportunistically, taking earthworms, Tipulidae (leatherjackets) or adult insects according to their availability (Galbraith 1989). Since the Coleoptera (beetles) are the most species-rich of all orders and are abundant in all habitats, they are likely to form a large part of the diet of such opportunistic feeders. The Carabidae (ground beetles), in particular, are preyed upon by at least 203 species of bird in the Palaearctic region (Larochelle 1980), including 12 of the 26 scarce species mentioned above (13 of the remaining 14 being waterfowl). Nidifugous charadriiform chicks have been shown to rely heavily upon carabids during the early part of their lives, when foraging time may be limited by weather conditions

See Glossary, p. 305, for explanation of technical terms.

European Wet Grasslands: Biodiversity, Management and Restoration. Edited by Chris B. Joyce and P. Max Wade.
© 1998 John Wiley & Sons Ltd.

(Beintema *et al.* 1990). Grassland management practices are likely to affect both the species composition and the abundance of the carabid fauna, with possible consequent effects on the energy budgets of the birds. In particular, it is of interest to consider the effects of management on the body size distribution of the beetles, since their food value to predators is positively correlated with their size.

Both foraging and handling time are reduced when a predator takes fewer, but larger, prey items. Weight increases approximately as the cube of body length, the relationship for adult carabids being weight $= 0.03069 \times \mathrm{length}^{2.63885}$ (Jarošík 1989). All beetles possess a thick cuticle of chitin, an indigestible polysaccharide that birds must eliminate before or after ingestion of the prey, reducing its nutritive value and possibly increasing handling time. However, the proportion of chitin is negatively correlated with body size (Kaspari and Joern 1993), so that larger beetles are both heavier and more nutritious than smaller ones with the same total length. Beintema (1991) has estimated that a *Tringa totanus* (redshank) chick near fledging would require an hourly intake of about 100 insects 9 mm in length, rising to almost 700 insects per hour if the insects' body size fell to 4 mm.

This chapter examines the effect of management intensification on the body-size distribution of carabid beetles in both wet and dry agricultural and semi-natural grasslands in Scotland and north-east England.

METHODS

Data on the carabid assemblages of 55 grassland sites in central and southern Scotland were collected by pitfall trappings between 1989 and 1994. At each location, traps consisted of two replicate sets of nine plastic cups, 8.5 cm diameter and 10 cm deep, partly filled with ethylene glycol (commercial antifreeze) and set flush with the ground surface about 2 m apart. Each trap was covered by a wire net to keep out birds and mammals. The traps were emptied and reset at approximately monthly intervals throughout the season, usually from late March to early October. Monthly catches were pooled to give two replicate annual totals for each site, a total of 110 sets of data.

In addition, data were analysed from pitfall trapping projects in north-east England between 1985 and 1992 (Blake *et al.* 1994). These sites comprised 39 coniferous and deciduous woodland, nine moorland and 50 improved and unimproved grassland. Grassland sites were assigned a management rating on a scale of 1 to 5 (1 = least intensively managed), taking account of the age of the pasture, the intensity and duration of grazing and the level of fertilizer and pesticide inputs (Luff *et al.* 1990). Environmental data available for grassland sites included volumetric soil water content, the proportion of organic matter in the soil and vegetation height.

For each of the Scottish sites, data were collected on sward type, sward age, cutting, grazing, and inorganic fertilization and organic inputs. Each of these aspects of the management regime was scored on a four-point scale from 0 to 3 in ascending order of intensity (Table 11.1). Natural or semi-natural vegetation scored 0, sown broad-leaved mixtures 1, sown grass and clover mixtures 2, and sown grasses 3. Uncultivated swards scored 0, permanent pastures more than 10 years old 1, pastures between 5 and 10 years old 2, and young grass or arable land 3. Defoliation and fertilization regimes were scored: none 0, low 1, moderate 2 and high 3. These scores were summed to arrive at a total

Table 11.1 Management intensity scores for four levels of six management parameters

Score	Sward type	Sward age	Cutting intensity	Grazing intensity	Inorganic inputs	Organic inputs
0	Natural or semi-natural	Uncultivated	None	None	None	None
1	Sown broad-leaved mixtures	Pasture >10 years old	Low	Low	Low	Low
2	Sown grass and clover mixtures	Pasture 5–10 years old	Moderate	Moderate	Moderate	Moderate
3	Sown grasses	Pasture <5 years old	High	High	High	High

management score lying between 0 and 18 for each site. Sites were then assigned to one of five levels of Grassland Management Intensity as follows: score 0–2, level 1; score 3–6, level 2; score 7–10, level 3; score 11–14, level 4; score 15–18, level 5. This scheme was based on qualitative rather than quantitative data and placed sites into fairly broad categories. This was felt to be desirable, given the often incomplete information available from farmers and other land managers. The categories were similar to those used by Luff et al. (1990) to describe the management of grassland sites in north-east England.

The average body size of the carabid fauna of each site was expressed as the Weight Median Length (WML) as described in Blake et al. (1994). This is the median point of the biomass distribution and is sensitive to the presence of larger individuals, which are more important ecologically, especially in terms of their value as prey items. The relationships between WML and habitat type, management intensity and environmental variables were examined using non-parametric statistical tests and generalized linear interactive modelling.

RESULTS

The mean carabid WML was greatest in moorland at 18.8 mm and woodland at 17.7 mm (Table 11.2, Figure 11.1). The overall mean values for grasslands were 13.4 mm for 50 north-east England sites and 13.2 mm for 110 Scottish sites. The differences in WML between grassland and woodland and between grassland and moorland were statistically significant (Mann–Whitney test: $P<0.001$ and $P<0.05$ respectively). There was no significant difference in WML between moorland and woodland. In grassland, average beetle body size decreased as Management Intensity level increased (Tables 11.2 and 11.3, Figures 11.1 and 11.2). The differences in WML according to Management Intensity level were significant for both the English and the Scottish data sets (Kruskal–Wallis One-way ANOVA: $K=22.28$, $P<0.001$ and $K=23.51$, $P<0.001$ respectively).

Carabid WML in the north-east England data set was also negatively correlated with volumetric soil water content and organic matter content of the soil, and positively correlated with vegetation height (Table 11.4). However, in this data set Management Intensity level was positively correlated with soil water and negatively with vegetation height. The mathematical model of the response of WML to management and environ-

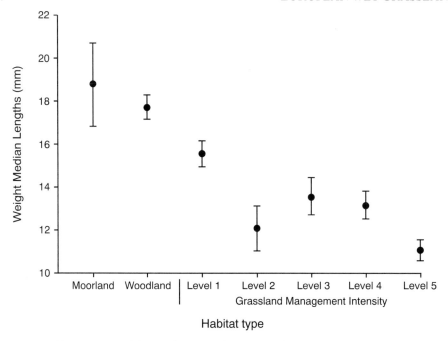

Figure 11.1   Mean (±SE) Weight Median Length (WML) of carabid (ground beetle) assemblages of nine moorland, 39 woodland and 50 grassland sites in north-east England. The grassland sites are classified according to Grassland Management Intensity: level 1 the least and level 5 the most intensively managed

mental variables interactively suggested that the principal influence on body size was that of Management Intensity level. Within each management level, the additional effect of soil water content was insignificant (Blake *et al.* 1994). In the Scottish grassland sites, WML responded negatively to intensification of all the components of management, but especially to changes in sward type and age (Blake *et al.* 1996).

DISCUSSION

These results suggest that all types of grassland, even when relatively undisturbed, support a carabid fauna of smaller body size than is supported by either moorland or woodland. Within grasslands, body size is greatly reduced by intensification of management. The reduction in WML from 15.6 to 11.1 mm (Table 11.2) translates to a reduction in dry weight from 45.9 to 18.4 mg. The average beetle in the most intensively managed grassland is only 40% as heavy as its counterpart under the least intensive management.

Soil water content is known to exert an influence on the carabid fauna (Luff *et al.* 1989; Eyre *et al.* 1990; Rushton *et al.* 1991). While average beetle size may be smaller in wetlands than in other grassland habitats, the most important factor affecting size within any wetland site will be intensity of management. Because of cutting and grazing, management results in an overall reduction in vegetation height, which may in itself be unfavourable to certain bird species such as *Gallinago gallinago* (snipe) (José and Self 1994). Increased management of wet grasslands normally involves increased drainage,

Table 11.2   Mean Weight Median Length (WML) of carabid (ground beetle) catches in different habitat types in north-east England

	Moorland	Woodland	Level of Grassland Management Intensity				
			1	2	3	4	5
Mean WML (mm)	18.8	17.7	15.6	12.1	13.6	13.2	11.1
SE	1.89	0.54	0.67	1.00	0.92	0.63	0.49
n	9	39	18	5	7	6	14

Intensity of management increases from level 1 to level 5 of Grassland Management Intensity. SE, standard error; n, number of sites of each habitat type.

Table 11.3   Mean Weight Median Lengths (WML) of carabid (ground beetle) catches in grasslands of different Management Intensity levels in Scotland

	Level of Grassland Management Intensity				
	1	2	3	4	5
Mean WML (mm)	15.3	13.3	12.8	11.8	11.2
SE	0.90	0.53	0.62	0.79	0.66
n	28	26	22	18	16

Intensity of management increases from level 1 to level 5 of Grassland Management Intensity. SE, standard error; n, number of sites in each level.

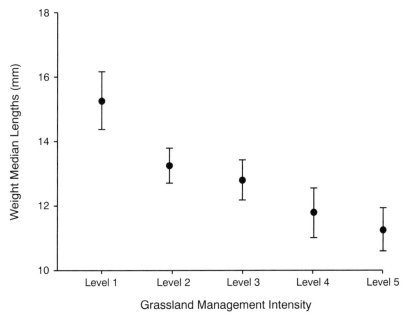

Figure 11.2   Mean (±SE) Weight Median Length (WML) of carabid (ground beetle) assemblages of 110 grassland sites in Scotland, classified according to Grassland Management Intensity: level 1 the least and level 5 the most intensively managed

Table 11.4 Results of Spearman's rank order tests for correlations between Weight Median Lengths (WML) and environmental variables at 50 grassland sites in north-east England

	r_s	n	P
WML: Management	−0.629	50	<0.001
WML: Soil water	−0.434	50	<0.01
WML: Organic matter	−0.314	50	<0.05
WML: Sward height	0.464	46	<0.01
Management: Soil water	0.293	50	<0.05
Management: Sward height	−0.734	46	<0.001

earlier and heavier stocking with livestock, earlier and more numerous cuts for fodder, and a reduction in both species and structural diversity of the vegetation. The detrimental effects on ground-nesting birds of trampling by livestock and of early cutting are well known, but the indirect effects of management practices through effects on the invertebrate fauna should also be considered. Early and repeated cutting for silage was associated in this study with especially low values of WML, usually below 10 mm, and this lack of larger prey would place the bird populations under greater pressure.

The relationship between WML and sward type and age reflects the requirement of larger carabid species for a more stable environment with less disturbance and fewer fluctuations in energy and nutrient levels. The larger species are predominately autumn breeders with slow-growing overwintering larvae (Kegel 1990) and most are flightless, while smaller species tend to be spring breeders with rapidly growing larvae and are more likely to be full-winged. Short-lived, fluctuating energy sources are more readily exploited by small-sized, dispersive species able to colonize and increase rapidly in abundance (Schoener and Janzen 1968; Gaston and Lawton 1988). The addition of nutrients, both from artificial fertilizer and from grazing animals, and the rapid removal of plant biomass by cutting and grazing, serve to create a fluctuating, unstable environment unfavourable to larger carabid species.

José and Self (1994) recommend that wet grassland management for the benefit of birds should be of low intensity, with no more than one cut, late in the season, a minimum of spring grazing and the maximum possible botanical diversity. All of these practices are likely further to benefit the birds by favouring a carabid fauna of relatively large body size.

ACKNOWLEDGEMENTS

This analysis was conducted while S. Blake was in receipt of a William Stewart Scholarship from the University of Glasgow. Facilities were provided by the Environmental Sciences Department of the Scottish Agricultural College (SAC), Auchincruive. SAC receives financial support from the Scottish Office Agriculture, Environment and Fisheries Department.

SUMMARY

The health of British lowland wet grasslands can be assessed by the populations of breeding or overwintering birds they support. It is important to consider the effects of management practices on the invertebrates on which the birds depend for prey, especially the effects on body size. Data on catches of Carabidae (ground beetles) in 110 sets of pitfall traps in Scotland and 50 sets in north-east England were analysed, estimating average body size by the Weight Median Length (WML), the median point of the biomass distribution. Carabid body size was negatively correlated with soil moisture, but the most important factor influencing the WML was intensity of management, with a reduction of almost 30% as management increased from the lowest to the highest level of intensity. Larger carabid body size will be favoured by low-intensity management, with reduced fertilization, cutting and grazing, and a more diverse sward.

Keywords: Body size, Carabidae (ground beetles), Grassland management, Prey, Wading birds.

REFERENCES

Beintema, A.J. (1991) Insect fauna and grassland birds. In: Curtis, D.J., Bignal, E.M. and Curtis, M.A. (eds), *Birds and pastoral agriculture in Europe*, pp. 97–101. Scottish Chough Study Group, Joint Nature Conservation Committee, Peterborough.

Beintema, A.J., Thissen, J.B., Tensen, D. and Visser, G.H. (1990) Feeding ecology of charadriiform chicks in agricultural grassland. *Ardea*, **79**, 321–344.

Blake, S., Foster, G.N., Eyre, M.D. and Luff, M.L. (1994) Effects of habitat type and grassland management practices on the body size distribution of carabid beetles. *Pedobiologia*, **38**, 502–512.

Blake, S., Foster, G.N., Fisher, G.E.J. and Ligertwood, G.E.L. (1996) Effects of management practices on the carabid faunas of newly-established wildflower meadows in southern Scotland. *Annales Zoologici Fennici*, **33**, 139–147.

Eyre, M.D., Luff, M.L. and Rushton, S.P. (1990) The ground beetle (Coleoptera, Carabidae) fauna of intensively managed agricultural grasslands in northern England and southern Scotland. *Pedobiologia*, **34**, 11–18.

Galbraith, H. (1989) The diet of lapwing *Vanellus vanellus* chicks on Scottish farmland. *Ibis*, **131**, 80–84.

Gaston, K.J. and Lawton, J.H. (1988) Patterns in the distribution and abundance of insect populations. *Nature*, **331**, 709–712.

Jarošík, V. (1989) Mass vs length relationship for carabid beetles (Col., Carabidae). *Pedobiologia*, **33**, 87–90.

José, P. and Self, M. (1994) The management of lowland wet grassland for birds. In: Crofts, A. and Jefferson, R.G. (eds), *The lowland grassland management handbook*, pp. 10.1–10.13. English Nature, Peterborough/The Wildlife Trusts, Lincoln.

Kaspari, M. and Joern, A. (1993) Prey choice by three insectivorous grassland birds: reevaluating opportunism. *Oikos*, **68**, 414–430.

Kegel, B. (1990) Diurnal activity of carabid beetles living on arable land. In: Stork, N.E. (ed.), *The role of ground beetles in ecological and environmental studies*, pp. 65–76. Intercept, Andover.

Larochelle, A. (1980) A list of birds of Europe and Asia as predators of carabid beetles including Cicindelini (Coleoptera: Carabidae). *Cordulia*, **6**, 1–19.

Luff, M.L., Eyre, M.D. and Rushton, S.P. (1989) Classification and ordination of habitats of ground beetles (Coleoptera, Carabidae) in north-east England. *Journal of Biogeography*, **16**, 121–130.

Luff, M.L., Eyre, M.D. and Rushton, S.P. (1990) Grassland management practices and the ground beetle fauna. *Proceedings of the 2nd Research Conference of the British Grassland Society*. British Grassland Society, Reading.

Rushton, S.P., Luff, M.L. and Eyre, M.D. (1991) Habitat characteristics of grassland *Pterostichus* species (Coleoptera, Carabidae). *Ecological Entomology*, **16**, 91–104.

Schoener, T.W. and Janzen, D.H. (1968) Notes on environmental determinants of tropical versus temperate insect size patterns. *American Naturalist*, **102**, 207–224.

Smith, K.W. (1983) The status and distribution of waders breeding on wet lowland grasslands in England and Wales. *Bird Study*, **30**, 177–192.

Part Three

MANAGEMENT

12 Plant Community Dynamics of Managed and Unmanaged Floodplain Grasslands: An Ordination Analysis

CHRIS B. JOYCE

International Centre of Landscape Ecology, Loughborough University, UK

INTRODUCTION

There has been a widespread loss of habitat diversity in European river floodplains over the past 200 years (Petts *et al*. 1989) and a marked reduction in the extent and biodiversity of semi-natural grasslands in Europe, particularly in the last 50 years (Fuller 1987; International Union for the Conservation of Nature and Natural Resources 1991; van Dijk 1991). Consequently, semi-natural floodplain grasslands (also termed inundation grasslands, alluvial meadows or flood meadows) have been similarly degraded and lost (Wells and Sheail 1988; Dargie 1993; Joyce 1994). Such grasslands are often of high nature conservation value, especially as they support rare plant species and vegetation types (Rodwell 1992; Jefferson and Robertson 1996; Jefferson and Grice 1998) and internationally important concentrations of wintering wildfowl and breeding wading birds (Hötker 1991; Jefferson and Robertson 1996).

The distinctive and ecologically valuable plant communities of European semi-natural floodplain grasslands are characterized by periodic inundation and are maintained by regular appropriate management, usually mowing and/or grazing. Indeed, past degradation and losses have been due largely to alterations in these factors through the implementation of flood defence, land drainage and agricultural intensification (Wells and Sheail 1988).

In contrast, there are many other remaining floodplain grasslands that are threatened by a lack of management. In northern and western Europe, for example, agricultural overproduction and policy reforms are leading to a reduction in the number of livestock in lowland areas and the withdrawal of marginal areas from agriculture (Bignal and McCracken 1992). In addition, following the dissolution of the Soviet Union, many central and eastern European countries have undergone reform in their agricultural as well as political systems. This has led to not only a shortage of public funds for agricultural subsidies and the abandonment of many areas, but also a lack of finance for grassland management for nature conservation objectives (Straškrabová *et al*. 1996). At the same time, reprivatization of land ownership is creating uncertainty in a number of

---

See Glossary, p. 305, for explanation of technical terms. Scientific names of vascular plants follow Tutin, T. G. *et al*. (1964–80) *Flora Europaea* Volumes 1–5. Cambridge University Press. See p. 319.

European Wet Grasslands: Biodiversity, Management and Restoration. Edited by Chris B. Joyce and P. Max Wade.

these countries (Baldock 1994), since many owners who have had land returned, or allocated, to them often have little incentive to manage it.

A number of studies have investigated the effects of management on the botanical composition of semi-natural grasslands (e.g. Baker 1937; Hopkins 1986; Bakker 1989; Smith and Jones 1991) and the changes brought about by neglect have been summarized (e.g. Ellenberg 1988, pp. 642–643; Wells 1989). However, the rate of community change can vary considerably between different grassland types (Fossati and Pautou 1989; Wells 1989; Gibson and Brown 1992). Little is known about the dynamics of semi-natural floodplain grassland plant communities in relation to management, or their sensitivity to the cessation of management. This chapter therefore seeks to explore the effects of management and abandonment on floodplain grassland plant community dynamics.

The analytical method selected for this study is ordination. Ordination techniques provide an effective means of describing temporal or spatial variation in vegetation composition. They have often been used to summarize or expose major features of the variation in plant community composition over time and to generate hypotheses about the underlying environmental factors, particularly from large observational data sets (Austin 1977; ter Braak 1995). Indeed, the study of grassland plant community dynamics in relation to management and succession has provided a focus for ordination. For example, Gibson and Brown (1992) assessed vegetation change in an early successional calcicolous grassland during six years of grazing and concluded that ordination analysis clearly separated grazing and temporal effects at the community scale. A study of succession in an abandoned damp meadow also found that ordination clearly differentiated the effects of cessation of grazing (Regnéll 1980).

This study aims to use ordination as a framework upon which to:

- elucidate the influence of particular management activities on floodplain grassland plant communities
- describe the short-term responses of different floodplain grassland communities to the cessation of management
- examine factors that influence floristic variation

The chapter concludes by indicating how an appreciation of such information may promote effective conservation management and rehabilitation of European floodplain grasslands.

STUDY SITES

The study was undertaken at two sites along the River Trent in the English Midlands (Figure 12.1). The Trent floodplain in this area is generally underlain by sand and gravel with a 1–2 m overburden of alluvial silts and clays. The history of the Trent is characterized by human modification, particularly systematic attempts at river regulation and land-use changes, often accompanied by a marked reduction in ecological diversity (Petts et al. 1992; Large et al. 1994). Pollen analysis has indicated that grassland utilization for agriculture was well established on the floodplain by 3200 BP (Lillie and Grattan 1995). Subsequently, the rich alluvial soils supported productive grazing for cattle and sheep, and rough grassland cut for hay, into the twentieth century (Edwards 1944), although the area was prone to periodic severe inundation (Marshall 1955). However, increasingly

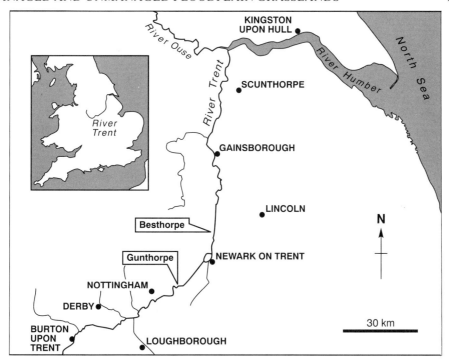

Figure 12.1 Location of the Gunthorpe and Besthorpe study sites

effective flood control facilitated land drainage and more intensive agriculture on the floodplain. Today, the river is regulated for most of its length, being channelized and with long reaches impounded by weirs and sluices. Flooding is largely controlled within washlands, and arable agriculture and intensively exploited grasslands predominate. Nevertheless, Dargie (1993) indicated that the Trent floodplain is a major source of the remaining wet grassland of potential nature conservation value in eastern England.

The study sites represent the two types of floodplain grassland plant communities that are of particular nature conservation interest along the Trent. One site was situated near the village of Gunthorpe (52° 58′ N, 1° 00′ W; National Grid Reference SK 665431) and was located in an approximately level pasture of 11.5 ha grazed by cattle and horses (Table 12.1). The pasture is underlain by loamy alluvial soils of the Wharfe series (Soil Survey of England and Wales 1983) with pH in the top 10 cm averaging 6.7. The grassland is regularly inundated and supports a plant community characteristic of wet circumneutral soils. This is recognized as the *Agrostis stolonifera–Alopecurus geniculatus* MG13 association in the National Vegetation Classification (Rodwell 1992), and as CORINE biotope type C37.242, *Agrostis stolonifera* and *Festuca arundinacea* swards (Devillers *et al.* 1991). This type of vegetation is a valuable forage resource for wildfowl (Burgess *et al.* 1990).

The second study site was situated in a 5.2-ha field near to Besthorpe village (53° 10′ N, 0° 46′ W; National Grid Reference SK 817641). The field is generally level with soils belonging to the Fladbury 2 series, being clayey and of alluvial origin (Soil Survey of England and Wales 1983) with an average pH in the top 10 cm of 6.2. The plant

Table 12.1 Summary of management and flooding at Gunthorpe and Besthorpe over the study period

Characteristic plant species	Gunthorpe (11.5 ha)			Besthorpe (5.2 ha)		
	Agrostis stolonifera, Alopecurus geniculatus, Poa trivialis			*Agrostis capillaris, Alopecurus pratensis, Festuca rubra, Holcus lanatus, Sanguisorba officinalis*		
	1993 (Mar–Dec)	1994 (Jan–Dec)	1995 (Jan–Sept)	1993 (Mar–Dec)	1994 (Jan–Dec)	1995 (Jan–Sept)
Management*						
Cutting	None	None	None	2 July	26 June	29 June
Grazing	1 Apr–7 Oct	10 Apr–16 Nov	1 Apr–1Dec	13 Sept–3 Oct	3 Oct–21 Oct	30 Sept (–Oct)
	8 horses	8 horses	6 horses	250 sheep	220 sheep	180 sheep
	44 cattle	24 cattle	36 cattle			
Duration of flooding†	59 days, mostly Oct–Dec	110 days, mostly Jan–Mar and Dec	65 days, Jan–Mar	None	None	10 days, Jan–Feb

*Outside the experiment enclosure only; † Inside and outside the experiment enclosure.

community present at Besthorpe is described by the National Vegetation Classification as MG4, the *Alopecurus pratensis–Sanguisorba officinalis* association (Rodwell 1992), and by CORINE biotope type C38.2, lowland hay meadows (Devillers *et al*. 1991). This community type is of considerable nature conservation value because of its botanical diversity and limited European distribution, being effectively confined to England (Council of the European Communities 1992; Rodwell 1992). The vegetation type is characteristic of areas where traditional, less intensive, hay-meadow management has been applied to seasonally flooded land with alluvial soils. At Besthorpe, there is a long history of late hay-cutting, with the regrowth grazed by sheep (Table 12.1). There is no history of fertilizer use in the area other than occasional small applications in the 1980s. Prior to flood defence works being completed in the 1970s, the area was likely to have been inundated annually (Marshall 1955), although since then it has flooded less frequently (N. Lewis, pers. comm. 1994).

METHODS

FIELD STUDY

The period of field study was March 1993–September 1995. In March 1993 an enclosure was established at each of the two study sites, excluding a section of the floodplain grassland from subsequent management, but not from flooding (Table 12.1). Within the enclosure, no management activity (including cutting and/or grazing) took place during the entire study period. Over the rest of the floodplain grassland, outside the enclosure, management continued as described in Table 12.1. There was also evidence of grazing by rabbits but only outside the enclosure. Flooding patterns at both sites are also summarized in Table 12.1.

The vegetation at both Gunthorpe and Besthorpe was monitored, usually monthly, between March and September in 1993, 1994 and 1995. Plant species presence and percentage cover were recorded in permanently marked 1-m^2 quadrats at both sites, four quadrats located at random within the enclosure (unmanaged) and either four or six located randomly in the floodplain grassland outside the enclosure (managed) in order to facilitate comparison with other experiments at the sites. Quadrats of 1 m^2 are recommended for botanical monitoring of grasslands (Smith *et al*. 1985) and were confirmed by examination at the beginning of the study to be appropriate to the spatial heterogeneity of the species composition of the swards.

DATA ANALYSIS AND PRESENTATION

Descriptive ordination analysis was most appropriate to investigate the effects of management regime and abandonment on plant community dynamics at the two different sites, particularly as statistical analysis was restricted by practical limitations in the field experimental design (see Hurlbert 1984). At both study sites, quadrats located within the enclosed unmanaged section of the floodplain grassland were segregated from quadrats in the managed remainder of the site. The study was therefore not controlled for the possibility that any small initial dissimilarities between the two parts of the same study site, or a chance event during the study, could have had an influence on plant community composition. Testing between such spatially segregated quadrats using inferential statistics would therefore only enable examination of the relationship between different parts of the floodplain rather than the effects of management or its cessation.

Ordination methods are used to represent visually the arrangement of units (species or samples) in two or more spatial dimensions. The position of each unit in the ordination will be uniquely determined by its particular combination of observed values. Units with similar sets of values will therefore occupy adjacent positions, or points, on the ordination diagram. The axes of the ordination diagram are either constrained, expressing measured environmental variables, or unconstrained, representing theoretical variables that explain point dispersion such that the points on each axis can subsequently be interpreted in relation to underlying environmental variables.

The unconstrained technique Detrended Correspondence Analysis (DCA) (Hill and Gauch 1980) was used to ordinate data sets from both study sites. DCA is appropriate for normally distributed species data with a long gradient of variation (ter Braak 1995). It is based on the observation that, within their range, species tend to be most abundant around their particular environmental optimum, i.e. they are unimodal (ter Braak and Prentice 1988). The method estimates the optima of the species response by a process known as reciprocal averaging (Hill 1973). This orders samples according to their species complement, and species by the samples in which they occur, by weighted averaging. With abundance data (e.g. percentage cover), weighted averaging applies weights proportional to species abundance. These orderings form the basis of the DCA axes, which each consists of a set of species scores derived from weighted averages and a corresponding set of sample scores that are weighted averages of the species scores. The first axis is constructed to maximize the dispersion of the species values; a second and further axes that maximize species dispersion can also be extracted by the same process such that they are uncorrelated with previous axes to ensure that additional information is expressed.

DCA yields species and sample scores that can be plotted against each other to fix the species or samples as points on a diagram. The species points are approximately the optima of their response to the environmental variables described by the DCA axes. Hence, the abundance or probability of occurrence of a species tends to decrease with distance from the location of its point on the diagram. Species plot closest to those samples in which they attain maximum abundance, and species near the centre of the diagram may be ubiquitous or they may be genuinely specific with an optimum near the centre of the sample range. Each sample score is the weighted average score of the species that occur in it. Hence, the species composition of a particular sample can be inferred from nearby species points, and samples that lie close to the point of a species are likely to have a high abundance of that species. Thus, samples that are most similar plot closest together and the degree of community change is positively related to distance on the DCA diagram.

Hill and Gauch (1980) noted that the variance of the optima of the species present in a sample is an estimate of the average response curve breadth of those species. They used the standard deviation as a measure of this breadth and defined the length of the DCA axis in standard deviation units (SD). This can aid interpretation of the DCA output because a species may be expected to appear, rise to its optimum, and disappear again in approximately 4 SD; a 50% change in sample composition will occur in approximately 1 SD (Hill and Gauch 1980).

Ordination was performed using the computer package CANOCO (ter Braak 1988).

An exploratory Correspondence Analysis (CA) on each of the Gunthorpe and Besthorpe data sets revealed an 'arch effect' distortion in the arrangement of points, indicating that the second axis of each diagram was systematically related to the first in a way that could impede interpretation (Hill and Gauch 1980). However, CA confirmed that variation within the sample data was sufficient to allow effective ordination using DCA. Hence, DCA was performed on both data sets incorporating detrending by polynomials to remove the arch effect (ter Braak 1995).

Plant community composition in the analysis was expressed as mean percentage cover for every species recorded from the managed and unmanaged sets of quadrats on each sampling date. Two other important variables that help define community composition and vegetation change were also included, namely bare ground and litter (defined as dead plant material of small size lying on the ground). These were also treated as response variables expressed as mean percentage cover, giving a total data set of 32 samples × 30 'species' for Gunthorpe and 33 samples × 29 'species' for Besthorpe. Cover data were transformed using $1n\,(y+1)$ so that data were approximately normally distributed.

Separate diagrams for Gunthorpe and Besthorpe were constructed from DCA by plotting scores for samples and species together against the first two axes of the ordination to summarize the main gradients of variation in the data.

RESULTS

Figures 12.2 and 12.3 illustrate the relative positions of the species and samples derived from DCA for the Gunthorpe and Besthorpe study sites respectively. The sample range along axis 1 was 2 SD for Gunthorpe and 1.5 SD for Besthorpe, indicating relatively subtle changes in community composition during the study.

GUNTHORPE

The first two axes of the DCA for Gunthorpe represent an important proportion of the total variation in the data set, together explaining 42% of this variation. Initially, scores for both managed and unmanaged sets of quadrats were similarly located along axis 1 of the DCA (Figure 12.2), indicating very little difference in community composition. Over the study generally however, grazed plots, and their characteristic species, had lower scores on axis 1 than had ungrazed plots and their associated species. Axis 1 therefore described a managed to unmanaged gradient. The pattern became increasingly evident after 1993, with progressively higher scores for managed plots on axis 1 related to ongoing grazing management in 1994 and 1995. This was typified by the enhanced presence of annual and ruderal species of low physical stature, such as *Chamomilla recutita, Plantago major* and *Polygonum aviculare*. In addition, several species newly colonized the managed plots (but not the unmanaged ones) during the study, including *Stellaria media, Taraxacum officinale* and *Juncus bufonius*. Only one species, *Rumex obtusifolius*, was eliminated from the grazed plots during the study.

Points representing unmanaged plots diverged from points for managed plots during the study period with declining scores along axis 1 (Figure 12.2). The cover of tall ruderal species such as *Rumex crispus* and *R. obtusifolius* increased in unmanaged plots over the period (e.g. mean total cover of these two species was 0.7% in June 1993, compared to

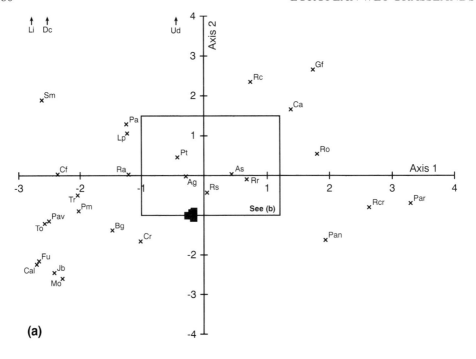

(a)

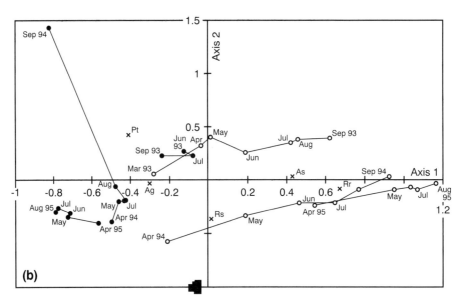

(b)

8.1% in June 1995). *R. crispus* and *Phalaris arundinacea* newly established themselves in the unmanaged plots, but not in the managed plots, during the study period. Three species, *Lolium perenne, Poa annua* and *Trifolium repens*, were lost from the unmanaged plots during the study, despite maintaining their presence in the managed plots.

Axis 2 in Figure 12.2 corresponded to the seasonal development of the vegetation. With one exception (September 1994; see below), sample dispersion along this axis was relatively narrow, indicating that community composition and species abundance in managed and unmanaged plots generally changed little over a season. Nevertheless, some temporal succession was evident. Species that were prominent earlier in the season, and study plots in which these species were relatively abundant, had low values along axis 2. In contrast, species that tended to achieve maximal abundance later in the season, and their corresponding plots, displayed higher axis 2 scores. Such species tended to be later successional ones, such as *R. obtusifolius* and *R. crispus*, that increased cover in a single season in the unmanaged plots (e.g. from a mean total cover value of 1.5% in April 1994 to 10.5% in September) and also increased cover in unmanaged plots over the three-year study period as a whole.

An important factor influencing the distribution of sample points for Gunthorpe was the extent of bare ground. This parameter had relatively low scores on both axes (Figure 12.2), indicating that it was prevalent early in the season but throughout the season only in managed plots. Patches of bare ground at Gunthorpe were often created by the deposition of silt during flooding, particularly in winter (Table 12.1), and were also developed and maintained through grazing and trampling by livestock. Thus, managed plots retained a mean value of at least 10% bare ground throughout the season, which partly accounts for the narrow range these points typically display on axis 2 in Figure 12.2. However, managed plots tended to exhibit decreasing scores along axis 1, where species that are able to withstand grazing and/or trampling were optimally located (e.g.

Figure 12.2   Positions of species and sample scores (standard deviation units) on the first two axes of a Detrended Correspondence Analysis (DCA) of species abundance data (mean % cover) from Gunthorpe 1993–95.

Species are signified by a cross, which represents the location of their optimal abundance in relation to the environmental variables described by the axes. Hence, species that respond similarly to the variables are positioned close to each other.

Samples are signified by filled circles for managed plots and open circles for unmanaged plots. Sample points represent weighted average abundance values of species recorded in those plots. The degree of community change is therefore indicated by distance on the diagram, i.e. the closer the sample points, the greater their similarity.

(a) Distribution of species scores (and other measured community variables). Positions of outliers are indicated by arrows.

(b) Centre of (a) enlarged to display sample scores. Lines linking samples recorded on different dates have been added to clarify temporal relationships.

Species codes: Ag, *Alopecurus geniculatus*; As, *Agrostis stolonifera*; Ca, *Cirsium arvense*; Cal, *Chenopodium album*; Cf, *Cardamine flexuosa*; Cr, *Chamomilla recutita*; Dc, *Deschampsia cespitosa*; Fu, *Filaginella uliginosa*; Gf, *Glyceria fluitans*; Jb, *Juncus bufonius*; Lp, *Lolium perenne*; Mo, Moss; Pa, *Poa annua*; Pan, *Potentilla anserina*; Par, *Phalaris arundinacea*; Pav, *Polygonum aviculare*; Pm, *Plantago major*; Pt, *Poa trivialis*; Ra, *Rumex acetosa*; Rc, *Rumex conglomeratus*; Rcr, *Rumex crispus*; Ro, *Rumex obtusifolius*; Rr, *Ranunculus repens*; Rs, *Rorippa sylvestris*; Sm, *Stellaria media*; To, *Taraxacum officinale*; Tr, *Trifolium repens*; Ud, *Urtica dioica*.

Other measured variables: Bg, bare ground; Li, litter

L. perenne, P. annua, S. media and *Rumex acetosa*). Managed plots therefore contrasted with the more dynamic unmanaged plots in which bare ground tended to become overgrown during the season.

The impact of flooding on vegetation dynamics at Gunthorpe was not restricted to its role in the generation of bare ground. Plant litter deposited by flooding was responsible for the outlying DCA score for managed plots in September 1994 (Figure 12.2), which was due to an unusually high mean litter cover of 18.7%. Nevertheless, the botanical composition of the affected plots had recovered by the next sampling date (April 1995) relative to records for the same plots in April 1994.

Several species were constant and relatively abundant components of both the managed and unmanaged plots at Gunthorpe throughout the three seasons of study, and could therefore be considered as describers of the community. These species, whose points were distributed close to the intersection of the axes in Figure 12.2, were *Agrostis stolonifera, Alopecurus geniculatus, Poa trivialis, Rorippa sylvestris* and *Ranunculus repens*. Although the species were all tolerant of temporary inundation and remained frequent in both managed and unmanaged plots, *A. geniculatus* and *P. trivialis* seemed to be slightly favoured by the grazing management, whilst a weak preference for the ungrazed treatment was shown by *A. stolonifera* and *R. repens*.

BESTHORPE

Twenty-one per cent of the variation in the Besthorpe data set was contained in the first two DCA axes. This was less than for Gunthorpe, indicating that the Besthorpe community was generally not as responsive to management and seasonal factors. Nevertheless, managed and unmanaged treatments at Besthorpe were clearly separated along axis 1 of the DCA (Figure 12.3). Plots that received annual cutting and grazing management

---

Figure 12.3   Positions of species and sample scores (standard deviation units) on the first two axes of a Detrended Correspondence Analysis (DCA) of species abundance data (mean % cover) from Besthorpe 1993–95.

Species are signified by a cross, which represents the location of their optimal abundance in relation to the environmental variables described by the axes. Hence, species that respond similarly to the variables are positioned close to each other.

Samples are signified by filled circles for managed plots and open circles for unmanaged plots. Sample points represent weighted average abundance values of species recorded in those plots. The degree of community change is therefore indicated by distance on the diagram, i.e. the closer the sample points, the greater their similarity.

(a) Distribution of species scores (and other measured community variables). Positions of outliers are indicated by arrows.

(b) Centre of (a) enlarged to display sample scores. Lines linking samples recorded on different dates have been added to clarify temporal relationships.

Species codes: Ac, *Agrostis capillaris*; Ae, *Arrhenatherum elatius*; Ao, *Anthoxanthum odoratum*; Ap, *Alopecurus pratensis*; Asy, *Anthriscus sylvestris*; Bh, *Bromus hordeaceus*; Cfo, *Cerastium fontanum*; Cn, *Centaurea nigra*; Cp, *Cardamine pratensis*; Dg, *Dactylis glomerata*; Fr, *Festuca rubra*; Gv, *Galium verum*; Hl, *Holcus lanatus*; Hs, *Heracleum sphondylium*; Hse, *Hordeum secalinum*; Lp, *Lolium perenne*; Lpr, *Lathyrus pratensis*; Mo, Moss; Pt, *Poa trivialis*; Ra, *Rumex acetosa*; Rac, *Ranunculus acris*; Rb, *Ranunculus bulbosus*; Sg, *Stellaria graminea*; So, *Sanguisorba officinalis*; Tf, *Trisetum flavescens*; To, *Taraxacum officinale*; Tp, *Tragopogon pratensis*.

Other measured variables: Bg, bare ground; Li, litter

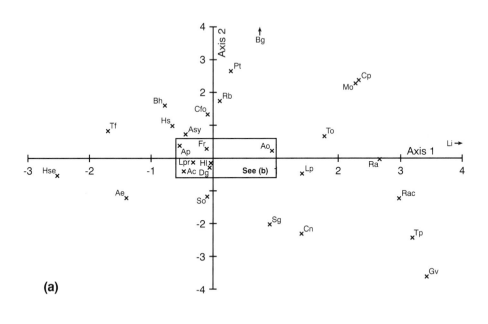

(a)

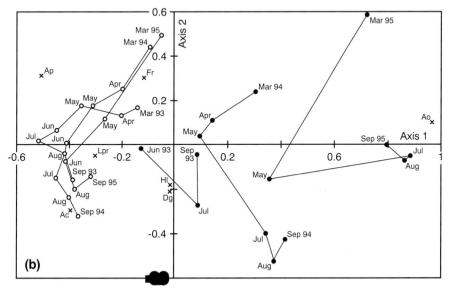

(b)

usually had higher scores along this axis than had unmanaged plots. Species and other measured parameters that reflected these treatments were similarly distributed. Thus, managed plots were characterized by the presence of litter following mowing for hay (e.g. mean of 6.5% litter cover in managed plots in August 1995, compared to none in unmanaged plots at the same time), probably due to ineffective removal of the cut material. Earlier in the season, managed plots also exhibited a greater cover of *Cardamine pratensis* and moss (mean moss cover of 9.5% in managed plots in March 1995 compared to 2.5% in unmanaged plots). In contrast, unmanaged plots were described by the enhanced presence of grasses that flower later in the season, most notably *Hordeum secalinum*, and by a greater cover of *Arrhenatherum elatius* and *Agrostis capillaris* than cut and grazed plots. *A. capillaris*, in particular, was present in abundance in unmanaged plots late in the season, achieving a mean cover of 45% in September 1995. The robust forbs *Anthriscus sylvestris* and *Heracleum sphondylium* also typified unmanaged plots. No species either was eliminated from, or newly established itself in, the managed or unmanaged plots during the three-year study.

Axis 2 in Figure 12.3 described the seasonal succession of the vegetation at Besthorpe. Higher scores on this axis correspond to earlier in the season, with later in the season being represented by lower scores. Cover of *Poa trivialis* peaked in spring in both managed and unmanaged plots, and *Ranunculus bulbosus, Cerastium fontanum, Bromus hordeaceus, C. pratensis* and moss were also conspicuous early in the season. Bare ground rarely featured in the experimental plots at Besthorpe, but tended to be associated with managed plots in March–April.

Unmanaged plots at Besthorpe followed a similar successional pattern each year, with very little between-year variation, as shown by the clustering of points in Figure 12.3. Managed plots exhibited a more variable response, both within and between seasons. Cutting, in particular, induced a pronounced shift in the sample scores each year, generally resulting in increased scores along axis 1. Cutting appeared to constrain the seasonal development of *A. capillaris* such that the increase in its cover observed later in the season in unmanaged plots was restricted. This seemed to enable other species, including *Anthoxanthum odoratum* and *Lolium perenne,* to maintain a presence in the managed plots later in the season. An explanation for the incremental yearly shift of managed plots along axis 1 in the diagram was not possible. Further study would have elucidated whether this pattern was either directionless variability introduced by cutting and grazing or represented the initial stages of a sustained trend of vegetation change.

It could be expected that managed and unmanaged plots at Besthorpe sampled in June 1993 would be similar in plant community composition, and hence in DCA scores, since management during the study did not commence until hay-cutting in July. However, DCA differentiated slightly between the two treatments (Figure 12.3) even at this early period of the experiment. These initial differences may have been due to the effects of rabbits grazing the managed plots outside the enclosure, as access to the unmanaged plots by rabbits was prevented following erection of the enclosure in March 1993. Alternatively, or in addition, they could have been caused by environmental heterogeneity within the study site, e.g. soil variability.

A number of plant species were relatively constant constituents of the community at Besthorpe during the study, irrespective of the implementation or absence of management. These included *A. capillaris, Alopecurus pratensis, Festuca rubra, Holcus lanatus, Sanguisorba officinalis* and, at low cover, *Dactylis glomerata* and *Lathyrus pratensis*.

Typically, points for these species are distributed around the axes intersection on the DCA diagram (Figure 12.3).

DISCUSSION

Ordination analyses described impacts of management and its cessation on floodplain grassland plant communities. Species composition and abundance appeared to follow a similar pattern to competition models presented by Grime (1979), which indicate that disturbance intensity and resource availability are important determinants of the competitive abilities of plants.

MANAGEMENT

The Gunthorpe environment was variable, being disturbed by grazing, trampling and episodic inundation. In such productive habitats exposed to repeated and severe disturbance, highly competitive plant species are suppressed and plants with a more ruderal strategy prevail (Grime 1979), often featuring short life cycles able to exploit environments only intermittently favourable for rapid plant growth. At Gunthorpe, recurrent flooding, and grazing and trampling facilitated the availability of moist bare mud and silt that was rich in available mineral nutrients, and therefore able to support rapid growth. However, the period available for growth was often relatively short, favouring annuals that germinated early and had a high potential growth rate, such as *Poa annua*, *Polygonum aviculare* and *Stellaria media*. Annual ruderal species were also prominent in particularly heavily grazed patches and trampled ground. Species such as *P. annua*, *Chamomilla recutita*, *Plantago major* and *Trifolium repens* could withstand intensive grazing and trampling because of their small size, prostrate habit, tough and elastic tissues, and rapid regeneration.

In contrast to Gunthorpe, the Besthorpe plant community was characterized by competitive stress-tolerant species. These were adapted to the relatively unproductive habitat and greater stability facilitated by the long history of extensive management comprising annual cutting for hay and grazing of the regrowth. Grime (1979) asserted that low intensities of disturbance function as a modifier of competition by favouring those species which tend to maintain their competitive ability either by avoidance of damage or by rapid recovery from its effects. At Besthorpe, the former were represented by rosette species (e.g. *Taraxacum officinale*) and species able to exploit the period of the year prior to hay-cutting (e.g. *Poa trivialis* and *Cardamine pratensis*). Species able to recover from cutting included *Sanguisorba officinalis*, *Festuca rubra* and *Agrostis capillaris*.

Regular cutting and grazing are likely to have a major impact on shorter-lived species unless they can recover vegetatively or have a persistent seedbank. Indeed, very few species at Besthorpe were short-lived with transient seedbanks (a notable exception being *Bromus hordeaceus*), suggesting that hay-making generally took place before many annuals shed seed. However, although most species at Besthorpe were perennials able to tolerate regular defoliation and reproduce vegetatively, they could also reproduce by seed production. In this context, the sheep that grazed Besthorpe were likely to be an important factor influencing species composition, for example by dispersing seeds in dung and wool (Bakker 1989; Fischer *et al*. 1996).

Floristic variation at Gunthorpe was not only related to grazing management, but was also associated with flooding. Regular inundation contributed to the preponderance of moisture-loving plants in the community and also created regeneration niches through sediment movement and deposition. Flooding can also provide a supply of nutrients so that damaged plants can regenerate strongly and seedlings develop. It can also aid dispersal by transporting diaspores (Ellenberg 1988). Many species of flooded swards can disperse by means of water-transported seeds or shoots, for example *Alopecurus geniculatus, Rumex crispus* and *Ranunculus repens,* which were all prominent species at Gunthorpe. Regular flooding at Gunthorpe also helped maintain the wet conditions that many characteristic species of the site require for germination, for example *Plantago major* and *Polygonum aviculare*.

CESSATION OF MANAGEMENT

Three years without management at the Gunthorpe and Besthorpe study sites induced changes in their plant communities that are largely consistent with the successional patterns widely described for neglected grasslands in Europe (Ellenberg 1988; Bakker 1989; Wells 1989; Rychnovská *et al*. 1994). It is generally accepted that following abandonment, grasses increase in leaf area, and above-ground biomass accumulates as standing dead material and litter. Productive, robust herbaceous species with a strong competitive ability prevail, excluding lower-growing, less competitive plants. Hence, plant species diversity is reduced and just one or two species, often grasses, become dominant. In relation to Grime's (1979) models of competition, this initial sequence suggests that more competitive plant species displace ruderal and stress-tolerant strategists with decreasing disturbance. Eventually shrubs invade and the grassland succeeds to scrub and woodland, although this can be slowed by the presence of accumulated litter and the dense competitive plant cover (Facelli and Pickett 1991; Prach 1994).

The cessation of grazing management at Gunthorpe removed some of the regeneration opportunities that the smaller ruderal plants had been able to exploit, although disturbance through inundation continued. A group of competitive ruderal strategists of taller stature were therefore favoured, particularly *Rumex obtusifolius* and *R. crispus*. In the absence of management, these species are able to complete their life cycle and maximize seed production to produce large reserves, and they are also able to tolerate flooding (van der Sman *et al*. 1993). Bakker (1989) found that *R. obtusifolius* increased its cover from 0 to 15% over 8 years in an abandoned wet grassland in The Netherlands, and Rychnovská *et al*. (1994) observed that *R. obtusifolius* and *R. crispus* were characteristic invaders of flooded meadows in central Europe. Although the subsequent course of succession at Gunthorpe cannot be predicted accurately, continued neglect would probably enable further expansion of the Rumices and the establishment and increase of other tall competitive ruderal species.

At Besthorpe, species of later phenological development that were previously suppressed by the cutting and grazing regime, notably *Arrhenatherum elatius* and *Agrostis capillaris*, increased in the unmanaged sward. Thus, removal of management favoured particularly competitive species able to tap any surplus resources and rapidly maximize production. Both of the above species are relatively coarse grasses. *A. elatius* quickly develops a tall, tussocky physiognomy that facilitates lateral expansion (Grime 1979; Rodwell 1992). *A. capillaris* is also capable of rapid growth and is moderately competitive when conditions are favourable (Grime *et al.* 1988). At Besthorpe, both species achieved a dense canopy of leaves from mid-summer when conditions for high productivity were suitable. Prach (1992) reported a similar succession in neglected floodplain grasslands in central Europe, where competitive and robust grasses such as *A. elatius, Alopecurus pratensis* and *Phalaris arundinacea* first increased and then dominated. At Besthorpe, further neglect would probably encourage the continued expansion of coarse grasses and the possible formation of an *Arrhenatheretum elatioris* community (Rodwell 1992).

A difference between the typical scheme of grassland succession and those described by this three-year study was the lack of litter accumulation in neglected grassland at both study sites. It is likely that flooding removed litter at Gunthorpe, although its potential importance was highlighted by the deposition of floodborne litter in September 1994, which had a temporary effect on species abundance. However, litter may begin to accumulate in these unmanaged floodplain grasslands after more than three years, particularly in the absence of flooding.

COMMUNITY DYNAMICS

Changes in management generally induced a greater community response at Gunthorpe than Besthorpe, as described by ordination. This suggests that variation is more likely to be expressed through the shorter-lived plant species in the short term, and that impact on the longer-lived perennials is delayed. The community at Gunthorpe, containing a large proportion of annual and biennial species able to tolerate unpredictable disturbance, was dynamic and sensitive to environmental influences. Yearly species turnover was observed, particularly in the managed vegetation where grazing and trampling removed plant biomass and also maintained gaps suitable for plant colonization. In contrast, the Besthorpe plant community was composed almost entirely of perennial species able to regenerate vegetatively and was comparatively resistant to the cessation of management with respect to species composition. This may have been a function of the long history of consistent hay-meadow management at the site, as the community contained many competitive strategists adapted to a taller dense canopy, which is a feature of hay meadows in the period before cutting as well as a characteristic of abandoned grasslands.

The results of ordination in this study have clearly illustrated changes in floodplain grassland plant communities in response to management activities and abandonment. However, there is a limit on the extrapolation of the results, partly because some floristic variation was not explained by ordination, some of which could have been due to spatial heterogeneity (e.g. within-site soil differences), but particularly as the study period of three years was insufficient to allow the grassland plant communities to respond fully to the cessation of management. Although Rychnovská *et al.* (1994) reported that derelict flooded meadows in Europe deteriorate relatively rapidly and that visible quantitative and incipient qualitative changes in the vegetation can occur after two years of neglect, abandoned wet grasslands in The Netherlands are characterized by the occurrence of *Rumex obtusifolius* after five years and *Heracleum sphondylium* after eight years (Bakker 1989). Also, Fossati and Pautou (1989) observed that it was nine years before *Sanguisorba officinalis*, a key component of the community at Besthorpe, was eliminated from a French wet grassland following the cessation of mowing. Furthermore, variations in management tend to have a more immediate effect on the structure and dominance of grassland vegetation than on species composition (Parr and Way 1988; Rychnovská *et al.* 1994). Indeed, many grassland plant species, including *Festuca rubra, Agrostis capillaris* and *Ranunculus repens,* which were all prominent in this study, exhibit large population fluctuations, even in successive years (Bakker 1989).

CONCLUSIONS: IMPLICATIONS FOR CONSERVATION

For nature conservation reasons, it is important that the dynamic effects of management and neglect of floodplain grasslands are better understood. This study has illustrated the role of regular management in maintaining floristic variation and characteristic species. Impacts on communities were mediated particularly through their effects on plant competition and regeneration. Thus, disturbance from grazing and trampling, and periodic inundation, created variable conditions suitable for a dynamic community characterized by short-lived ruderal species that responded rapidly to perturbations. A more stable environment featuring a long history of low-intensity management was expressed in a community that was composed mostly of perennial competitive stress-tolerant species and exhibited some inertia to management cessation. However, an absence of management in both types of grassland encouraged the expansion of robust competitive species, although the rate of successional change was greater in the community of short-lived plants from the unpredictable floodplain environment.

In Europe there are substantial areas of neglected floodplain grasslands that were dependent on appropriate management to maintain their nature conservation value. Reinstatement of regular management at an appropriate intensity could restore recently abandoned floodplain grasslands to their former community type, as cutting and grazing alters the competitive relationships in floodplain grassland vegetation by reducing the vigour of the taller-growing coarse species that tend to dominate derelict sites. Straškrabová and Prach (1998) reintroduced regular cutting to a floodplain meadow in the Czech Republic abandoned for 20 years. After five years, species diversity and composition were restored to a quality comparable with local floodplain meadows that had received uninterrupted management. However, this relatively rapid rehabilitation was supported by the seedbank of the neglected meadow, in which species typical of cut grasslands persisted, and by flooding, which transported diaspores from nearby sources (K. Prach, pers. comm. 1996). Rehabilitation of floodplain grasslands abandoned for longer than this example may be more problematic, particularly where a viable seed source and flooding are lacking. Cutting may lead to gaps appearing between the tall tussocks that characterize derelict grasslands, enabling undesirable herbaceous and woody species to colonize (Rychnovská *et al.* 1994). Time will therefore be needed for the desired grassland species, with their specialized ecological niches, to establish themselves naturally. As this study has indicated, however, although the superficial response of different floodplain grasslands to management change is similar, impact at the community level can vary greatly and individual communities will respond differently to the resumption of management. Proposals for rehabilitation and conservation of floodplain grasslands should not overlook their inherent heterogeneity and should therefore focus on developing management prescriptions for specific floodplain grassland ecosystems.

ACKNOWLEDGEMENTS

The financial support of European Union Contract No. EV5 vct 920100 is gratefully acknowledged. I thank the landowners and managers for kindly allowing permission to use the two study sites: Crown Estates Commissioners and G.S. Chatterton Ltd at Gunthorpe and Redland Aggregates Ltd and Nottinghamshire Wildlife Trust at Besthorpe. Thanks also to Val Black, Max Wade, Jane Reed, Jane Hallam, John Taylor and Erica Milwain.

SUMMARY

Ecologically valuable floodplain grasslands in Europe are threatened both by changes in management and by abandonment. In this study, the effects of management and its cessation on floodplain grassland plant community dynamics are examined using ordination. The study was undertaken at two grasslands in the English Midlands: one a grazed inundation community (Gunthorpe) and the other characteristic of traditional hay-meadow management (Besthorpe). Plant species cover at both sites was monitored in three successive growing seasons in one set of permanent quadrats subject to ongoing management and in another set in which no management activity took place. Data from Gunthorpe and Besthorpe were ordinated separately using Detrended Correspondence Analysis (DCA). At both sites, DCA separated managed from unmanaged treatments and described the seasonal succession of the vegetation. Grazing, trampling and episodic inundation at Gunthorpe maintained bare ground and favoured a dynamic community characterized by short-lived small ruderal species that responded rapidly to perturbations. At Besthorpe, the long history of low-intensity management was expressed in a community composed mostly of perennial competitive stress-tolerant species that were resistant to the cessation of management. Abandonment at both sites encouraged the expansion of robust competitive species that tended towards phenological development later in the season. It is recommended that the different responses by specific plant communities to management implementation and change should be considered in proposals for conservation and rehabilitation of floodplain grasslands.

Keywords: Competition, Conservation, Floodplain grasslands, Grassland management, Ordination, Plant community dynamics, Rehabilitation.

REFERENCES

Austin, M.P. (1977) Use of ordination and other multivariate descriptive methods to study succession. *Vegetatio*, **35**, 165–175.

Baker, H. (1937) Alluvial meadows: a comparative study of grazed and mown meadows. *Journal of Ecology*, **25**, 408–420.

Bakker, J.P. (1989) *Nature management by grazing and cutting*. Kluwer Academic, Dordrecht.

Baldock, D. (1994) Possible policy options and their implications for conservation. In: Haggar, R.J. and Peel, S. (eds), *Grassland management and nature conservation*, pp. 167–176. British Grassland Society, British Grassland Society Occasional Symposium No. 28. Reading.

Bignal, E. and McCracken, D. (1992) *Prospects for nature conservation in European pastoral farming systems*. Joint Nature Conservation Committee, Peterborough.

Burgess, N.D., Evans, C.E. and Thomas, G.J. (1990) Vegetation change on the Ouse Washes wetland, England, 1972–88 and effects on their conservation importance. *Biological Conservation*, **53**, 173–189.

Council of the European Communities (1992) Council Directive 92/43/EEC of 2 April 1992 on the conservation of natural habitats and of wild fauna and flora. *Official Journal of the European Communities*, No. L206.

Dargie, T.C. (1993) *The distribution of lowland wet grassland in England*. English Nature Research Report No. 49. English Nature, Peterborough.

Devillers, P., Devillers-Terschuren, J. and Ledant, J.P. (1991) *Habitats of the European Community. (CORINE Biotopes Manual.)* Office for Official Publications of the European Communities, Luxembourg.

Edwards, K.C. (1944) *Land utilization survey, part 60 – Nottinghamshire*. Geographical Publications, London.

Ellenberg, H. (1988) *Vegetation ecology of central Europe*. Cambridge University Press, Cambridge.

Facelli, J.M. and Pickett, S.T.A. (1991) Plant litter: its dynamics and effects on plant community structure. *The Botanical Review*, **57**, 1–32.

Fischer, S.F., Poschlod, P. and Beinlich, B. (1996) Experimental studies on the dispersal of plants and animals on sheep in calcareous grasslands. *Journal of Applied Ecology*, **23**, 1206–1222.

Fossati, J. and Pautou, G. (1989) Vegetation dynamics in the fens of Chautagne (Savoie, France) after the cessation of mowing. *Vegetatio*, **85**, 71–81.

Fuller, R.M. (1987) The changing extent and conservation interest of lowland grasslands in England and Wales: a review of grassland surveys 1930–84. *Biological Conservation*, **40**, 281–300.

Gibson, C.W.D. and Brown, V.K. (1992) Grazing and vegetation change: deflected or modified succession? *Journal of Applied Ecology*, **29**, 120–131.

Grime, J.P. (1979) *Plant strategies and vegetation processes*. Wiley, Chichester.

Grime, J.P., Hodgson, J.G. and Hunt, R. (1988) *Comparative plant ecology*. Unwin Hyman, London.

Hill, M.O. (1973) Reciprocal averaging: an eigenvector method of ordination. *Journal of Ecology*, **61**, 237–249.

Hill, M.O. and Gauch, H.G. (1980) Detrended correspondence analysis: an improved ordination technique. *Vegetatio*, **42**, 47–58.

Hopkins, A. (1986) Botanical composition of permanent grassland in England and Wales in relation to soil, environment and management factors. *Grass and Forage Science,*, **41**, 237–246.

Hötker, H. (ed.) (1991) *Waders breeding on wet grasslands*. Wader Study Group Bulletin No. 61. Wader Study Group, Tring.

Hurlbert, S.H. (1984) Pseudoreplication and the design of ecological field experiments. *Ecological Monographs*, **54**, 187–211.

International Union for the Conservation of Nature and Natural Resources (1991) *The lowland grasslands of central and eastern Europe*. Information Press, Oxford.

Jefferson, R.G. and Grice, P.V. (1998) The conservation of lowland wet grassland in England. In: Joyce, C.B. and Wade, P.M. (eds), *European wet grasslands: biodiversity, management and restoration*, pp. 31–48. John Wiley, Chichester.

Jefferson, R.G. and Robertson, H.J. (1996) *Lowland grassland: wildlife value and conservation status*. English Nature Research Report No. 169. English Nature, Peterborough.

Joyce, C.B. (1994) Effects of land-use changes on the Luznice floodplain grasslands in the Czech Republic. In: Haggar, R.J. and Peel, S. (eds), *Grassland management and nature conservation*, pp. 299–301. British Grassland Society Occasional Symposium No. 28. British Grassland Society, Reading.

Large, A.R.G., Prach, K., Bickerton, M.A. and Wade, P.M. (1994) Alteration of patch boundaries on the floodplain of the regulated River Trent, UK. *Regulated Rivers: Research and Management*, **9**, 71–78.

Lillie, M.C. and Grattan, J.P. (1995) Geomorphological and palaeoenvironmental investigations in the lower Trent valley. *The East Midland Geographer*, **18**, 12–24.

Marshall, J.D. (1955) A history of River Trent flooding. *Nottinghamshire Countryside,* **16**, 3–6.

Parr, T.W. and Way, J.M. (1988) Management of roadside vegetation: the long-term effects of cutting. *Journal of Applied Ecology*, **25**, 1073–1087.

Petts, G.E., Möller, H. and Roux, A.L. (eds) (1989) *Historical change of large alluvial rivers*. Wiley, Chichester.

Petts, G.E., Large, A.R.G., Greenwood, M.T. and Bickerton, M.A. (1992) Floodplain assessment for restoration and conservation: linking hydrogeomorphology and ecology. In: Carling, P.A.

and Petts, G.E. (eds), *Lowland floodplain rivers: geomorphological perspectives*, pp. 217–234. Wiley, Chichester.

Prach, K. (1992) Vegetation, microtopography and water table in the Luznice River floodplain, South Bohemia, Czechoslovakia. *Preslia*, **64**, 357–367.

Prach, K. (1994) Succession of woody species in derelict sites in central Europe. *Ecological Engineering*, **3**, 49–56.

Regnéll, G. (1980) A numerical study of successions in an abandoned, damp calcareous meadow in S. Sweden. *Vegetatio*, **43**, 123–130.

Rodwell, J.S. (ed.) (1992) *British plant communities*, Vol. 3, *Grasslands and montane communities*. Cambridge University Press, Cambridge.

Rychnovská, M., Blažková, D. and Hrabé, F. (1994) Conservation and development of floristically diverse grasslands in central Europe. In: 't Mannetje, L. and Frame, J. (eds), *Grassland and society*, pp. 266–277. Wageningen Pers, Wageningen.

Smith, I.R., Wells, D.A. and Welsh, P. (1985) *Botanical survey and monitoring methods for grasslands*. Focus on Nature Conservation No. 10. Nature Conservancy Council, Peterborough.

Smith, R.S. and Jones, L. (1991) The phenology of mesotrophic grassland in the Pennine Dales, northern England : historic hay cutting dates, vegetation variation and plant species phenologies. *Journal of Applied Ecology*, **28**, 42–59.

Soil Survey of England and Wales (1983) *1:250,000 Map of England and Wales, Sheet 4, Eastern England*. Soil Survey of England and Wales, Harpenden.

Straškrabová, J. and Prach, K. (1998) Five years of restoration of alluvial meadows: a case study from central Europe. In: Joyce, C.B. and Wade, P.M. (eds), European wet grasslands: biodiversity, management and restoration, pp. 295–303. John Wiley & Sons, Chichester.

Straškrabová, J., Prach, K., Joyce, C. and Wade, M. (eds) (1996) Floodplain meadows – ecological functions, contemporary state and possibilities for restoration. *Priroda*, **4**, 1–176.

ter Braak, C.J.F. (1988) *CANOCO – a FORTRAN program for canonical community ordination by (partial) (detrended) (canonical) correspondence analysis, principal component analysis and redundancy analysis (version 2.1)*. Agricultural Mathematics Group, Wageningen.

ter Braak, C.J.F. (1995) Ordination. In: Jongman, R.H.G., ter Braak, C.J.F. and van Tongeren, O.F.R. (eds), *Data analysis in community and landscape ecology*, pp. 91–173. Cambridge University Press, Cambridge.

ter Braak, C.J.F. and Prentice, I.C. (1988) A theory of gradient analysis. *Advances in Ecological Research*, **18**, 271–317.

van der Sman, A.J.M., Joosten, N.N. and Blom, C.W.P.M. (1993) Flooding regimes and life-history characteristics of short-lived species in river forelands. *Journal of Ecology*, **81**, 121–130.

van Dijk, G. (1991) The status of semi-natural grasslands in Europe. In: Goriup, P.D., Batten, L.A. and Norton, J.A. (eds), *The conservation of lowland dry grassland birds in Europe*, pp. 15–36. Joint Nature Conservation Committee, Peterborough.

Wells, T.C.E. (1989) Responsible management for botanical diversity. In: British Grassland Society (ed.), *Environmentally responsible grassland management*, pp. 4.4–4.16. British Grassland Society, Hurley.

Wells, T.C.E. and Sheail, J. (1988) The effects of agricultural change on the wildlife interest of lowland grasslands. In: Park, J. (ed.), *Environmental management in agriculture : European perspectives*, pp. 186–201. Belhaven Press, London.

13 Dynamics of Plant Litter in Riparian Meadows: Setting a Reference for Management

SHAOJUN XIONG and CHRISTER NILSSON
Riparian Ecology Group, Umeå University, Sweden

INTRODUCTION

The ongoing deterioration of the world's ecosystems has emphasized the need for conservation action both to maintain the remaining biodiversity and to allow a sustainable use of ecosystems (Robinson 1993). To succeed in this matter, a reference base and guidelines need to be developed. It is necessary to decide what we want our damaged ecosystems to become and look like in the future, to understand what sorts of processes operate in them under pristine conditions, and to what extent these processes can be restored.

Riparian meadows represent an example of an ecosystem that is in urgent need of both restoration and management in order to conserve it in a more original state. Riparian meadows occur naturally along some river stretches in the far north, where strong ice action during the spring flood prevents trees and shrubs from establishing in the upper riparian zone (Julin 1963; Arnqvist and Dynesius 1987). The resulting ecosystem often has a species-rich flora and fauna, because ice disturbance produces shifting mosaics of communities, and because the riparian zone is an ecotone between the upland and the water.

Besides the natural ice-governed meadows, the majority of riparian corridors in non-alpine and non-tundra regions in the northern hemisphere were covered by forest and shrubland in their natural states. Hundreds of years ago, these riparian forests were cleared following an expansion of the agricultural human population, and the areal extent of riparian meadows in northern regions such as northern and central Sweden was increased enormously. The resulting meadows were highly productive because of the seasonal fertilization by floods, and they have been used for hay-making over the centuries. However, during the twentieth century, their management successively stopped. Many of these meadows have been permanently inundated or damaged by flow regulation following hydroelectric operations and can no longer be recovered. Others are still regularly flooded but are being recolonized by riparian forest and shrubland. However, there are still some remnants of meadow patches created by humans along northern rivers that can be used as references in restoration and management of other

See Glossary, p. 305, for explanation of technical terms. Scientific names of vascular plants follow Tutin, T. G. *et al.* (1964–80) *Flora Europaea* Volumes 1–5. Cambridge University Press. See p. 319.

European Wet Grasslands: Biodiversity, Management and Restoration. Edited by Chris B. Joyce and P. Max Wade.
© 1998 John Wiley & Sons Ltd.

riparian meadows, not only in the north but elsewhere. This chapter focuses on those meadows which have been developed through human activity.

We have studied remnants of meadows in northern Sweden. The vegetation in these remnants often shows a typical vertical zonation and within the zones a mosaic structure (Nilsson 1983) that is species-rich and dynamic (Nilsson *et al.* 1989). It is largely governed by seasonal floods and related factors such as erosion, transport and deposition of sediments, litter and waterborne plant propagules. In this chapter we focus on the redistribution of plant litter and discuss this process in the context of management and restoration of riparian meadows. Litter is defined as dead plant material of small size lying loose on the ground (Facelli and Pickett 1991a).

THE ROLE OF PLANT LITTER IN RIPARIAN CORRIDORS

Riparian corridors belong to an ecosystem in which plant litter plays an important role for a number of reasons.

1.   The riparian corridor is highly dynamic in terms of how plant litter is redistributed (Nilsson *et al.* 1993). Much of the plant litter in the riparian corridor, including wind-borne plant litter from the adjacent uplands, is eroded during flood periods, transported by the river, decomposed, and deposited along other sections of the riparian corridor. Nutrients released from litter during decomposition in the river are also likely to return to the riparian corridor adsorbed to sediment particles (Mitsch *et al.* 1979). Both erosion and deposition of plant litter vary spatially and temporally. Some riparian areas export plant litter whereas others receive plant litter, depending on flood regimes, current velocity, and riparian structure. Redistribution of plant litter both within and between riparian corridors may contribute to the heterogeneity of riparian habitats and their vegetation by providing variation in nutrient conditions and physical geography. The microtopography of the riparian corridor resulting from scouring and deposition at least partly governs the distribution of species (Hardin and Wistendahl 1983).

2.   It is suggested that the decomposition of plant litter is faster in the riparian zone than in most uplands within the same climatic region (Merritt and Lawson 1980; Malanson 1993). High rates of decomposition imply high rates of mineral recycling, which favours primary productivity.

3.   Rivers are important natural corridors for the flows of species (Nilsson *et al.* 1991a,b; Malanson 1993; Naiman *et al.* 1993), and waterborne plant litter has an important role in these flows. Drifting plant litter not only brings plant diaspores that are part of the litter itself, but also forms floating or semi-floating islands that trap drifting plant diaspores and carry invertebrates. Furthermore, it creates bare ground for plant colonization by burying and suppressing established vegetation and leaving bare soil after decomposition (Nilsson *et al.* 1993). For example, Nilsson and Grelsson (1990) documented a maximum of 189 000 seeds m^{-2} in stranded plant litter along the Vindel River in northern Sweden, and Nilsson and co-workers (unpublished observations) found that artificial seed replicas used in experiments were often trapped in packs of plant litter.

4.   Production of plant litter in riparian areas is higher than that in uplands (e.g. Brinson 1990; Malanson 1993). This abundant plant litter provides a high potential for influencing both riparian and aquatic communities. For example, riparian plant litter is the major source of food for aquatic invertebrates (e.g. Cummins *et al.* 1989).

Not surprisingly, being a carrier of materials and energy, plant litter is an essential element in concepts of the river continuum (Vannote *et al.* 1980), the nutrient spiralling (Webster 1975), the riparian filtering (Ward 1989) and the flood pulse (Junk *et al.* 1989). It has also been demonstrated that species richness and species composition of riparian vegetation may vary along the gradient of litter accumulation (Nilsson and Grelsson 1990).

POTENTIAL IMPACTS OF PLANT LITTER ON RIPARIAN MEADOWS

Riparian meadows in northern regions both lose and receive plant litter during seasonal floods. The losses mainly consist of leaf litter, but the deposits include both wood and leaf litter; however, the former component has successively become less important since riparian forests and river channels began to be cleared of woods and trees (Maser and Sedell 1994). The deposits of plant litter may range from a few hundred grams to several kilograms per square metre (Nilsson and co-workers, unpublished observations).

The accumulation of plant litter has physical, chemical and biological impacts on vegetation (Facelli and Pickett 1991a; Nilsson *et al.* 1993). The physical and chemical impacts of plant litter have been studied from several aspects (Facelli and Pickett 1991a, and references therein), whereas the biological impact is little known (Nilsson *et al.* 1993).

The physical impact concerns the burial of plants and subsequent changes in light, temperature and moisture on the soil surface. This burial may also lead to floristic changes, e.g. the exclusion of small plants such as *Selaginella* and *Euphrasia* (Nilsson and co-workers, unpublished observations). Plant litter also protects seeds and young plants from erosion and animals, and insulates the soil, thus reducing fluctuations in temperature and moisture.

The chemical impact concerns the nutrients and phytotoxins that might be added and changes in soil acidity following decomposition of plant litter. Plant litter may also indirectly increase soil fertility by intercepting and trapping suspended nutrient particles from the floodwater (Jordan *et al.* 1989).

The biological impact involves both a direct effect, in that plant diaspores are carried by plant litter and deposited on the riverbank together with the litter, and indirect effects, either through the creation of patches for plant colonization or through the attraction of animals and micro-organisms that may change plant communities.

Many studies on plant litter in a variety of non-forested ecosystems have shown that plant litter may affect germination and establishment, species richness and biomass production of plant communities (Table 13.1), mostly due to the physical and chemical impacts of plant litter (Facelli and Pickett 1991a). Although study sites range from dry deserts to lacustrine meadows, significant effects of leaf litter accumulation on plant communities were found in most cases and most of them were negative.

DYNAMICS OF PLANT LITTER IN THE RESTORATION AND MANAGEMENT OF RIPARIAN MEADOWS

The redistribution of plant litter in riparian meadows is intimately linked to flooding regimes. Flooding prevents the litter from accumulating and suppressing vegetation over

Table 13.1 Examples of leaf-litter effects on plant communities in non-forest ecosystems. The studies are arranged from dry (top) to wet (bottom) conditions

Ecosystem type and location	Germination and establishment	Species richness	Biomass production	Reference
Desert, California, USA	−			Sheps (1973)
Arid grassland, Texas, USA	±[b]			Fowler (1986)
Arid grassland, Texas, USA	−			Fowler (1988)
Chalk grassland, UK	−	−		Watt (1974)
Highland grassland, Australia	−			Williams and Ashton (1987)
Grassland, Argentina		−		Facelli et al. (1988)
Annual grassland, California, USA			+	Heady (1956)
Annual grassland, Washington, USA	−			Bergelson (1990)
Prairie grassland, California, USA			−	Weaver and Rowland (1952)
Prairie grassland, Oklahoma, USA			−	Knapp and Seastedt (1986)
Prairie, Kansas, USA			−	Hulbert (1969)
Pasture, Durham, UK	±[b]			Lee and Cooke (1989)
Old field, Michigan, USA	−			Goldberg and Werner (1983)
Old field, New Jersey, USA	±[b]	0	−	Facelli and Pickett (1991b), Facelli (1994)
Old field, New Jersey, USA		−	0	Carson and Peterson (1990)
Old field, Michigan, USA	−			Werner (1975)
Old field, South California, USA		−		Monk and Gabrielson (1985)
Lacustrine wetlands, Canada	−	−		van der Valk (1986)
Lake shore, Sweden			−	Granéli (1989)
Lake and saltmarsh, UK	−	−		Haslam (1971a, b)

[a]+, − and 0 represent positive, negative and no significant responses respectively.
[b]The effects of litter vary with litter amount and/or the species studied.

large areas. Instead, litter is eroded from some places and accumulated on others, and this contributes to the development and maintenance of species-rich and productive meadows. Flooding also leads to litter decomposition in the river channel, and is therefore also responsible for feedback responses such as redeposition in the riparian corridor of nutrients from litter decomposed in the river. Flooding also enhances rates of litter decomposition in the riparian corridor, and dilutes the phytotoxin compounds that are considered to harm the development of plant communities in uplands (Sheps 1973; van der Valk 1986; Chapin et al. 1994). Therefore, to restore the dynamics of plant-litter redistribution, recovery of flooding regimes has to be incorporated into programmes for restoration and management of riparian meadows. At present, the hydrological regime in the majority of rivers in the northern hemisphere is to some extent modified by dams, diversions and withdrawals, and riparian ecosystems cannot be maintained (Dynesius and Nilsson 1994). Therefore, reconnecting the rivers with their floodplains has become a major goal in many river restoration schemes (e.g. Large and Petts 1994; Dahm *et al.* 1995).

Plant litter redistribution operates not only on local but also on regional scales, in that plant litter and associated plant diaspores are transported downstream and thus connect upstream with downstream sites. This transport is hampered by the fragmentation and regulation of rivers by dams. Programmes for full-scale restoration of litter redistribution in riparian meadows therefore have also to include dam removal. Such measures have been put into practice, for example in North America (Shuman 1995), but the economic costs can be high, especially if large amounts of accumulated sediments have to be removed to re-establish the former channel morphology and pre-regulation flooding regimes in the riparian corridors (Wunderlich *et al.* 1994).

Although restoration schemes designed to restore flooding regimes and allow nature to do the remaining job are likely to be more successful, they might not always be feasible, for economic or other reasons. Such constraints provide a challenge to search for other, alternative measures that are not aimed at complete restoration but imply some remediation. However, to succeed in such efforts, whatever they be, we need a fuller understanding of the dynamics of litter redistribution in rivers, including their riparian corridors. For example, how large must deposits be to change the vegetation of riparian meadows from one year to the next and, conversely, how small can deposits be to leave vegetation unaffected? Which life histories of plants are the most successful in resisting large litter deposits and, also, which species are the most sensitive to litter deposition? Which of the physical, chemical and biological effects of deposited litter are most important, and does the relative importance of these effects vary among different types of rivers in different biogeographic regions?

Finally, besides the restoration and management of natural ecological processes, we also need a comprehensive knowledge of how hay-making procedures could be operated to maintain the value of riparian meadows as products of a successful and long-term co-operation between nature and humans.

ACKNOWLEDGEMENTS

We thank Gunnel Grelsson, Roland Jansson, Mats Johansson and Elisabet Nilsson for valuable comments on the manuscript. This work was funded by the Swedish Institute, and the European Commission Environment Programme (through the Swedish Environmental Protection Agency) under Research Grant EV5V-CT92-0100 (ERMAS). The authors are grateful to Dr H. Barth for his interest in the project.

SUMMARY

The redistribution of plant litter in riparian meadows is reviewed in the context of the management and restoration of this habitat. Such redistribution is a dynamic process with implications for nutrient release and export, mineral cycling, the transportation of plant propagules and the creation of bare ground, and influencing both riparian and aquatic plant and animal communities. The accumulation of plant litter has physical, chemical and biological impacts on vegetation. Flooding is a key factor with litter being eroded from some places and accumulating in others. Although restoration schemes designed to restore flooding regimes and allow nature to take its course are likely to be most successful, they might not always be feasible.

Keywords: Germination, Litter, Mineral cycling, Nutrient release, Restoration, Riparian meadows.

REFERENCES

Arnqvist, G. and Dynesius, M. (1987) *Råneälven Naturinventering och bedömning av vetenskapliga naturvärden*, Länsstyrelsen i Norrbottens län, Luleå.

Bergelson, J. (1990) Life after death: site pre-emption by the remains of *Poa annua. Ecology*, **71**, 2157–2165.

Brinson, M.M. (1990) Riverine forests. In: Lugo, A.E., Brinson, M. and Brown, S. (eds), *Forested wetlands. Ecosystems of the world*, Vol. 15, pp. 87–141. Elsevier, Amsterdam.

Carson, W.P. and Peterson, C.J. (1990) The role of litter in an old field community: impact of litter quantity in different seasons on plant species richness and abundance. *Oecologia*, **85**, 8–13.

Chapin, F.S., Walker, L.R., Fastie, C.L. and Sharman, L.C. (1994) Mechanisms of primary succession following deglaciation at Glacier Bay, Alaska. *Ecological Monographs*, **64**, 149–175.

Cummins, K.W., Wilzbach, M.A., Gates, D.M., Perry, J.B. and Taliaferro, W.B. (1989) Shredders and riparian vegetation. *BioScience*, **39**, 24–30.

Dahm, C.N., Cummins, K.W., Valett, H.M. and Coleman, R.L. (1995) An ecosystem view of the restoration of the Kissimmee River. *Restoration Ecology*, **3**, 225–238.

Dynesius, M. and Nilsson, C. (1994) Fragmentation and flow regulation of river systems in the northern third of the world. *Science*, **266**, 753–762.

Facelli, J.M. (1994) Multiple indirect effects of plant litter affect the establishment of woody seedlings in old fields. *Ecology*, **75**, 1727–1735.

Facelli, J.M. and Pickett, S.T.A. (1991a) Plant litter: its dynamics and effects on plant community structure. *Botanical Review*, **57**, 1–32.

Facelli, J.M. and Pickett, S.T.A. (1991b) Plant litter: light interception and effects on an old field plant community. *Ecology*, **72**, 1024–1031.

Facelli, J.M., Montero, C.M. and León, R.J.C. (1988) Effect of different disturbance regimes on seminatural grasslands from the subhumid pampa. *Flora*, **180**, 241–249.

Fowler, N.L. (1986) Microsite requirements for germination and establishment of tree grass species. *American Midland Naturalist*, **115**, 131–145.

Fowler, N.L. (1988) What is a safe site?: neighbor, litter, germination date, and patch effects. *Ecology*, **69**, 947–961.

Goldberg, D.E. and Werner, P.A. (1983) The effects of size of opening in vegetation and litter cover on seedling establishment of goldenrod (*Solidago* spp.). *Oecologia*, **60**, 149–155.

Granéli, W. (1989) Influence of standing litter on shoot production in reed, *Phragmites australis* (Cav.) Trin. ex. Steudel. *Aquatic Botany*, **35**, 99–109.

Hardin, E.D. and Wistendahl, W.A. (1983) The effects of floodplain trees on herbaceous vegetation patterns, microtopography and litter. *Bulletin of the Torrey Botanical Club*, **110**, 23–30.

Haslam, S.M. (1971a) Community regulation in *Phragmites communis* Trin. I. Monodominant stands. *Journal of Ecology*, **59** , 65–73.

Haslam, S.M. (1971b) Community regulation in *Phragmites communis* Trin. II. Mixed stands. *Journal of Ecology*, **59**, 75–88.

Heady, H.F. (1956) Changes in the central California annual plant community induced by the manipulation of natural mulch. *Ecology*, **37**, 798–811.

Hulbert, L.C. (1969) Fire and litter effects in undisturbed bluestem prairie in Kansas. *Ecology*, **50**, 874–877.

Jordan, T.E., Whigham, D.F. and Correll, D.L. (1989) The role of litter in nutrient cycling in a brackish tidal marsh. *Ecology*, **70**, 1906–1915.

Julin, E. (1963) Den isgångsbetingade ängen vid Sundholmen i Torne älvs mynningsområde. *Svensk Botanisk Tidskrift*, **57**, 19–38.

Junk, W.J., Bayley, P.R. and Spark, R.E. (1989) The flood pulse concept in river floodplain systems. *Canadian Special Publication of Fisheries and Aquatic Sciences*, **106**, 110–127.

Knapp, A.K. and Seastedt, T.R. (1986) Detritus accumulation limits productivity of tallgrass prairie. *BioScience*, **36**, 662–668.

Large, A.R.G. and Petts, G.E. (1994) Rehabilitation of river margins. In: Calow, P. and Petts, G.E. (eds), *The rivers handbook*, pp. 401–418. Blackwell Scientific, Oxford.

Lee, H.C. and Cooke, J.A. (1989) Effects of bracken litter and frond canopy on emergence and survival of ryegrass and white clover seedlings. *Journal of Agricultural Science, Cambridge*, **113**, 397–400.

Malanson, G.P. (1993) *Riparian landscapes*. Cambridge University Press, Cambridge.

Maser, C. and Sedell, J.R. (1994) *From the forest to the sea. The ecology of wood in streams, rivers, estuaries and oceans*. St Lucie Press, Delray Beach, Florida.

Merritt, R.W. and Lawson, D.L. (1980) *Leaf litter processing in floodplain and stream communities*. US Department of Agriculture, Forest Service, General Technical Report WO-12, 93–105. Washington, DC.

Mitsch, W.J., Dorge, C.L. and Wiemhoff, J.R. (1979) Ecosystem dynamics and a phosphorus budget of an alluvial cypress swamp in southern Illinois. *Ecology*, **60**, 1116–1124.

Monk, C.D. and Gabrielson, F.C. (1985) Effects of shade, litter and root competition on old-field vegetation in South Carolina. *Bulletin of the Torrey Botanical Club*, **112**, 383–392.

Naiman, R.J., Décamps, H. and Pollock, M. (1993) The role of riparian corridors in maintaining regional biodiversity. *Ecological Applications*, **3**, 209–212.

Nilsson, C. (1983) Frequency distributions of vascular plants in the geolittoral vegetation along two rivers in northern Sweden. *Journal of Biogeography*, **10**, 351–369.

Nilsson, C. and Grelsson, G. (1990) The effects of litter displacement on riverbank vegetation. *Canadian Journal of Botany*, **68**, 735–741.

Nilsson, C., Grelsson, G., Johansson, M. and Sperens, U. (1989) Patterns of plant species richness along riverbanks. *Ecology*, **70**, 77–84.

Nilsson, C., Gardfjell, M. and Grelsson, G. (1991a) Importance of hydrochory in structuring plant communities along rivers. *Canadian Journal of Botany*, **69**, 2631–2633.

Nilsson, C., Grelsson, G., Dynesius, M., Johansson, M.E. and Sperens, U. (1991b) Small rivers behave like large rivers: effects of postglacial history on plant species richness along riverbanks. *Journal of Biogeography*, **18**, 533–541.

Nilsson, C., Nilsson, E., Johansson, M.E., Dynesius, M., Grelsson, G., Xiong, S., Jansson, R. and Danvind, M. (1993) Processes structuring riparian vegetation. In: Menon, J. (ed.), *Current topics in botanical research*, pp. 419–431. Council of Scientific Research Integration, Trivandrum.

Robinson, N.A. (1993) *Agenda 21: Earth's action plan*. Oceana, New York.

Sheps, L.O. (1973) Survival of *Larrea tridentata* S. & M. seedlings in Death Valley national monument, California. *Israel Journal of Botany*, **22**, 8–17.

Shuman, J.R. (1995) Environmental considerations for assessing dam removal alternatives for river restoration. *Regulated Rivers*, **11**, 249–261.

van der Valk, A.G. (1986) The impact of litter and annual plants on recruitment from the seed bank of a lacustrine wetland. *Aquatic Botany*, **24**, 13–26.

Vannote, R.L., Minshall, G.W., Cummins, K.W., Sedell, J.R. and Cushing, C.E. (1980) The river continuum concept. *Canadian Journal of Fisheries and Aquatic Sciences*, **37**, 130–137.

Ward, J.V. (1989) Riverine–wetland interactions. In: Sharitz, R.R. and Gibbons, J.W. (eds), *Freshwater wetlands and wildlife*, pp. 385–400. Office of Scientific and Technical Information, US Department of Energy, Oak Ridge.

Watt, A.S. (1974) Senescence and rejuvenation in ungrazed chalk grassland in Breckland: the significance of litter and moles. *Journal of Applied Ecology*, **11**, 1157–1171.

Weaver, J.E. and Rowland, N.W. (1952) Effect of excessive natural mulch on the development, yield and structure of a native grassland. *Botanical Gazette*, **114**, 1–19.

Webster, J.R. (1975) *Analysis of potassium and calcium dynamics in stream ecosystems on three southern Appalachian watersheds of contrasting vegetation.* Unpublished PhD Dissertation. University of Georgia, Atlanta.

Werner, P.A. (1975) The effect of plant litter on germination in teasel, *Dipsacus sylvestris. American Midland Naturalist,* **94**, 470–476.

Williams, R.J. and Ashton, D.H. (1987) Effect of disturbance and grazing by cattle on the dynamics of heathlands and grasslands communities on the Bogong High Plains, Victoria. *Australian Journal of Botany,* **35**, 413–431.

Wunderlich, R.C., Winter, B.D. and Meyer, J.H. (1994) Restoration of the Elwha River ecosystem. *Fisheries,* **19**, 11–19.

14   Managing Water for Wetland Ecosystems: A Case Study

ADRIAN ARMSTRONG and STEVE ROSE
ADAS Land Research Centre, Gleadthorpe, UK

INTRODUCTION

Restoration of wetland status to land that has been improved for agricultural usage involves much more than the simple cessation of drainage. It also requires active management of both the hydrology and the land. Although target land management and water regimes can be established for some major species, these have to be implemented within the requirements of the continued agricultural usage of the land, and the institutional and economic constraints (Armstrong *et al.* 1995). Importation of the water volumes necessary to achieve the desired objectives may be a major component of the cost of the restoration scheme. This chapter illustrates some of these themes with reference to the Broads Environmentally Sensitive Area (ESA), reporting a detailed monitoring and modelling exercise in the Halvergate area.

THE STUDY AREA: THE BROADS ENVIRONMENTALLY SENSITIVE AREA (ESA)

The Broads ESA consists of a series of low-lying river valleys, marshes and fens in Norfolk and north Suffolk in eastern England (Figure 14.1). The water bodies that give the area the name, the Broads, are shallow lakes, believed to be flooded medieval peat workings, concentrated in the northern part. The whole area of river valley, broad, fen and marsh forms an interconnected wetland system, unique in Europe, and the largest area of lowland grazing marsh in eastern England (Ministry of Agriculture, Fisheries and Food 1991).

In the late 1970s and early 1980s the grazing marshes of the Norfolk Broads appeared to be under threat from arable agriculture, following proposed improvements to the arterial drainage. Although the farmers of the area were not hostile to the environmental case, they were not in the business of subsidizing other peoples' scientific interests, nor of acting as curators of a historic landscape. The debate over the renewal of pump drainage in the early 1980s (reported in ADAS 1983), which resulted in a compromise in which grant aid on one pump drainage scheme was denied on conservation grounds, established

See Glossary, p. 305, for explanation of technical terms.

European Wet Grasslands: Biodiversity, Management and Restoration. Edited by Chris B. Joyce and P. Max Wade.
© 1998 John Wiley & Sons Ltd.

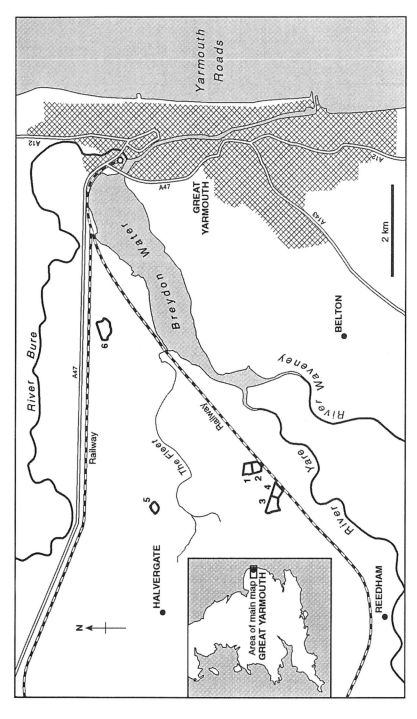

Figure 14.1 Location of the study fields (1–6) in the Halvergate area of the Broads Environmentally Sensitive Area

an important precedent. Since then, public policy has been concerned to preserve the landscape, and so various institutional opportunities have been introduced to encourage the retention of grassland, notably the Broads Grazing Marshes Conservation Scheme, set up in 1984, a joint Ministry of Agriculture, Fisheries and Food (MAFF)/Countryside Commission initiative. This established the principle of voluntary agreement under which farmers were paid to maintain land in grass and to manage their grass in a traditional way.

The Broads ESA was designated in March 1987, and since January 1988 farmers have been able to enter into agreements with MAFF, following the provisions of the Agriculture Act 1986 (Ministry of Agriculture, Fisheries and Food 1989). The objectives of the ESA are:

- to maintain and support traditional livestock farming on permanent grassland
- to conserve the landscape, wildlife, archaeological and other features that are responsible for the character of the traditional grazing marshes and meadows
- to encourage the continuation of grassland farming, especially where the threat of abandonment is greatest
- to extend the areas covered by traditional grazing marshes on wet meadows by encouraging the adoption of traditional farming practices on land now managed more intensively

In return for payments, the farmers agree to maintain traditional grassland management (Payment Tier 1), and may also raise water levels to a more ecologically sympathetic high level in the spring (Tier 2). Initial uptake for the ESA scheme was good and, in 1991, 89% of the eligible area was in the first designation of 13 780 ha (Ministry of Agriculture, Fisheries and Food 1991). In 1992 a third tier (Tier 3) was introduced, which required the participating farmers to flood land in spring, and thereby further improve the ecological value of the land. As anticipated, the uptake of this third tier has not been large. A further tier, for the restoration of grassland on previously cropped land, has also been introduced. Glaves (1998) gives further information on the ESA scheme in general.

THE HALVERGATE MONITORING SITES

The Halvergate area in the centre of the Broads ESA was chosen as the site for detailed monitoring studies. The wide expanse of floodplain centred on Halvergate marshes, which was traditionally summer grazing pasture, forms an area of considerable ornithological interest. Since 1989, water levels in six fields have been monitored (Figure 14.1) to identify the effectiveness of the management regimes imposed in response to the ESA scheme. Four fields were chosen within one block of land that was subject to a water management scheme for the retention of spring and summer water levels (ESA Tier 2). Two reference fields outside this area, which had the normal summer water levels, were identified and also monitored (ESA Tier 1). These fields were chosen to represent a variety of water management, but also to reflect areas that were known to have both higher and lower field surface levels, based on previous survey information (Table 14.1).

Instrumentation was installed in December 1989, and has been maintained for a core set of fields. For agricultural reasons, it was not possible to maintain the continuous records in field 5 beyond June 1991, and in June 1992 the instrumentation was withdrawn

Table 14.1   Fields monitored including payment tier, height of field surface and monitoring period

Field	Tier	Mean field surface height (m AOD)	Monitoring period
1	2	−0.70	Dec 89–Dec 94
2	2	−0.65	Dec 89–Jun 92
3	2	−0.52	Dec 89–Jun 92
4	2	−0.40	Dec 89–Dec 94
5	1	−0.53	Dec 89–Jun 91
6	1	−0.29	Dec 89–Dec 94

AOD, Above Ordnance Datum.

from fields 2 and 3. The dataset from field 4 was also incomplete. The data reported in this chapter thus concentrate on the two fields with the longest records, fields 1 and 6. These afford a contrast between the two tiers of management.

In these fields, water tables have been monitored by a combination of open auger holes (dipwells) and continuous-recording water-level meters. The dipwells formed a transect from ditch to ditch, and so recorded the shape of the water table, and were read on average every three to four weeks. Continuous records of the water levels were maintained in the field centres and in the ditches. Additionally, whenever the auger holes were read, soil surface strength was recorded using the Royal Society for the Protection of Birds (RSPB) penetrometer (Green 1986), which records the force necessary to insert a rod 20 cm into the soil, and so simulates soil penetration by a bird beak. On each occasion eight replicate measurements of soil strength were taken at three locations within each field. Parallel data were also collected using a conventional field shear vane inserted 2–3 cm into the soil.

Meteorological data were obtained from the synoptic station at Hemsby, about 10 km to the north. Some local rainfall information, gathered by RSPB staff at their Berney Arms site on the south-east edge of Halvergate, showed that there was little systematic difference between the two sets, and it was considered that the Hemsby data could be used as a good estimate of the conditions at Halvergate. Estimates of potential evapotranspiration for the site were provided using these data and the MORECS system (Thompson *et al.* 1981).

The soils of the area are alluvial clays of the Newchurch series (Clayden and Hollis 1984). These soils are very well structured in the topsoil, where they are rich in organic matter, but rapidly become structureless with depth, and at 2 m depth are 'buttery' and anaerobic. The hydraulic conductivity of the soil reflects this structural development, and varies between values in excess of 100 m day^{-1} close to the surface and values less than 0.1 m day^{-1} at depths below 1 m.

WATER REGIMES RESULTS

Results from two of the long-term monitoring locations, fields 1 and 6 (Figure 14.2), show the contrast between the water levels established under the two regimes. This figure shows the rainfall data, the calculated soil moisture deficit, and the depth of the water table in both of the fields and the adjacent ditches. All the water-table values are shown

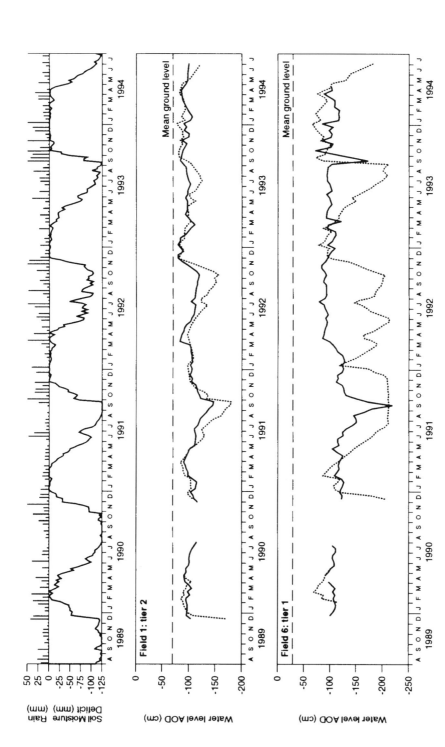

Figure 14.2 Rainfall, potential soil moisture deficit, in-field water regimes (solid lines) and ditch levels (dotted lines) for fields 1 and 6 during the monitoring period, December 1989 to June 1994. AOD, Above Ordnance Datum

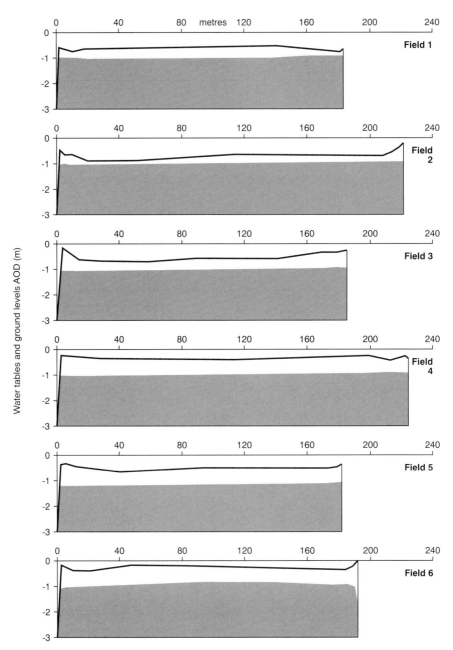

Figure 14.3 Example of water-table shapes during the winter period. These measurements were taken on 26 February 1992. AOD, Above Ordnance Datum

relative to mean sea level in metres Above Ordnance Datum (AOD), to facilitate comparison between the fields (Table 14.1).

In field 1, the ditch regimes remain high, and the field water levels remain close to the surface for all except the driest summer months. In the summer of 1991, severe water shortage prevented the maintenance of high water levels in the ditches, and the field water level fell accordingly. In every summer, the water level in the field centres falls below that of the ditch, indicating that recharge of water from the ditch to the field (subirrigation) is not sufficient to meet the evaporative demand from the vegetation.

The water regimes in field 6 are dramatically different. This is an area in which the ditch water levels are maintained at a low level to facilitate agricultural production. The ditch that controls the water level in this field is a main channel, and is controlled by the pump that drains the whole of the area. The water level is generally between −1.2 and −1.5 m AOD and falls even lower in the summer. However, the soil of the site has the same low conductivity as the other two fields, and so the water table in the field rises in the middle of the field, in response to the incident rainfall. The result is that the water table in the field is much less closely coupled to the ditch regime than that for field 1. The winter period is thus dominated by the development of a classic drainage condition, whilst in the summer there is only limited recharge. Because the land levels are higher than for field 1, the water table is a long way from the surface in the summer, and the site thus appears to be relatively very much drier.

The shape of the water tables throughout the area is shown by Figure 14.3, which demonstrates that the water table is essentially flat, and held at the ditch level for all except field 6, in which there is the classic domed water table. All the fields exhibit the 'bowl-shaped' topography that is characteristic of grazing marshes, as the repeated excavation and clearance of ditches lead to the deposition of material round the field margins. This variation in field level imposes a variation in water-table depth, so that the water table is further from the surface at the margins, but close to the surface near the field centre. This result demonstrates the overriding importance of topography in defining the water-table conditions at any point. It also shows that even where the water table can be controlled accurately, there will be variation in the hydrological regime within a field, reflecting the topographic variation.

SOIL-STRENGTH RESULTS

The soil-strength measurements typified by the data from field 1 (Figure 14.4) showed only a slight correlation between soil strength, as measured using either the RSPB penetrometer or the conventional shear vane. The penetrometer measurements all indicated the high resistance of the soil surface. Nearly all the observations are for values greater than a 6-kg force required to penetrate the surface, and this level of resistance was maintained even when the soil was flooded. Such values are greater than the estimated maximum of 3 kg for penetration by *Gallinago gallinago* (snipe) (Green 1986). The results contrast with those obtained at other sites, for example in the Somerset Levels and Moors ESA, where a clear correlation between water-table depth and soil surface strength has been observed (Armstrong *et al.* 1996).

It is considered that this behaviour is the consequence of the grazing management regime of the marshes, which are subject to short periods of high-intensity stocking. The

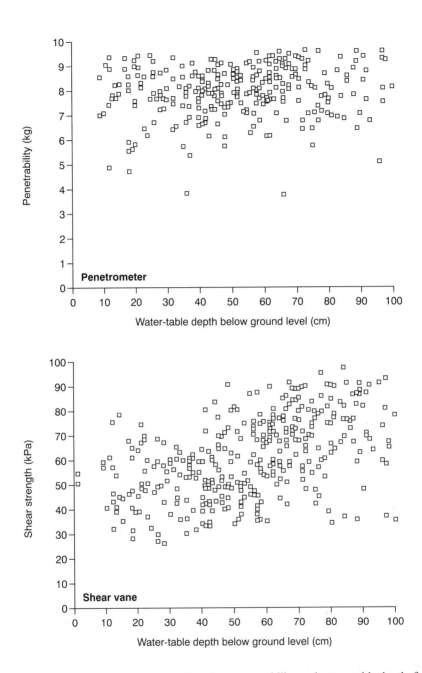

Figure 14.4   Relationship between soil surface penetrability and water-table depth, field 1

result is the formation of a hard surface mat of vegetation and compacted soil. It is suggested that this capping behaviour is one of the characteristics of clay marshes, such as the Halvergate marshes, compared to peat marshes. The results also serve to emphasize the importance of management practices in restricting the value of land for ecological purposes.

A MODEL OF WATER MOVEMENT: DITCH

Although the techniques for water management appear technically simple, involving the manipulation of inflow and outflow water levels, it is by no means clear that such actions will always lead to the desired effects. The results from the Broads ESA have shown that water tables in the field may be considerably lower in the centre of the field than in the controlling ditches. It is thus uncertain that the procedure of setting ditch levels will always be translated into the required soil water regimes. A model, called DITCH (Drain InTeraction with Channel Hydrology), was therefore developed that enabled the examination of the consequences of various ditch management regimes (Armstrong 1993; Armstrong *et al.* 1993a).

MODEL APPROACH

For simple situations, water levels in the field can be calculated from a consideration of the water balance:

$$M_t = M_{t-1} + (R - ET - Q_d)/f \qquad [1]$$

where the water elevation in the field on day t is M_t, R is the rainfall, ET is evapotranspiration, Q_d is the discharge through the drainage systems and f is the relevant porosity. Normally f, defined as the specific yield, is approximated by the drainable porosity of the soil, but where the water level is above the soil surface, then f becomes unity. The model thus assumes that evaporation taken from the profile results in a direct fall of the water table. This is a simplification, as the actual effect is to remove water from the whole profile. However, solutions to the equation of unsaturated flow (the Richards' equation) are notoriously difficult, and impose excessive demands on computing resources. For this reason the simplified representation is adopted.

Where sequences of rainfall and evaporation data are available from meteorological sources, the prediction of the soil water regime requires estimates of the flux through the drainage system, which can be calculated for soils of uniform hydraulic conductivity K, drained by parallel ditches at spacing L, using the Donnan drainage equation (International Institute for Land Reclamation and Improvement 1973):

$$Q_d = 4K(M_t^2 - D_t^2)/L^2 \qquad [2]$$

where D_t is the level in the ditches on day t. The flux between ditch and field, Q_d, can be in either direction, and therefore includes both drainage (Q_d is positive) and recharge (Q_d is negative). The model can therefore represent both the winter and summer phases of operation.

As implemented for this study, the level in the ditches is input into the model as an externally constrained set of values, so the water balance can be solved directly. The

Table 14.2 Mean annual fluxes predicted by the DITCH model

Soil	Rain (mm)	Evapotranspiration (mm)	Recharge (mm)	Drainage (mm)
Peat	616	446	134	235
Clay	616	356	7	10

model, however, also has the facility to predict the water level in the ditches by a second water-balance calculation (Armstrong *et al.* 1993a).

The use of the DITCH model to examine the ability of ditch management regimes to generate suitable within-field regimes was reported by Armstrong (1993). A theoretical analysis for uniform soils showed that, where the hydraulic conductivity of the soil is high (as in many peat soils), it is possible to control in-field water levels by setting the ditch water levels high. The model, however, also demonstrated for the UK that these high water levels require the import of significant quantities of water to meet the evaporative demand. By contrast, where soils have a low conductivity (as in some alluvial clays), it is much more difficult to maintain high water tables in the field, so that the water regime in such soils is dominated by the summer evaporative demand, and recharge through the ditch system is only very small. Table 14.2 gives the mean fluxes predicted in a 20-year run of the model using meteorological data from Hemsby, and two theoretical uniform soils, clay and peat, drained by ditches 70 m apart (Armstrong *et al.* 1993b). The model shows that where the soil is impermeable, then the recharge function is insufficient to maintain the water table close to the surface during the summer.

NON-UNIFORM SOIL PARAMETERS

The assumption of a vertically homogeneous soil, used for simple drainage models, has severe limitations, particularly for the Broads ESA area. Hydraulic conductivity data from the site, collected using the auger-hole method, were observed to decline rapidly with depth, following an exponential decline. If this is written as:

$$K(z) = K_0 e \beta^z \tag{3}$$

where z is height above some reference point (the base of the profile), K_0 is the hydraulic conductivity of the saturated soil at $z = 0$, and β is a constant, then the analysis of Youngs (1965) can be used to give an estimate of the flux between drain and field, q, as:

$$Qd = 2K_0[e^{\beta H_m} - e^{\beta H_w} - \beta H_m + \beta H_w]/\beta^2 S^2 \tag{4}$$

where the ditch drains are spaced $2S$ apart, H_w is the height of water in the ditches, and H_m is the maximum water-table height at mid-drain spacing. This can then be included into equation [1] to calculate the sequence of water-table heights that constitute the soil water regime in the centre of the field (Armstrong *et al.* 1993a).

TEST RESULTS

The DITCH model was adapted to match the conditions in the Broads ESA area. Values for the parameters K_0 and β required for equation [4] were calculated from the observed

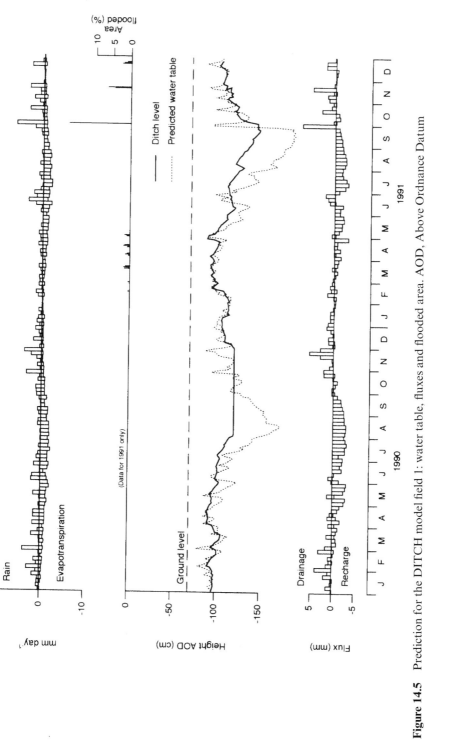

Figure 14.5 Prediction for the DITCH model field 1: water table, fluxes and flooded area. AOD, Above Ordnance Datum

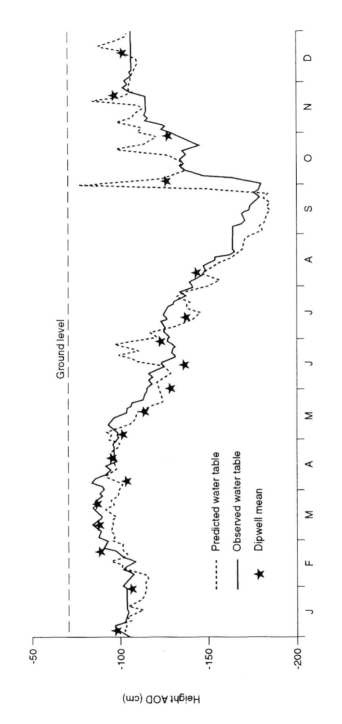

Figure 14.6 Modelled vs observed water-table depths, field 1, 1991. Observed water tables are plotted for both continuous recorders and as means of all dipwells in the field. AOD, Above Ordnance Datum

hydraulic conductivity data. The observed meteorological data for 1991 and the observed daily ditch values for field 1 were used to predict the water tables in the centre of the same field, the fluxes through the drainage system, and the percentage area flooded (Figure 14.5). Comparison between the predicted and observed water tables (Figure 14.6) show excellent agreement between the model and the observation. The only major discrepancy is with the large rainfall event at the end of September. Field observations suggested that much of this rain ran off the fields because of infiltration limitations, and did not enter the soil. This process is not considered in the model, and the short period of mismatch between model and observation was considered to be insufficient reason to reject the model.

On the same set of fields, measurements of microtopography had been undertaken using a 1-m-spaced transect from corner to corner to give detailed information. The cumulative distribution of height values was then used to infer the degree of surface flooding at any time. This is a significant variable because of the need for some flooded areas for the successful breeding of some important bird species (Tickner and Evans 1991). From Figure 14.5 it can be seen that some areas remained flooded until the end of May, but that apart from during the single large rainfall event in September, flooding was never more than 10% of the area of the land. This observation was in accord with the subjective estimates of flooded area made by the site operator.

CONCLUSIONS

In the study area in the Broads ESA, high water levels have been imposed, and these have, over the four-year study period, resulted in high water tables in the field centres. However, the study has also shown that maintenance of high water levels in clay soils requires very high ditch water levels, so that water can move through the higher conductivity surface layers, or even over the surface. Surface drainage channels that facilitate water movement from the ditch to the field centre should thus be maintained.

The management of water levels for wetland restoration thus requires close attention to the issue of land levels. Controlled water tables are essentially flat in the winter period and, as a consequence, topographic variation imposes a significant variation in water regime, as the depth between the water table and the surface varies. This topographic variation will thus impose the variation in conditions that appear to be the ecological ideal.

Maintenance of high water levels in the summer months requires the import of significant quantities of water. This places a demand on the water resources of the region and a cost on the restoration process. During periods of particular water shortage, it may not then be possible to maintain target water levels throughout the Broads ESA area.

The results from this study indicate the difficulties in maintaining wetland status with clay lowlands. The situation is very different in more conductive peat soils, where the rate of transfer between ditch and field centre is much greater (Armstrong 1993; Armstrong *et al.* 1996).

ACKNOWLEDGEMENTS

The financial support of the Ministry of Agriculture, Fisheries and Food is gratefully acknowledged. We are grateful to the many colleagues who have helped collect the data and who provided the background information, notably Colin Dennis, David Thomas, Kate Jewson and Frank Broughton; to Ian Barrie for the calculation of the PET data, and to the Royal Society for the Protection of Birds for access to their rainfall data. Especial thanks are due to the landowners who have allowed us access to their land to record this information.

SUMMARY

Active management of wetlands may involve the control of ditch water levels. The effectiveness of this strategy requires consideration of the interaction between managed ditches and the centres of the fields they surround, and the estimation of the volumes of water necessary to maintain desired levels of wetness. The DITCH hydrological model has been developed to predict the water table within a field surrounded by managed ditches, and the potential water flux necessary to maintain the desired level of wetness. It is shown that for low-conductivity soils, the use of ditches to maintain high water tables is unlikely to be successful. These themes are illustrated by examples from the Broads Environmentally Sensitive Area.

Keywords: Broads Environmentally Sensitive Area, Modelling, Water management.

REFERENCES

ADAS (1983) Conservation – the Halvergate story. In: *ADAS Annual Report 1983*, pp. 73–78. HMSO, London.

Armstrong, A.C. (1993) Modelling the response of in-field water tables to ditch levels imposed for ecological aims: a theoretical analysis. *Agriculture, Ecosystems and Environment*, **43**, 345–351.

Armstrong, A.C., Portwood, A. and Castle, D.A. (1993a) Simple models to predict field soil water regimes in the presence of ditch water levels managed for environmental aims. *Transactions, Workshop on Subsurface Drainage Simulation Models*, pp. 147–157. 15th International Congress on Irrigation and Drainage, The Hague.

Armstrong, A.C., Rose, S.C. and Treweek, J. (1993b) Water use requirements for managing wetland reserves for ecological aims. *Fourth National Hydrology Symposium*, pp. 2.23–2.28. September 1993, University of Wales, Cardiff.

Armstrong, A.C., Caldow, R., Hodge, I.D. and Treweek, J. (1995) Restoring wetlands: the hydrological, ecological and socio-economic dimensions. In: Hughes, J. and Heathwaite, L. (eds), *Hydrology and hydrochemistry of British wetlands*, pp. 445–466. Wiley, Chichester.

Armstrong, A.C., Rose, S.C. and Miles, D.R. (1996) *Effects of managing water levels to maintain or enhance ecological diversity within discrete catchments.* ADAS Report to Ministry of Agriculture, Fisheries and Food on Project BD0205. ADAS Land Research Centre, Gleadthorpe.

Clayden, B. and Hollis, J.M. (1984) *Criteria for differentiating soil series.* Soil Survey Technical Monograph 17. Soil Survey of England and Wales, Harpenden.

Glaves, D. (1998) Environmental monitoring of grassland management in the Somerset Levels and Moors Environmentally Sensitive Area, England. In: Joyce, C.B. and Wade, P.M. (eds), *European wet grasslands: biodiversity, management and restoration*, pp. 73–94. John Wiley, Chichester.

Green, R.E. (1986) *The management of lowland wet grassland for breeding waders.* Unpublished report. Royal Society for the Protection of Birds, Sandy.

International Institute for Land Reclamation and Improvement (1973) *Drainage principles and applications.* ILRI Publication No. 16. Wageningen, The Netherlands.

Ministry of Agriculture, Fisheries and Food (1989) *Environmentally Sensitive Areas.* HMSO, London.

Ministry of Agriculture, Fisheries and Food (1991) *The Broads Environmentally Sensitive Area. Report of monitoring 1991.* HMSO, London.

Thompson, N., Barrie, I.A. and Ayles, M. (1981) *The Meteorological Office rainfall and evaporation calculation system: MORECS (July 1981).* Meteorological Office Hydrological Memorandum No. 45. The Meteorological Office, Bracknell.

Tickner, M.B. and Evans, C.E. (1991) *The management of lowland wet grasslands on RSPB reserves.* Royal Society for the Protection of Birds, Sandy.

Youngs, E.G. (1965) Horizontal seepage through unconfined aquifers with hydraulic conductivity varying with depth. *Journal of Hydrology*, **3**, 283–296.

15 The Influence of Minor Variations in Hydrological Regime on Grassland Plant Communities: Implications for Water Management

DAVID J. G. GOWING[1], GORDON SPOOR[1] and
OWEN MOUNTFORD[2]
[1]*Silsoe College, Cranfield University, Silsoe, UK*
[2]*Institute of Terrestrial Ecology, Monks Wood, UK*

INTRODUCTION

The type of lowland wet grassland plant community, its floristics and its extent are usually determined to a large degree by the hydrology of the site. Other factors such as nutrient availability and vegetation management also play an important role, but these too are often a function of the underlying hydrology. As lowland wet grasslands are often groundwater systems, the depth of the water table below the surface is arguably the most important environmental factor when considering the distribution of the grassland flora. This has been widely appreciated by Dutch workers (Grootjans and Ten Klooster 1980; Gremmen *et al.* 1990), but there has been little attention paid to quantifying plant water-regime requirements in the British context.

Effective water management in wetland areas requires quantitative information on species and community preferences and tolerances. Qualitative autecological data (Ellenberg 1988; Grime *et al.* 1988; Mountford and Chapman 1993) using ranking systems on arbitrary scales are available, but no comprehensive set of quantitative data exists. Whilst the former are of value to ecologists, they are of limited use for managers. This chapter describes a method by which the water-regime requirements of some native species and communities have been quantified.

The traditional approach to quantifying water requirements in agricultural research has been to use lysimeters in which single species are grown above a range of controlled water-table depths. This method could, in principle, be paralleled with native species. Initial trials using native wet grassland species grown individually in lysimeters did not provide useful information however. This was due to species showing a much broader physiological tolerance than their expected ecological niche would suggest (Ellenberg 1953). It was therefore concluded that any experimental approach that did not take

See Glossary, p. 305, for explanation of technical terms. Scientific names of vascular plants follow Tutin, T. G. *et al.* (1964–80) *Flora Europaea* Volumes 1–5. Cambridge University Press. See p. 319.

account of interspecific competition was fundamentally flawed. To reproduce 'natural' conditions of competition in a lysimeter would be prohibitively time-consuming, and therefore the collection of data from the field was recognized as a more appropriate technique. By selecting field sites where plant communities have reached a near-equilibrium position with respect to water regime, collection of useful data could begin immediately.

Ditch-drained fields have a range of water regimes within them, due to variations in microtopography and seepage potentials. The latter are a function of field shape and proximity to ditches (Youngs 1992). These variations give rise to different plant communities and, given historical hydrological data such as pumping-station records are available, past water regimes at any point in the field can be estimated and related to the observed community.

In effect, an unperturbed lowland wet grassland site may be regarded as an infinite set of long-standing 'natural lysimeters', each with its own water-regime history and its own subset of the vegetation, which has apparently proved itself to be best adapted to that particular regime. Information on plant water-regime tolerances may thus be derived from an analysis of the correlation between water-table behaviour and plant species distribution.

A CASE STUDY

Hydrological and botanical information has been collected from a deep-peat area of the Somerset Moors, England, providing an extensive data set composed of 600 microsites, randomly distributed over 25 ha of traditionally managed lowland wet grassland. This information has formed the basis of a case study designed to reveal the quantitative tolerance limits of a range of plant species and communities. The site at Tadham Moor is a Site of Special Scientific Interest on which detailed studies of vegetation (Mountford *et al*. 1993) and soil nutrient status (Kirkham and Wilkins 1994) have been undertaken.

HYDROLOGICAL MODELLING

For each microsite, weekly water-table depths over a 15-year period (1976–1990) were generated using the hydrological model of Youngs *et al*. (1989). Land drainage theory is used to model water-table depth as a continuous succession of steady states. It enables water tables to be considered in three dimensions using the analysis of Childs and Youngs (1961). The inputs for this model are set out in Table 15.1. Physical survey data (spot elevations and positions relative to ditches) were recorded at the time of the botanical surveys. Soil parameters were estimated from field measurements and the output of the model was validated against dipwell readings taken at fortnightly intervals on the site.

The model has been successfully validated for relatively flat ditch-drained land on peat (Youngs *et al*. 1989). Other soil types are currently being investigated. Ditch banks and drainage grips were excluded from the surveys.

THRESHOLD APPROACH

The hydrological model in this instance produced 780 weekly values of water-table depth (15-year run) for each microsite. A number of methods were explored for expressing these

Table 15.1 The inputs required by the hydrological model to retrodict the weekly water-table depths for a single quadrat location within a wetland, with typical values

Input variables	Values used in case study
Physical survey of site	
Surface elevation	3.0 m AOD
Position relative to field centre	30 m long axis, 15 m short axis
Field dimensions	150 m long axis, 80 m short axis
Elevation of ditch base	2.0 m AOD
Soil parameters	
Hydraulic conductivity of each soil layer	1.5 and 0.7 m day^{-1}
Specific yield of each soil layer	15%
Depth of boundaries between soil layers	0.3 m
Depth to impermeable layer	3 m
Unsaturated hydraulic conductivity	
exponential coefficient	6.4 m^{-1}
Weekly inputs	
Rainfall data from nearest gauge	10 mm (mean for whole year)
Potential evapotranspiration	
from reference crop (Penman)	20 mm (mean for summer)
Ditch stage level (mean)	2.7 m AOD

Field data were collected individually for each of the 1000 microsites.
AOD, Above Ordnance Datum.

data as a single variable characteristic of an individual microsite. Initially mean water-table depth during the growing season was used to explain the patterns of plant distribution (Spoor *et al.* 1990), following the approach of Gremmen *et al.* (1990), who estimated aeration stress by considering the mean water-table depth in spring. Subsequent work, however, has shown the most discriminating variable to be one that involves the duration and the extent to which the water table falls below, or rises above, a threshold depth (e.g. Noest 1994). The former quantifies the likely exposure of the vegetation at a given microsite to drought conditions; the latter quantifies the potential stress due to anoxia occurring at a site, caused by high water tables or flooding. This 'peak-over-threshold' approach was based on an earlier method of water-regime evaluation from The Netherlands (Sieben 1965). Figure 15.1 illustrates the approach schematically. The critical depths for setting the two thresholds were based upon the soil properties and the

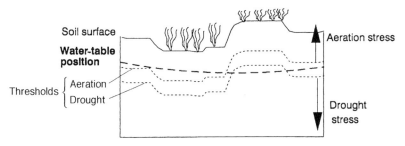

Figure 15.1 The threshold approach to quantifying potential stress conditions. When the water table is located as shown, the plants on the right are potentially drought-stressed whilst those on the left are exposed to potential aeration stress

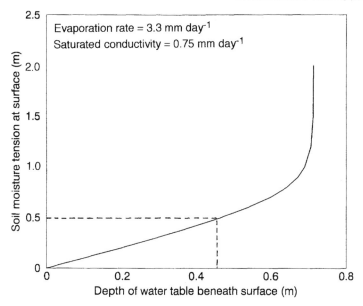

Figure 15.2   The relationship between water-table depth and the availability of water at the soil surface (after Gardner 1957), illustrating the calculation of the 'drought threshold' as the depth at which plants may first respond to soil drying. All evaporation (3.3 mm) is assumed to occur in a 10-h period and the unsaturated hydraulic conductivity exponential coefficient is taken as 6.4 m^{-1} (Youngs *et al.* 1989)

prevailing climatic conditions at the site. The 'drought threshold' was calculated as the water-table depth beneath which the soil moisture tension at the surface would exceed 0.5 m (a level of soil-drying detectable by plants; Henson *et al.* 1989), at a time when the potential evaporative demand is a typical summer value. At the study site, the long-term mean daily evapotranspiration rate in June from grass is 3.3 mm d^{-1} and an example of a surface moisture tension/water-table depth relationship (after Gardner 1957) is illustrated in Figure 15.2. The 'aeration threshold' was the water-table depth above which the air-filled porosity of the surface soil layers would be insufficient to allow ready diffusion of oxygen to the roots.

The degree to which these thresholds were breached was calculated for each microsite independently, from a plot of water-table depth against time. The area of the graph beyond the threshold line is termed the Sum Exceedence Value (SEV), and takes the unit of metre.weeks (Figure 15.3). For each threshold, this value was calculated for each year and a mean taken over a 15-year period, to reduce the effect of climatic variation. The mean SEVs were then used to characterize the water regime of that microsite.

CORRELATION BETWEEN WATER REGIME AND SPECIES DISTRIBUTION

To correlate species distribution with potential stress, the microsites were ranked according to their mean SEV score and a plot was made for each species, showing relative frequency of occurrence against SEV (Figure 15.4). The plot in the figure represents a rolling average generated by samples of 50 quadrats. The relative frequency is defined as

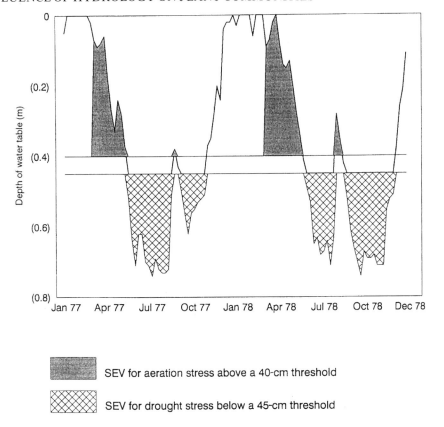

SEV for aeration stress above a 40-cm threshold

SEV for drought stress below a 45-cm threshold

Figure 15.3   The model's output in terms of weekly water-table depths over a two-year period for a typical quadrat location on a Somerset moor in a field with deep-peat soil and 60-m drain spacing. The horizontal line at 0.4 m represents the aeration threshold depth, and that at 0.45 m the drought threshold depth. The calculation of Sum Exceedence Values (SEVs) is illustrated schematically

the number of occurrences of the species within the 50 quadrats, divided by the expected frequency, which is based on the total number of its occurrences in the full sample of 600. If the species distribution were unaffected by water regime, relative frequency would be unity over the whole range. Where relative frequencies differ significantly from this value, the species is considered to either 'prefer' or 'avoid' a particular regime. The sharp transitions displayed by many of the species plots, such as in Figure 15.4, suggest that SEV is indeed quantifying a real variable in terms of a plant's physiological requirements.

For the site in question, the preferences and tolerances of a number of species have been summarized graphically (Figure 15.5), and compared to the F-value rankings assigned to them by Ellenberg (1988). A similar analysis has been repeated using National Vegetation Classification (NVC) community types instead of individual species (Rodwell 1992). To achieve this, the quadrat data at each microsite were subjected to TABLEFIT analysis to assign the most appropriate community (Hill 1991). A summary of the results from this type of analysis is shown in Figure 15.6.

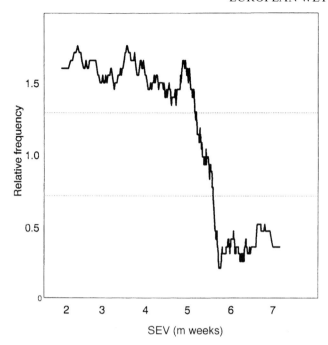

Figure 15.4 A plot of relative frequency of *Carex nigra* against the Sum Exceedence Value (SEV) of potential drought stress below a 45-cm threshold. The dotted horizontal lines define the 95% confidence interval for the 'expected' value of unity. Where the relative frequency lies outside these limits, the water regime is having a significant effect on the distribution of the species

DISCUSSION

APPLICABILITY OF THE INFORMATION

The approach has been tried on four other wet grassland sites in Britain, all managed for hay. They are: West Sedgemoor, Somerset (south-west England), which is similar to Tadham Moor, having deep-peat soils, but generally higher water tables; Cricklade North Meadow, Wiltshire (central England), which has a clayey alluvial soil overlying a shallow gravel aquifer; Wicken Fen, Cambridgeshire (east England), a deep-peat site with no aftermath grazing; and Upwood Meadows, Huntingdonshire (east England), which has a well-structured clay soil with ridge and furrow topography. From experience of applying the approach to these sites the information produced should be suitable for any European lowland wet grassland sites that fulfil the following criteria:

1. There is a true water table and its depth is the dominant environmental factor determining the pattern of vegetation on the site.
2. The nutrient status of the soil and the vegetation management of the site in question are comparable to those pertaining on one of the sites from which the reference data were obtained.
3. Historical records of ditch stage levels and climate exist, such that the current water regime may be assessed and any predicted perturbation quantified.

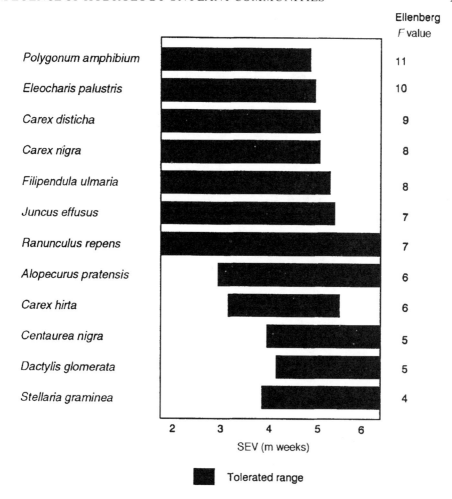

Figure 15.5 Summary of tolerated ranges for a number of species found in the wet meadows of the Somerset Moors, with respect to potential drought stress below a 45-cm threshold. They are listed in order of descending Ellenberg *F* value. SEV, Sum Exceedence Value

4. The site's vegetation has attained or will be allowed to attain a near-equilibrium position with respect to the prevailing water regime. There is no firm guide as to the length of time it may take a grassland community to achieve a near-equilibrium position with respect to the water regime; in practice we use 20 years.

If a change in hydrological regime can be estimated quantitatively, then the type of data presented here would indicate directly which species and communities were likely to benefit or suffer from a given change in the water regime.

The same hydrological model (Youngs *et al.* 1989) has been used to describe all the sites with peat soils, but separate models (Gowing and Youngs 1996; Youngs, unpublished observations) have been developed for those with mineral soils.

The range of sites from which information has been gathered is as yet limited.

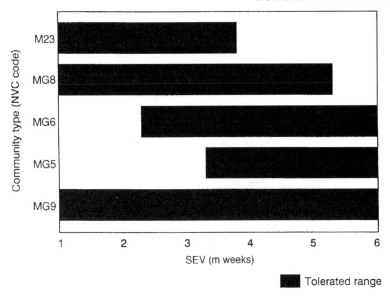

Figure 15.6 Summary of Sum Exceedence Value (SEV) ranges tolerated by the five different communities identified on the test site, with respect to potential drought stress below a 45-cm threshold. The communities are labelled according to the British National Vegetation Classification (NVC) (Rodwell, 1992): M23 (*Juncus effusus–Galium palustre* rush pasture); MG8 (*Cynosurus cristatus–Caltha palustris* grassland); MG6 (*Lolium perenne–Cynosurus cristatus* grassland); MG5 (*Centaurea nigra–Cynosurus cristatus* grassland); MG9 (*Holcus lanatus–Deschampsia cespitosa* grassland)

Continuing field work is gradually expanding this base and making the results more widely applicable, but a much larger data set needs to be amassed before generalizations can be made on the effects of soil nutrient status and vegetation management on water-regime requirements.

It should be possible, during the course of this current research programme, to derive tolerance limits for the large majority of species found on English lowland wet grasslands, most of which are also widespread throughout northern and central Europe. Such limits are more difficult to define for species that are very sparse in distribution within the studied sites because other environmental factors may be preventing them from occupying their full ecological niche with respect to water regime. Quantifying their requirements directly is often not possible. An alternative approach, which may have to be used in this situation, is a reliance on autecological rankings, such as those of Ellenberg (1988) and Grime *et al.* (1988), to group these species with others believed to possess similar water-regime tolerances and whose requirements have been positively quantified by the procedure described in this chapter.

MANAGEMENT OF GRASSLAND HYDROLOGY

Many, if not most, lowland wet grasslands of ecological interest, in the UK and much of Europe, have their hydrology managed artificially. Indeed, the features for which they are renowned and hence conserved have often evolved in response to the managed water regime. This is particularly true of the plant communities. The very fact that they are classed as grassland, rather than mire, can be the result of water-table control. Therefore, when considering the conservation or restoration of such habitats, the management of water is a vital component. Achieving the appropriate water regime on a site has three prerequisites:

1. A clear objective in terms of the plant community that is the target of the management or restoration effort.
2. A knowledge of the range of water regimes over which that community will be best suited to the site
3. The necessary infrastructure to control the water regime effectively.

The first of these has to be a subjective decision made by ecologists with local knowledge of the site. The second is the subject of the methodology and case study described above. The third can only be identified by consultation with water engineers, who would ideally be able to formulate a model for the hydrology of the site. They could thereby calculate the volumes of water that need to be supplied or removed during the year, and even more crucially, the rates of water movement within the site, based on the density of water control conduits, be they ditches, pipes or natural water courses.

Once these three conditions have been met, the components may be combined to formulate a water-management plan for the site. Success of the scheme should then be assessed relative to the objectives set. The floristics of the plant community may be used as an indicator of the water-regime behaviour, using information such as that presented in Figure 15.5. It is important to remember, however, that conclusions should only be drawn from floristic changes if the monitoring programme has been sufficiently long term to avoid inaccuracies introduced by the variation in annual weather patterns.

The actual values of threshold depth for the two potential stresses (drought and aeration) will vary between sites, depending upon soil character and climate (summer evaporative demand). It can be seen from the example that the two thresholds are close together, suggesting that as soon as aeration stress has been relieved in the upper soil layers of the soil, drought stress begins to become a factor. Plant species tend to be adapted to cope with one or other of these stresses, possessing aerenchyma to avoid aeration stress or sunken stomata to deal with drought stress. We would submit that all sites expose plants to one, or more often, in the case of wet grasslands, both, of these stresses and it is the relative degree of the two that largely determines the plant community composition. What may appear to be very minor variations in water-table regime between two sites may nevertheless be sufficient to explain differences in plant community. Small alterations to a site's hydrology, therefore, may cause significant alterations to its vegetation. To be able to predict such changes, a method for understanding the tolerances of species in a quantifiable way is required. The work described here offers one such approach.

ACKNOWLEDGEMENTS

The authors wish to thank Flood Defence and Conservation Policy Divisions of the Ministry for Agriculture, Fisheries and Food, the National Rivers Authority (now the Environment Agency) and the Royal Society for the Protection of Birds for funding the research described in this paper. They also acknowledge the co-operation of English

Nature for allowing the use of Tadham Moor in this project and of ADAS for collecting the dipwell records used for validating the model.

SUMMARY

Water-table regime is a dominant environmental factor determining the distribution of plant species and communities within lowland wet grassland. To be able to manage such grasslands effectively, a quantitative knowledge of plant water-regime requirements is necessary. A ditch-drained wetland provides an infinite set of 'natural lysimeters' from which information of this kind may be obtained. A hydrological model may be used to characterize the individual water regimes of these lysimeters. The impact on plants can then be derived in terms of the potential stresses that the water regime may impose on the vegetation. A peak-over-threshold analysis technique is used to quantify these stresses. Plant distribution patterns have been found to correlate strongly with water regimes, quantified in this manner, across a trial site in the Somerset Moors, England. Tolerance limits for both species and communities have been derived, and a summary of this information is presented. The results indicate that very small perturbations in water regime, if sustained over a number of years, would lead to changes in the grassland community growing above. The possibility of actively managing a site's water regime in order to achieve specific objectives in terms of the plant community present is discussed.

Keywords: Biodiversity, Drought stress, Hay meadow, Hydrological model, Water tables, Water-logging.

REFERENCES

Childs, E.C. and Youngs, E.G. (1961) A study of some three-dimensional field-drainage problems. *Soil Science*, **92**, 15–24.

Ellenberg, H. (1953) Physiologisches und okologisches Verhalten derselben Pflanzenarten. *Berliner Deutsche Botanische Gesturn*, **65**, 351–362.

Ellenberg, H. (1988) *Vegetation ecology of central Europe*. Cambridge University Press, Cambridge.

Gardner, W.R. (1957) Some steady state solutions of the unsaturated moisture flow equation with application to evaporation from a water table. *Soil Science*, **85**, 228–232.

Gowing, D.J.G. and Youngs, E.G. (1997) The effect of the hydrology of a Thames flood meadow on its vegetation pattern. In: *Floodplain rivers: hydrological processes and ecological significance*. Occasional Paper No. 8, British Hydrological Society, London.

Gremmen, N.J.M., Reijen, M.J.S.M., Wiertz, J. and van Wirdum, G. (1990) A model to predict and assess the effects of groundwater withdrawal on the vegetation in the Pleistocene areas of The Netherlands. *Journal of Environmental Management*, **31**, 143–155.

Grime, J.P., Hodgson, J.G. and Hunt, R. (1988) *Comparative plant ecology: a functional approach to common British species*. Unwin Hyman, London.

Grootjans, A.P. and Ten Klooster, W.P. (1980) Changes of ground water regime in wet meadows. *Acta Botanica Neerlandica,* **29**, 541–554.

Henson, I.E., Jenson, C.R. and Turner, N.C. (1989) Leaf gas exchange and water relations of lupins and wheat. I. Shoot responses to soil water deficits. *Australian Journal of Plant Physiology*, **16**, 401–413.

Hill, M.O. (1991) *TABLEFIT, Program manual (version 1)*. Institute of Terrestrial Ecology, Huntingdon.

Kirkham, F.W. and Wilkins, R.J. (1994) The productivity and response to inorganic fertilisers of species-rich wetland hay meadows on the Somerset Moors – nitrogen response under hay cutting and aftermath grazing. *Grass and Forage Science*, **49**, 152–162.

Mountford, J.O. and Chapman, J.M. (1993) Water regime requirements of British wetland vegetation: characterising vegetation using the moisture classifications of Ellenberg and Londo. *Journal of Environmental Management*, **38**, 275–288.

Mountford, J.O., Lakhani, K.H. and Kirkham, F.W. (1993) Experimental assessment of the effects of nitrogen addition under hay-cutting and aftermath grazing on the vegetation of meadows on a Somerset peat moor. *Journal of Applied Ecology*, **30**, 321–332.

Noest, V. (1994) A hydrology–vegetation interaction model for predicting the occurrence of plant species in dune slacks. *Journal of Environmental Management*, **40**, 119–128.

Rodwell, J.S. (1992) *British plant communities*, Vol. 3, *Grasslands and montane communities*. Cambridge University Press, Cambridge.

Sieben, W.H. (1965) Het verband tussen outwatering en obrengst bij de jonge zavelgranden in de Noordoostpolder. *Van Zee tot Land*, **40**, 1–117.

Spoor, G., Chapman, J.M. and Leeds-Harrison, P.B. (1990) Water regime requirements of wildlife habitats: research and guidelines. *Transactions, Ministry of Agriculture, Fisheries and Food conference of river and coastal engineers*, 1990, MAFF, Eastbury House, London.

Youngs, E.G. (1992) Patterns of steady groundwater movement in bounded unconfined aquifers. *Journal of Hydrology*, **131**, 239–253.

Youngs, E.G., Leeds-Harrison, P.B. and Chapman, J.M. (1989) Modelling water-table movement in flat low-lying lands. *Hydrological Processes*, **3**, 301–315.

16 A Comparison of the Management and Rehabilitation of Two Wet Grassland Nature Reserves: The Nene Washes and Pevensey Levels, England

BILL JENMAN[1] **and CHARLIE KITCHIN**[2]

[1]*Sussex Wildlife Trust, Henfield, UK*
[2]*Royal Society for the Protection of Birds, Peterborough, UK*

INTRODUCTION

Non-governmental organizations play a vital role in the conservation of wet grasslands. This role is illustrated by the Royal Society for the Protection of Birds (RSPB) and the Sussex Wildlife Trust (the Trust), two non-governmental charitable organizations in the UK responsible for managing wet grassland nature reserves in the Nene Washes, Cambridgeshire and the Pevensey Levels, Sussex respectively (Figure 16.1). The RSPB, a relatively large organization, manages a number of other lowland wet grassland nature reserves in the UK, including Old Hall Marshes, Essex; Ouse Washes, Cambridgeshire; West Sedgemoor, Somerset; and Loch Gruinart, Islay, Strathclyde. The Trust also manages other nature reserves in Sussex, two of which contain wet grasslands, Waltham Meadows and Amberley Wild Brooks, both of which are floodplain meadows.

An important objective for both organizations in acquiring the grasslands at Nene and Pevensey as nature reserves was to attempt to restore the wetland conditions that would enable wildfowl and wading bird (wader) populations to achieve their full potential. In recent decades, both sites had experienced marked declines in the numbers of these birds, which not only was a considerable conservation loss in its own right, but also prompted concern for the rest of the biological interest of the sites. The primary reason for these declines was improvements in drainage for agriculture benefit, which made the sites drier. This chapter describes and compares the approaches taken by the RSPB and the Trust in managing these two nature reserves to reverse this decline and improve the nature conservation value of the sites.

Although the two wet grassland areas are superficially similar and are being managed by conservation organizations to meet similar aims, a comparison of the two sites reveals a number of notable differences. These differences emphasize factors critical to conserv-

See Glossary, p. 305, for explanation of technical terms. Scientific names of vascular plants follow Tutin, T. G. *et al.* (1964–80) *Flora Europaea* Volumes 1–5. Cambridge University Press. See p. 319.

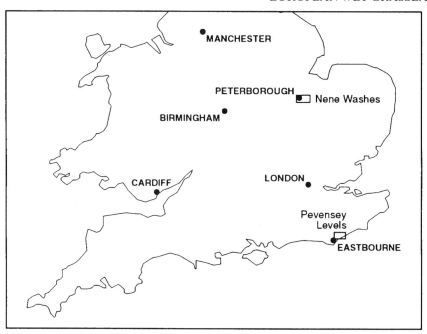

Figure 16.1   Location of the Royal Society for the Protection of Birds Nene Washes and Sussex Wildlife Trust Pevensey Levels nature reserves

ing biodiversity in this habitat, namely hydrology and winter flooding, soil characteristics, land ownership and the character of the managing organization itself.

NENE WASHES NATURE RESERVE

BACKGROUND

The Nene Washes (Figure 16.1) are part of the Fens, a large coastal basin in eastern England that has been progressively occupied and drained over several thousand years. The Washes came into existence in the 1700s after the completion of drainage by the Dutch engineer Vermuyden (Self *et al.* 1994). This involved the straightening of rivers and the construction of washlands adjacent to the rivers. These washlands stored floodwater from the rivers and released it back when the tide and river levels dropped. The development of pumping technology, through wind, steam power and, presently, electricity, allowed the fertile soils of the Fens to be drained and the area became dominated by arable farming. The contemporary drainage of the Nene Washes is the responsibility of the Nene Washlands Commissioners Internal Drainage Board (IDB), of which the RSPB has been a member since 1983.

Within the washlands, regular flooding preserved the traditional fenland agricultural system of summer grazing and hay-cutting. This combination of flooding and grassland management has in turn preserved some of the wildlife of the Fens. However, drainage improvements to the River Nene in the 1950s and 1960s reduced flooding on the Nene

Table 16.1 Characteristics of Nene Washes and Pevensey Levels and their nature conservation appellations

	Nene Washes	Pevensey Levels
Area (ha)	1400	3501
County	Cambridgeshire	East Sussex
Altitude	1–2 m AOD	0–3 m AOD
National Nature Reserve (ha)	–	Designated 1985 (52)
SSSI (ha)	Designated 1959 (1400)	Designated 1977 (3501) (renotified in 1990)
Ramsar site	Designated 1993	Proposed
SPA	Designated 1993	Proposed
Nature reserve (ha)	RSPB (286)	Sussex Wildlife Trust (129)

AOD, Above Ordnance Datum; SSSI, Site of Special Scientific Interest; SPA, Special Protection Area under European Union Directive on Conservation of Wild Birds (79/409/EEC); RSPB, Royal Society for the Protection of Birds.

Washes such that farmers could plough the grassland for arable crops. Records indicate that by the late 1970s over 75% of the area that is now nature reserve was intensively cropped (Self *et al.* 1994). This destroyed most of the botanical and bird interest, including the loss of the regular wintering flock of 5000 *Anser brachyrhynchus* (pink-footed geese). The reduction of the water table and ploughing caused the peat soils to oxidize and shrink, lowering the ground level. This made the crops vulnerable to flooding, and since the 1970s much of the arable land has been abandoned and has reverted to grassland.

Prior to acquisition by the RSPB, the area still supported a variety of important bird species and was a priority habitat for nature reserve acquisition. The RSPB began buying land in 1982 and now owns 22% of the Washes (Figure 16.2).

Some of the characteristics of the Nene Washes and its nature conservation appellations are summarized in Table 16.1. The site is internationally important for wintering wildfowl and nationally important for its assemblage of breeding waders and ducks. The site supports internationally important numbers of wintering *Cygnus columbianus bewickii* (Bewick's swan) and *Anas strepera* (gadwall) and nationally important numbers of wintering *Anser albifrons albifrons* (European white-fronted goose), *Anas crecca* (teal), *A. penelope* (wigeon), *A. acuta* (pintail) and *A. clypeata* (shoveler) and breeding *A. strepera, A. querquedula* (garganey), *A. clypeata* and *Limosa limosa* (black-tailed godwit).

There are three main aspects to RSPB management of the Nene Washes nature reserve: the provision of refuge areas for wintering wildfowl, the control of water levels, and grassland management. However, provision of refuge areas for wildfowl is contingent upon appropriate hydrological and grassland management.

HYDROLOGICAL MANAGEMENT

Winter flooding and wildfowl refuges

Extensive winter flooding on the whole of the Nene Washes is required to attract large numbers of wildfowl when, for example, over 20 000 duck may be present. In these circumstances a variety of water depths are present as well as refuge areas with suitable

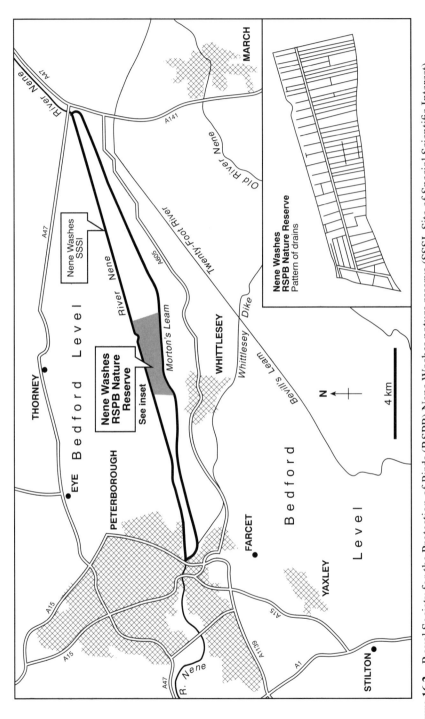

Figure 16.2 Royal Society for the Protection of Birds (RSPB) Nene Washes nature reserve (SSSI, Site of Special Scientific Interest)

habitat for birds that are disturbed from other parts of the Washes. The dabbling duck species, *Anas penelope, A. strepera, A. crecca, A. acuta* and *A. clypeata*, require shallow flooding, generally less than 30 cm deep, to feed in, while deeper water attracts the diving duck species, *Aythya ferina* (pochard) and *A. fuligula* (tufted duck). Open water also offers security, for example the floods in the nature reserve are especially important for roosting *Cygnus columbianus bewickii*, which feed by day on neighbouring arable land. Wintering and passage waders favour short grassland with shallow flooding for both safe roosting and feeding, and in recent winters large numbers of *Vanellus vanellus* (lapwing) (9000), *Pluvialis apricaria* (golden plover) (7500), *Limosa limosa* (1300) and *Philomachus pugnax* (ruff) (200) have been observed.

The decreased winter flooding experienced since river regulation in the 1950s led to a decline in the importance of the Nene Washes for wintering wildfowl. However, it is now possible to flood artificially over 100 ha of the reserve where the land level was lowered due to peat shrinkage during the arable era. Flooding begins in November and is maintained until the end of February, after which some areas are drained for breeding waders. In dry winters the largest block of artificial flooding, about 50 ha, becomes an important refuge, attracting most of the wildfowl on the Washes.

A second refuge area on the reserve is the tidal River Nene. An agreement with the farmer on the north side of the river to prohibit shooting makes this one of the most important areas on the washlands for *Anas penelope* and *A. crecca*. Since 1992, the RSPB has acquired a lease on shooting rights over other parts of the Nene Washes, away from the managed nature reserve.

Water for feeding and breeding

The important breeding birds on the Washes all require some degree of wetness for feeding and breeding. The soils of the Nene Washes are soft peats that are easily probed when wet, and contain an abundance of invertebrates able to tolerate flooding. Ideal conditions are produced by receding winter floods. However, in order to increase the flood storage capacity of the Washes the water levels in the ditches are lowered in winter, affecting some 65% of the reserve. They are not raised until the end of April, some time after the start of the breeding season. The RSPB, negotiating as a member of the IDB, has enabled levels in many ditches to be raised earlier in April.

The most important improvement in breeding conditions since RSPB involvement on the site was brought about in 1985 when the RSPB warden persuaded the IDB to raise the summer water levels on the reserve section of the Washes by 30 cm. However, the long-term solution to maintaining high water levels in the ditches is the hydrological isolation of all or parts of the reserve, allowing high levels to be established at the start of the breeding season independent of the lower water levels set by other landowners.

The fields on the Washes are typically long and narrow (Figure 16.2), allowing water from the ditches to percolate through the peat soils to most of the field. Many years of ditch slubbings (sediment dredged from the ditch and dumped on the top of the bank) have raised field edges, while oxidation has been shrinking the field centres, creating saucer-shaped profiles that retain winter floodwater into the spring. In extreme cases the middle of the field is at or lower than the water level in the surrounding ditches. As their concave shape closely resembles the summer profile of the water table, so parts of the field remain damp long enough for feeding waders to probe for food.

Large parts of the reserve remain shallowly flooded at the start of the bird breeding season. These areas are of little value for nesting but are important feeding sites. Food becomes concentrated in them as the water levels subside and they act as 'honey pots' to which waders bring their young. Maintaining this excessive wetness in the spring is important for attracting large numbers of breeding birds such as *Gallinago gallinago* (snipe) and *Anas querquedula*. By July, the land is dry enough to graze by livestock. The spring flooding encouraged on the nature reserve would be uneconomical to farmers and it is one of the few distinctions between the reserve and neighbouring washland. The reserve (22% of the whole Nene Washes area) usually holds 65% of the breeding waders, largely because of the spring floods.

The ditch system and its management

There are 43 km of ditches in the Nene Washes nature reserve, which support nationally important communities of plants and aquatic invertebrates. A survey in 1993 of 138 ditches (101 of which were in the nature reserve) recorded 86 aquatic and wetland species within the reserve (90 species for the Washes as a whole) (Jerram 1993). An earlier survey in 1984 identified four particularly notable species, *Nymphoides peltata, Potamogeton trichoides, Rumex palustris* and *Alisma lanceolatum* (Evans 1985), the first three of which are nationally scarce. Three of these species were still to be found in the nature reserve in 1993.

At the outset of the reserve, most of the ditches were in a derelict condition, but since 1984 the RSPB has been gradually opening them up using a mechanical bucket. The same machine is used to slub out (mild form of dredging) and reprofile the ditches on a five- to seven-year rotation. In 1984 a large proportion of the ditches were heavily silted and dominated by *Glyceria maxima* (Evans 1985), while a repeat survey in 1993 found a marked improvement in nature conservation value since regular management had been reinstated, with 90% of the ditches being open with a diverse flora (Jerram 1993).

The open water of the ditches and their edges are important for birds, particularly for breeding ducks, which nest in the fields but bring their young to the water to feed. Cattle that graze the fields have unrestricted access to the ditches, which act as a 'wet fence' and a source of drinking water. Their heavy bodies push in the ditch edges, creating a margin of marsh and mud around the fields, which provides excellent feeding sites for waders throughout the summer.

The water levels in the ditches have to be consistently high to be effectively stock-proof. This level is set by the IDB, but since 1991 the National Rivers Authority (which became the Environment Agency in April 1996) has taken over the responsibility of maintaining this. Problems of maintaining the agreed summer ditch levels have been exacerbated by the recent dry summers and over-abstraction of water from the River Nene once it reaches Peterborough and the Fens, mostly for drinking water for human consumption, maintaining navigation levels and irrigation in the Fens.

GRASSLAND MANAGEMENT

Lowland wet grassland is by nature conservation standards an intensively managed habitat. The aim on the reserve is to have produced a short sward over 85% of the area by

November. This is attractive to wintering wildfowl and subsequently to breeding birds for feeding. On 10% of the site a taller swamp community is encouraged and 3% is reed fen. Grazing is mostly by cattle to promote an uneven sward and some areas are mown for hay in summer and aftergrazed, either lightly to encourage botanical diversity or more heavily to create a short sward favoured by grazing *Anas penelope* (Self *et al.* 1994). The few trees present are managed by pollarding or coppicing to prevent them being used as nesting or vantage sites by corvids, which predate other birds' eggs.

Wintering wildfowl such as *Anas penelope* and *Cygnus columbianus bewickii* require a short sward that is both accessible and palatable. *Limosa limosa* and *Vanellus vanellus* also require a relatively short sward (10–15 cm) that allows them to feed whilst watching for predators. Other waders such as *Tringa totanus* (redshank) and *Gallinago gallinago*, and all of the duck species, conceal their nests in long vegetation. This can be produced by tussocks in a predominantly short sward or as patches of faster-growing grasses. A varied structure and species composition also increases the variety and quantity of plant and invertebrate food available to breeding and wintering birds.

The taller swamp habitat, when flooded, attracts wintering wildfowl and, when not flooded, is used by roosting *Asio flammeus* (short-eared owl) and *Circus* spp. (harriers). It is also the preferred nesting habitat for many ducks, rails, *G. gallinago* and certain passerines.

Tall vegetation along ditch margins occurs around hay fields and the ungrazed droves. It is an important habitat for Odonata (dragonflies and damselflies) and breeding birds such as *Gallinula chloropus* (moorhen), *Acrocephalus schoenobaenus* (sedge warbler) and *Emberiza schoeniclus* (reed bunting). The vegetation is mown in the autumn to maintain the open nature of the lowland wet grassland habitat and prevent woody plants from becoming established.

Grazing

The key breeding bird species on the reserve are ground-nesting and so, to reduce the losses of eggs and young to trampling, the start of the grazing season is delayed until mid-May. Losses are further reduced by first grazing the drier fields, which have fewer breeding birds, and then moving the livestock onto wetter fields gradually as the summer progresses. Thus some fields that remain flooded throughout the spring are not grazed until July.

The grazing season finishes at the end of October but can be extended slightly if palatable grass remains and other conditions such as ditch levels and weather are favourable. Grazing pressures vary between fields but on average are near to 200 livestock unit days ha^{-1}. Approximately 500 cattle graze during the summer on the reserve. Cattle are preferred to sheep or horses because they are easier to manage, tolerate wetter ground and eat a wide variety of grasses and sedges, producing a more varied sward structure.

This flexible grazing regime is possible because individual fields are not rented to farmers. Instead, the cattle are taken on a headage basis with RSPB staff responsible for the shepherding, earning the RSPB about £15 000 per annum. This requires staff to have training and experience in animal husbandry and the nature reserve to have facilities for handling livestock.

Mowing

Mowing grass for hay or silage is a simple way to achieve a short sward habitat and as much as 50 ha are usually mown every summer on the reserve. The hay is usually sold as a standing crop to the farmer who cuts it, thus saving the RSPB a large investment in time and machinery. Cutting does not start before 15 June each year in order to allow most breeding birds to nest and rear young. However, for late broods the operation at this time is usually fatal. In practice, mowing does not often begin until July and many fields are still being mown in August. Mowing produces a homogeneous vegetation structure, which can be accentuated by repeated cutting, although there is usually a sufficient diversity of grass species to produce some structural variety for nesting birds.

The timing of traditional hay-making allows plants to flower and set seed, encouraging or maintaining an interesting flora. Most of this had been eliminated on the Nene Washes by ploughing in the 1960s and 1970s but two fields on the reserve retain some botanical interest and these are cut for hay by the RSPB every year. The restoration of the grassland through traditional management is producing an increasingly semi-natural flora, aided by the introduction of diaspores in floodwater.

Following hay-making, the grass continues to grow through the remainder of the summer, so it is essential to bring in livestock to graze on this growth, i.e. aftermath graze, or, if necessary, take a second cut. Sheep are used to graze much of the aftermath, as they help control *Senecio aquaticus*, a toxic plant that can be fatal in hay crops fed to livestock.

As on many other grassland nature reserves, a tractor-mounted cutter (a 'topper') is used extensively to produce a sward of the desired height at the end of the growing season. From July onwards the topper is used to cut and spread the coarse vegetation that remains once the cattle have grazed the palatable plants. It is also used to control the seed dispersal and increase of *Cirsium arvense*, which is a serious nuisance in neighbouring arable land. The amount of grass growth varies from season to season and so in some years it may not be possible to arrange for extra hay-cutting or livestock. In such a case, topping can be used.

PEVENSEY LEVELS NATURE RESERVE

BACKGROUND

The Pevensey Levels are an extensive area of flat low-lying grazing marshes on the south coast of England (Table 16.1, Figures 16.1 and 16.3). The Levels were originally a shallow bay that developed first to saltmarsh and later to freshwater marsh as a result of the deposition of silt and land. Silt deposition was probably considerably enhanced by the natural development of a shingle spit along the present shoreline enclosing the area (English Nature 1990). The shingle beach now protects the Levels from sea-water inundation, and sluices at the end of Waller's Haven and Pevensey Haven (Figure 16.3) prevent saltwater intrusion into the drainage system. Soil pits dug for the Sussex Wildlife Trust showed that there is sand or silt at approximately 2 m below ground level (approximately modern sea level), which is probably marine in origin. This is overlain by alluvial clay with extensive lenses of peat. The Pevensey Levels were drained in medieval

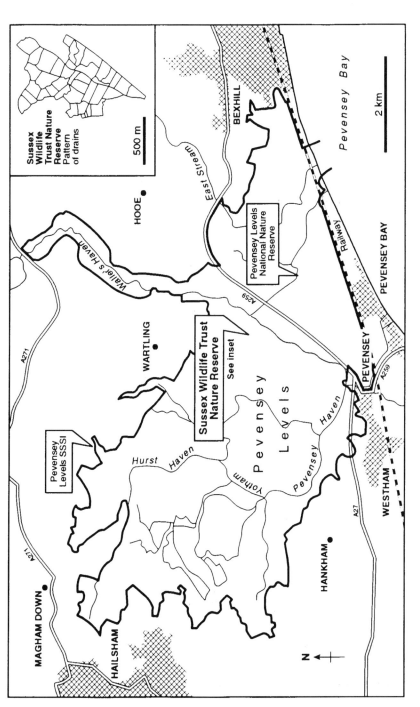

Figure 16.3 Sussex Wildlife Trust Pevensey Levels nature reserve (SSSI, Site of Special Scientific Interest)

times and converted to pasture, and have been gradually drying out over the centuries. The nature conservation value of the Levels was recognized in the 1960s, and in 1977 3501 ha of the Pevensey Levels were designated as a Site of Special Scientific Interest (SSSI) (Table 16.1). Since 1982 the Trust has acquired part of the Pevensey Levels as a nature reserve (currently 129 ha) (Figure 16.3).

The drainage of the Levels is the responsibility of the Environment Agency (the main river channels) and the Pevensey Levels IDB (other main channels), although the operating authority is actually the Environment Agency (previously the National Rivers Authority), a role taken on in the early 1970s. Pump drainage was introduced in the late 1960s and early 1970s, and now approximately two-thirds of the SSSI are pump-drained. Intensification of agriculture following more effective drainage means that about 20% of the land is now in arable production and virtually all the grassland is either improved or semi-improved. Livestock farming patterns have also changed and, where grazing was once seasonal because of winter flooding, farmers now generally rely on utilizing their land all year. Ownership patterns have not altered much, however, and landholdings remain relatively small and fragmented, with about 160 owners over the whole Levels area (National Rivers Authority 1992).

DECLINE IN WINTERING AND BREEDING BIRDS

The Trust's nature reserve supports an important assemblage of breeding birds characteristic of lowland wet grassland. These include *Cygnus olor* (mute swan), *Anas platyrhynchos* (mallard), *Vanellus vanellus, Gallinago gallinago, Tringa totanus, Motacilla flava* (yellow wagtail), *Acrocephalus schoenobaenus, A. scirpaceus* (reed warbler) and *Emberiza schoeniclus*. In winter the reserve is notable for large numbers of *V. vanellus* and *G. gallinago*.

Robust historical data on bird numbers are lacking but it is widely acknowledged that there has been a steep decline in the numbers of wintering and breeding birds on the Levels as a whole over the last 25 years since drainage was improved (Sussex Ornithological Society 1974–92; Royal Society for the Protection of Birds 1993). The RSPB carried out intensive and systematic winter bird counts in 1992/93 and 1993/94, which showed very low numbers, with no species reaching nationally important populations. Out of 16 sections (i.e. independent hydrological units), 11 supported less than one bird per hectare (Royal Society for the Protection of Birds 1993). Consequently, the Pevensey Levels have not been designated as a Special Protection Area. Comparisons with carefully managed sites like the RSPB reserve at Pulborough indicate that very much greater numbers could be accommodated.

The Trust originally owned 49 ha of land on the Levels. Although this is in the most ecologically diverse section of the SSSI that is still gravity-drained, management on the nature reserve has been little different from agricultural operations on the Levels as a whole. Lacking a discrete hydrological unit where water levels could be raised independently, the Trust confined its activities to summer grazing and maintaining the ditches. The Trust's land has therefore been as poor as the rest of the Levels from an ornithological perspective. As elsewhere, the soil surface has been too dry for wader chicks in spring and early summer, while winter flooding – extensive surface-water splashes rather than true seasonal lakes – has been limited to the occasional exceptionally wet period.

HYDROLOGICAL MANAGEMENT

The Nature Conservancy Council, later English Nature (EN), the Trust, the RSPB and others had long been concerned by the decline in the importance of the area for birds although the ditches and drains remained exceptionally valuable for their aquatic flora and invertebrate fauna. The continuing decline in avifauna led in 1991 to the selection of the SSSI as a pilot for EN's Wildlife Enhancement Scheme (WES), which paid farmers approximately £75 ha^{-1} to manage their ditches on a six-year cycle and to maintain high ditch water levels (30 cm below field level in the critical January to August period) (English Nature 1991). The WES was primarily aimed at maintaining the nature conservation value of the ditches and drains. The scheme has been extremely successful in terms of uptake (almost two-thirds of the SSSI is in the WES or similar schemes, e.g. Countryside Stewardship Scheme) (English Nature, pers. comm. 1997) and has had substantial benefits for the ditch flora and fauna. However, various bodies, notably the RSPB, have criticized the results of the scheme in relation to water-level management. Since no farmer owns an independent hydrological unit, no farmer can fulfil the water-level management requirements by themselves. To date, the WES has therefore had almost no effect on bird numbers.

In 1992 the National Rivers Authority (NRA), in response to requirements to produce a Water Level Management Plan (Ministry of Agriculture, Fisheries and Food 1991) and pressure from conservation bodies including EN and the Trust, began a Pevensey Levels study (Hart and Douglas in prep.). One of the objectives of this study was to 'identify the geographical components of the Levels system capable of independent water management and their water balance related to land use'. When considering raised water levels it is vital to ensure that no land is made wetter without the owner's consent: hence the need for detailed knowledge of the system. In 1993, the Trust was able to buy over 80 ha of additional land with the aim of creating a discrete hydrological unit in order to raise spring water levels.

In 1994, these various elements came together. Supported by data from the NRA study, the Trust was able to commission the Agricultural Development Advisory Service (ADAS) to provide detailed information on water levels and produce appropriate proposals to isolate hydrologically a large part of the Trust's land. At an early stage a joint NRA/EN/Trust working group was established, as all parties were keen to see the project on the nature reserve as a trial for an approach for the whole of the WES area. The Trust's reserve provided an ideal pilot study as the Trust had a vested interest in success while being prepared to tolerate some mistakes. As the Trust's holding is quite large by Pevensey Levels' standards only one other owner's land was affected, fortunately an owner sympathetic to the Trust's aims. Finally, as its aims are primarily concerned with wildlife conservation rather than agriculture, the Trust was able to commit approximately £3000 priming money and the staff time needed to initiate the trial.

Careful examination of the complex ditch system within the hydrological unit showed that just four stops (i.e. a sluice created by an earth dam with a flexible pipe running through to control the outflow of water) could hydrologically isolate the nature reserve. The key to maintaining high water levels at the site in spring is retaining winter rainfall. Unlike most other wetland systems in the UK, there is effectively no input of water into the Pevensey Levels system except from precipitation, itself in short supply in spring. Low flows in the rivers passing through the Pevensey Levels (Figure 16.3), and existing abstraction from upstream of the Levels, mean that it is not possible to extract water from these rivers to maintain ditch water levels during the critical spring period. From May to July, when evaporation exceeds rainfall by about 80 mm a month (National Rivers Authority 1994), only the reservoir of water in the soil and ditches is available, so winter-derived precipitation has to sustain the wildlife. Set against this, while extra surface water collected in winter would be invaluable for birds, actual flooding probably would not be, and would certainly be unacceptable to most farmers.

The Trust is now at the stage of building the stops. The NRA and EN have identified future priority areas within the Levels where all or most of the landowners are in the WES scheme and where further units of suitable land can be identified and stops built to enable the farmers to meet their WES commitments. In the longer term it is planned to develop the Trust's reserve as a core of especially wet land where nature conservation has priority. On the rest of the Levels, significant areas will be wet enough to sustain large numbers of waders in both winter and spring while becoming agriculturally more productive. WES payments will compensate farmers for any loss of income, but the continuation of the Levels as an economically viable grazing area is essential to the maintenance of the grassland habitat.

GRASSLAND MANAGEMENT

Most of the fields in the nature reserve were sown as *Lolium perenne* leys with some *Agrostis stolonifera* before the site was purchased by the Trust and they are of low floristic diversity. However, these pastures are of interest as they remain damp and the shingle ridges of former beaches can still be seen. The fields are grazed by cattle and sheep, with the cattle grazing in summer (April–October) and the sheep in winter (November–February), summer grazing by cattle being traditional on the Levels. At the time of acquiring the nature reserve it was ungrazed and rank, making it necessary for the Trust to implement a relatively high grazing pressure. These rates have since been reduced and in 1993 there were 40 livestock units of cattle in the summer and 50 livestock units of sheep in the winter. The livestock belong to farmers immediately adjacent to the nature reserve who rent the land by paying a grazing fee to the Trust.

No hay or silage cuts are taken from the nature reserve but some restoration cuts were taken during the first few years after it was bought by the Trust. No fertilizers are used on the site.

THE DITCH SYSTEM AND ITS MANAGEMENT

The Pevensey Levels have an outstandingly rich ditch flora and fauna, probably a remnant of their natural wetland origins. Ditches follow the former lines of creeks and streams, which in the medieval period formed the basis of the drainage network. In the whole of the Pevensey Levels there are 40 km of main river channel managed by the Environment Agency, 70 km of other main channel managed by the IDB, and landowners are responsible for a further 600 km of ditches. Over 10 km of ditches occur within the nature reserve. The Levels support approximately 110 species of aquatic plants (two-thirds of the British aquatic flora), 37 of them rare or local, e.g. *Alisma lanceolatum*. It has approximately 120 species of insect considered nationally or locally rare, including 21 Red Data Book species, e.g. *Hydrophilus piceus* (great silver water beetle). It is also one of only two UK sites for *Dolomedes plantarius* (fen raft spider), and is one of the best UK sites for freshwater Mollusca (snails), with four nationally rare species (English Nature 1990; Killeen 1994; D. Harvey, pers. comm.).

A study undertaken to investigate the impact of a small-scale change from pasture to cereal farming identified fundamental changes in ditch vegetation as a result (Palmer 1986). Some species such as *Hydrocharis morsus-ranae* were lost, but submerged species and algae increased. After five years there was no sign of reversion to the typical pasture ditch flora. However, aquatic invertebrates, apart from two rare molluscs, had satisfactorily re-established themselves. Nevertheless, when the pasture was ploughed, about 40% of the ditch length was filled in, which represents a considerable loss of wetland habitat (Palmer 1986).

Regular clearance of sections of ditch is required to prevent them from becoming overgrown and the ditches in the nature reserve are managed on a 7–10 year cycle, at the end of each cycle being slubbed out using a mechanical bucket. The cleaning-out process stops the hydroseral succession in the ditch and returns it to an earlier stage. This type of traditional practice maintains the channels in the reserve and across the rest of the Levels at different stages in the hydroseral succession, accounting in part for the high diversity of flora and fauna.

DISCUSSION

Although two apparently similar areas of lowland wet grasslands, the Nene Washes and the Pevensey Levels have certain fundamental differences that have influenced the way in which the nature reserves, established by the RSPB and the Sussex Wildlife Trust respectively, have been managed. Further differences are due to the nature of these two non-governmental organizations, one a national institution and the other more locally based.

CONSERVATION ACHIEVEMENTS

The Nene Washes and the Pevensey Levels have similar histories in terms of the pursuit of ever more efficient drainage of the agricultural land claimed from the original wetlands. The potential for the conservation of waterfowl and waders was recognized for both areas and the resolve made to reverse the decline. Both organizations have therefore established a significant landholding, which has been instrumental in general conservation improvements. For example, Table 16.2 compares the numbers of key breeding birds on the Nene Washes nature reserve in the three seasons 1983–85 with those in 1993–95, illustrating a considerable increase in numbers of key breeding birds, including the establishment of a *Limosa limosa* colony, now the largest in England.

Numbers of wintering wildfowl generally are heavily influenced by the extent of winter flooding and the degree of shooting disturbance. There are a number of other factors, such as the accuracy of counts and changes in population size of bird species, that make it difficult to demonstrate the influence of conservation management on numbers of wildfowl. However, Table 16.3 offers a comparison of wildfowl numbers on the Nene Washes reserve before and after RSPB involvement, in a series of winters when there was extensive natural flooding. The nature reserve now regularly supports internationally

Table 16.2 Numbers of some key breeding bird species on the Nene Washes nature reserve

Birds (pairs)	1983	1984	1985	1993	1994	1995
Cygnus olor	3	3	11	14	14	9
Anas strepera	6	3	8	27	13	24
A. platyrhynchos	151	49	41	71	75	113
A. querquedula	3	2	2	5	4	2
A. clypeata	57	19	15	31	39	41
Aythya fuligula	13	1	5	5	7	9
Vanellus vanellus	17	22	29	38	62	68
Limosa limosa	0	0	0	14	14	14
Gallinago gallinago (displaying)	27	25	44	75	69	70
Tringa totanus	8	6	17	43	65	86

Table 16.3 Numbers of some key wintering bird species on the Nene Washes nature reserve

Species	Pre-RSPB			RSPB		
	1976/7	1977/8	1978/9	1992/3	1993/4	1994/5
Cygnus columbianus bewickii	157	23	67	203	567	1538
Anas penelope	1500	21	2000	2441	3440	6650
A. strepera	1	0	0	239	296	115
A. crecca	400	200	300	785	1400	1574
A. acuta	500	7	600	768	577	1794
A. clypeata	41	6	10	309	147	390
Aythya ferina	40	48	89	353	110	206

RSPB, Royal Society for the Protection of Birds.

important numbers of *Cygnus columbianus bewickii* and *Anas acuta* and nationally important numbers of *A. penelope, A. strepera, A crecca* and *A. clypeata.*

HYDROLOGICAL MANAGEMENT

The Nene Washes were designed to be flooded on a seasonal basis. The drainage engineering effort on the Pevensey Levels was aimed at preventing flooding throughout the year. This included the routing of upland waters through the Levels via Waller's Haven and Pevensey Haven (Figure 16.3). A combination of the type of soil in the Nene Washes, with its high content of peat, and the earlier conversion of much of the area to arable agriculture (in the 1950s and 1960s) meant that the Washes experienced a shrink-age of soil volume and a lowering of the field surface. This made the fields wetter and more prone to carrying standing water in comparison to those in the Pevensey Levels, where the peat layer is in lenses beneath the alluvial clay approximately 1 m below the ground surface. Also, at Pevensey arable agriculture is not as widespread and is of relatively recent origin.

Such differences have had an impact on the scope for hydrological management of the two nature reserves. The Trust is unsure if it is possible to retain enough water on the Levels to last through spring without actually flooding the land in winter. Although winter flooding might be acceptable to a conservation body, it is not usually to farmers, at least not without substantial financial compensation. It is also possible that floodwater will move seawards laterally through the peat, making survey work to identify independent hydrological blocks irrelevant. Although the peat is covered by apparently impermeable clay, this nevertheless remains a concern.

There is a further critical limiting factor that has dictated the emergence of the present management policies and practice. This is the degree to which the nature reserves can be managed as hydrologically discrete areas. In the case of the RSPB, the land acquired was independent of other parts of the Washes from the outset and this conferred a degree of freedom on the manipulation of flooding. In contrast, the Sussex Wildlife Trust began with only 49 ha, which was part of a larger hydrological unit constraining possible management options.

CONSTRAINTS AND OPPORTUNITIES

As the two reserves are being managed primarily for nature conservation with agricultural productivity a secondary aim, there has been and will remain a monetary cost associated with their management. The RSPB is a large organization, with a membership of approximately one million, whereas the Trust is much smaller, with a membership of 10 050. Thus, support from other agencies apart (e.g. EN and the NRA), the scope for undertaking options with associated monetary costs is likely to be less in the case of the Trust.

The Trust has needed to work in close co-operation with such agencies as EN and the NRA, and with landowners. Such co-operation has been essential to the success of the WES on Pevensey Levels. The RSPB, in contrast, has been able to adopt a relatively independent approach, although it has not been able to produce its own Water Level Management Plan, instead making a major contribution to the IDB (Nene Washlands) Water Level Management Plan. The Trust found that most farmers were sympathetic to its aims but naturally concerned about their business, so a good grant scheme and, importantly, personal contact (in this case from EN) have been vital in building up mutual trust. The critical test for the WES will come in carrying the local farming community with the Trust, for without their consent and co-operation Pevensey Levels will never achieve its potential as a wetland of international importance for birds.

It was also important for the Pevensey Levels, with its complex system of ownership, drainage and little excess water, to ensure that a sound conservation strategy was developed, achieved in this case by the NRA. The time put into the preparation of this strategy was well spent, not least because of the culture change needed for drainage engineers to adapt to a new role. Well-planned beneficial publicity was also essential and it was vital that the affected landowners received the correct message directly.

There is no relationship between the diversity and numbers of birds using the two nature reserves and floristic diversity. Neither of the two nature reserves has a particularly diverse grassland flora, having been improved by past agricultural practices. However, the Nene Washes nature reserve has shown a general diversification of its grasslands under the present management regime, mainly through the assistance of floods depositing propagules. This facility is not available on the Pevensey Levels, which do not receive such allochthonous material. The development of a diverse flora at both nature reserves will require either specific management and/or the passage of time.

ACKNOWLEDGEMENTS

Annual reports from the Royal Society for the Protection of Birds (1983–94) provided information on the Nene Washes. The support of the RSPB wardens and other staff who have worked on the Nene Washes nature reserve and have contributed towards its successful management is acknowledged. Thanks to Paul José for his comments on the first draft of this chapter.

The work at Pevensey Levels is the result of the co-operation of both the National Rivers Authority and English Nature, and without their help the Sussex Wildlife Trust would not have been able to achieve any improvements to the management of our nature reserve or Pevensey Levels SSSI as a whole. Special thanks are due to Caryl Hart and Simon Taylor of the NRA and Basil Lindsey of EN. Dave Burgess of RSPB has provided valuable help and information about bird numbers and organized the recent winter counts. Finally, Mr J. Norris (Jnr), our grazier, has been very helpful with practical advice.

SUMMARY

Although the Nene Washes and Pevensey Levels wet grassland nature reserves are superficially similar and are being managed by non-governmental nature conservation organizations to meet similar aims, a comparison of the two sites reveals a number of notable differences. Both sites had experienced a reduction in numbers of overwintering birds and breeding wading birds prior to acquisition as nature reserves due to more effective drainage for agriculture and associated land-use change. Management strategies being implemented by the organizations for restoring this nature conservation resource are described. The differences between the two sites emphasize factors critical to conserving biodiversity in this habitat, namely hydrology (winter flooding in particular), soil characteristics, land ownership and the character of the non-governmental organization responsible for management.

Keywords: Grassland nature reserve, Hydrological management, Nene Washes, Pevensey Levels, Wading birds.

REFERENCES

English Nature (1990) *Pevensey Levels Site of Special Scientific Interest*. SSSI citation. English Nature, Lewes.

English Nature (1991) *Wildlife Enhancement Scheme for Pevensey Levels Site of Special Scientific Interest*. English Nature, Lewes.

Evans, C. (1985) *Ditch plants at Nene Washes reserve*. Unpublished report to Royal Society for the Protection of Birds, Sandy.

Hart, C. and Douglas, S. (in prep.) The Pevensey Levels study. In: Harpley, A.J. and Wade, P.M. (eds), *The ecology and management of drainage channels*. Wiley, Chichester.

Jerram, R. (1993) *The Nene Washes: a survey of the aquatic and riparian flora of field ditches*. Report to English Nature. English Nature, Peterborough.

Killeen, I.J. (1994) *A survey of the freshwater Mollusca of Pevensey Levels, East Sussex*. National Rivers Authority Southern Region, Worthing.

Ministry of Agriculture, Fisheries and Food (1991) *Conservation guidelines for drainage authorities*. MAFF, London.

National Rivers Authority (1992) *A land use survey of Pevensey Levels, East Sussex*. National

Rivers Authority Southern Region, Worthing.

National Rivers Authority (1994) *Pevensey Levels draft strategy – consultation document*. National Rivers Authority Southern Region, Worthing.

Palmer, M. (1986) The impact of a change from permanent pasture to cereal farming on the flora and invertebrate fauna of watercourses in the Pevensey Levels, Sussex. *Proceedings of the European Weed Research Society/Association of Applied Biologists 7th Symposium on Aquatic Weeds*, pp. 233–238.

Royal Society for the Protection of Birds (1983–94) *Nene Washes RSPB nature reserve*. Annual Reports, 1983–94. Royal Society for the Protection of Birds, Sandy.

Royal Society for the Protection of Birds (1993) *Pevensey Levels wintering bird survey 1992/93*. Royal Society for the Protection of Birds, Shoreham.

Self, M., O'Brien, M. and Hirons, G. (1994) *Hydrological management for waterfowl on RSPB lowland wet grassland reserves*. RSPB Conservation Review No. 8. Royal Society for the Protection of Birds, Sandy.

Sussex Ornithological Society (1974–92) *Sussex bird reports 1973–91*. Sussex Ornithological Society.

Part Four

RESTORATION

17 Residual Effects of Phosphorus Fertilization on the Restoration of Floristic Diversity to Wet Grassland

JERRY TALLOWIN[1]**, FRANCIS KIRKHAM**[2]**, ROGER SMITH**[1] **and OWEN MOUNTFORD**[3]

[1]*Institute of Grassland and Environmental Research, North Wyke, UK*
[2]*ADAS, Wolverhampton, UK*
[3]*Institute of Terrestrial Ecology, Monks Wood, UK*

INTRODUCTION

Intensification of grassland management throughout western Europe this century has resulted in enormous decreases in biological diversity within the farmed landscape (Baldock 1990). The application of fertilizer has been a key factor in reducing floristic diversity in species-rich grassland (Brenchley and Warington 1958; Tilman 1982; Mountford *et al.* 1993). Current policies, such as the creation of Environmentally Sensitive Areas (ESAs), which are designed to safeguard rural landscapes in the UK (Ministry of Agriculture, Fisheries and Food 1994; Glaves 1998), could provide significant opportunities for enhancing wildlife resources in extensive farming systems. However, experience has shown that in order to achieve a species-rich state, de-intensification will need to reduce primary production and standing crop mass in many circumstances (Al-Mufti *et al.* 1977; Grime 1979; Vermeer and Berendse 1983; Wheeler and Shaw 1991). Enhanced availability of macronutrients in soils that have been under intensive farming now limit the restoration of floristic diversity when extensification occurs (Bakker 1989). Farming practices that simply involve a reduction or cessation of fertilizer inputs and the adoption of traditional managements, such as hay-making, can be a slow and uncertain means to achieve a recovery in species richness in wet grassland (Berendse *et al.* 1992). Nutrient-availability depletion can be an unacceptably slow process if dependent upon crop offtake alone, particularly with regard to phosphorus (Marrs 1990).

Information that predicts, with any degree of precision, either the extent or the rate of decline in the availability of nitrogen (N), phosphorus (P) or potassium (K), when fertilizer inputs cease, is limited for wetland systems in Europe.

This chapter examines the impact of inorganic N, P and K fertilizer application on

---

See Glossary, p. 305, for explanation of technical terms. Scientific names of vascular plants follow Tutin, T. G. *et al.* (1964–80) *Flora Europaea* Volumes 1–5. Cambridge University Press. See p. 319.

European Wet Grasslands: Biodiversity, Management and Restoration. Edited by Chris B. Joyce and P. Max Wade.
© 1998 John Wiley & Sons Ltd.

floristic diversity and on the restoration of diversity following cessation of fertilizer use on wet hay meadows of the Somerset Levels, UK.

METHODS

EXPERIMENTAL SITE

The experimental site was within the Tadham and Tealham Moors Site of Special Scientific Interest in Somerset, UK, on peat soil with an average pH of 5.7 (Kirkham and Wilkins 1994a). Prior to the start of the project the following grassland types, in accordance with the National Vegetation Classification (Rodwell 1992), were represented on the site: *Cynosurus cristatus–Centaurea nigra* (MG5), *Cynosurus cristatus–Caltha palustris* (MG8) and, to a lesser extent, the *Alopecurus pratensis–Sanguisorba officinalis* (MG4) association; these grasslands are equivalent to C38.112, C37.214 and C38.2, respectively, of the CORINE biotope classification (Devillers *et al.* 1991). These semi-natural grassland types reflect previous extensive management practices that the site had been subjected to for an indefinite period; these involved late hay-cutting followed by aftermath grazing with no inorganic fertilizer inputs.

LARGE-SCALE EXPERIMENT

In a large-scale experiment, started in 1986, the following inorganic fertilizer nitrogen (N) amounts were applied annually: 0, 25, 50, 100 and 200 $kg\,ha^{-1}$, referred to later as the N0–N200 treatments, respectively. The fertilizer N applications were split, with half being applied in mid-April and the other half after the hay cut in July. Inorganic phosphorus (P) and potassium (K) fertilizers were applied to those plots that had received fertilizer N in amounts to replace that removed in the hay crop. Treatment plots were cut for hay after 1 July and the regrowth was grazed with beef cattle. Plot size, which ranged from 1.1 ha for the N0 plots to 0.6 ha for the N200 plots, was designed to support a minimum of two steers throughout the grazing period. A compressed sward height (rising plate) of 5.5–6.5 cm was maintained during grazing by adjusting cattle numbers per plot to give a continuous but variable stocking density (Tallowin *et al.* 1990). In 1990 the plots were split, with one half continuing to receive fertilizer inputs as previously until April 1993 (N+), whilst, on the other half, fertilizer treatments ceased (N−).

SMALL-SCALE EXPERIMENT

A small-scale experiment (7.5-m^2 plots) was also started in 1986, under a cutting-only management (Kirkham and Wilkins, 1994b). In addition to the same fertilizer N treatments with replacement amounts of P and K used in the large-scale experiment, treatments also included: 0N with P and K replaced; 100 or 200 $kg\,N\,ha^{-1}$ with 0P and K replaced; 0 or 100 or 200 $kg\,N\,ha^{-1}$ with 75 $kg\,P\,ha^{-1}$ and K replaced; and 200 $kg\,N\,ha^{-1}$, 75 $kg\,P\,ha^{-1}$ with 200 $kg\,K\,ha^{-1}$. Plots were cut at the same time as the hay in the large-scale experiment and again in the autumn with the cut herbage removed from the plots. All fertilizer inputs ceased after 1989 in this small-scale experiment.

   The amounts of P and K applied in the large- and small-scale experiments to replace that removed in the hay from the N-fertilized plots were, respectively, between 6 and 12 kg $P\,ha^{-1}$ and between 35 and 110 kg $K\,ha^{-1}$. In both the large- and small-scale experiments the dry-matter yields of herbage and of N, P and K were measured at each cut between 1986 and 1993. Vascular plant and bryophyte (moss) species were scored individually for percentage ground cover within randomly located quadrats on each plot in May each year. Nomenclature for bryophytes follows Smith (1978).

DATA ANALYSIS

Individual species abundance (cover) and the number of species per plot were analysed separately for the large- and small-scale experiments by analysis of variance (ANOVA) using GENSTAT (GENSTAT V Committee 1987) and treatment effects were tested for difference using the Fisher Least Significant Difference (LSD) test. Prior to carrying out the ANOVA the cover values of individual species were angular transformed to stabilize the variances. The relationship between community composition and N, P and K application was examined using the ordination programmes of Canonical Correspondence Analysis (CCA) and Detrended Correspondence Analysis (DCA) (ter Braak 1987–92). Harvested dry-matter and nutrient yield data were analysed for each year separately by ANOVA using GENSTAT (GENSTAT V Committee 1987) and treatment effects were tested for difference using the Fisher LSD test. A significance level of $P = 0.05$ was used throughout unless otherwise stated.

RESULTS

IMPACT OF FERTILIZERS ON BOTANICAL COMPOSITION

In the large-scale experiment there was a significant downward trend in species richness with time where fertilizer inputs were continued over eight years, whilst in the unfertilized control no such changes were observed (Figure 17.1) (Mountford et al. 1993). In the small-scale experiment highly significant depletions in species richness occurred, relative to the unfertilized control, in plots that had received a combination of high N and P inputs (Table 17.1) (Kirkham et al., 1996). A species–treatment biplot produced by CCA after four years of fertilizer input showed that P was much more influential in determining botanical composition in this grassland than either N or K (Figure 17.2). There was, however, close correspondence between the P and K gradients. The biplot shows close association between some competitive species and P input.

Four years after the cessation of fertilizer inputs significant differences existed in the small-scale experiment between treatment plots and the control in the percentage cover values of several species, particularly where high P had been applied (Figure 17.3). For ease of interpretation the actual cover values and not the angular transformed values are shown in Figure 17.3. The most obvious residual effects of previous high P fertilizer treatment were in the enhanced cover values of *Holcus lanatus* and *Rumex acetosa* four years after cessation of inputs. Formerly common species in the unfertilized grassland, such as *Plantago lanceolata*, *Leontodon hispidus* and *Centaurea nigra* remained at lower cover values than in the control. The species richness of plots that had received high inputs of N and P with replacement amounts of K over four years had been reduced by

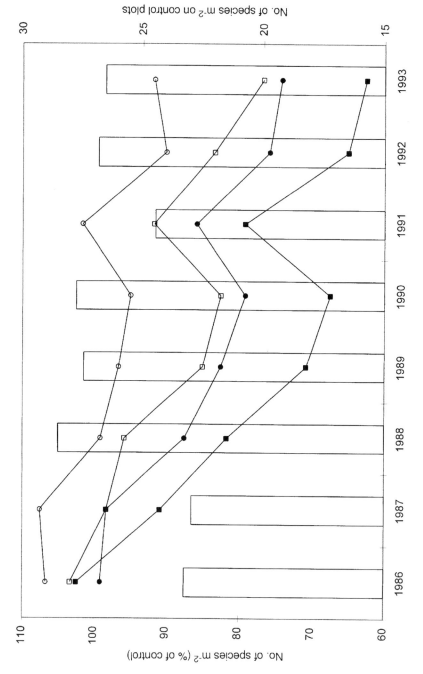

Figure 17.1 The influence of inorganic fertilizers on species richness recorded in May of each year, 1986–93, in the large-scale experiment. Bars represent the number of species m^{-2} of control plots receiving no fertilizer; lines represent numbers m^{-2} on plots receiving N at 25 kg ha^{-1} (open circles), 50 kg ha^{-1} (open squares), 100 kg ha^{-1} (filled circles) and 200 kg ha^{-1} (filled squares) per year, expressed as % of the control mean for each year

Table 17.1 Species richness (number of species per 7.5-m^2 plot) in the small-scale experiment under cutting management alone during the period of fertilizer application, 1986–89, and following the cessation of fertilizer application, 1990–93

No.	Treatment	1986	1987	1988	1989	1990	1991	1992	1993
T1	N0, P0, K0	33.3	37.3	35.3	33.0	31.7	31.3	26.7	24.7
T2	N0, Pr, Kr	32.0	36.7	34.7	32.7	30.3	31.3	29.7	28.0
T3	N25, Pr, Kr	29.7	37.0	34.0	29.7	29.7	25.3	27.7	27.0
T4	N50, Pr, Kr	30.7	37.0	32.0	29.7	30.7	27.3	27.0	24.7
T5	N100, Pr, Kr	29.3	33.7	28.3	27.3	27.3	26.7	26.7	25.3
T6	N200, Pr, Kr	30.0	36.0	29.7	28.7	28.3	22.7	25.7	23.7
T7	N100, P0, Kr	28.3	36.0	33.7	32.3	27.0	28.3	28.7	29.0
T8	N200, P0, Kr	31.3	35.0	30.0	31.3	26.3	28.3	27.0	25.0
T9	N0, P75, Kr	31.3	37.0	34.7	32.0	32.7	27.7	26.3	24.3
T10	N100, P75, Kr	28.7	34.3	27.0	23.3	26.0	23.7	19.7	26.0
T11	N200, P75, Kr	29.0	31.0	27.0	19.0	18.3	19.0	16.3	20.0
T12	N200, P75, K200	29.7	30.0	26.3	22.0	19.7	19.7	24.3	24.0
ESE		1.73	1.94	1.42	1.74	1.56	2.92	1.92	1.99
				**	***	***	***	**	

r, replacement: e.g. Kr = K applied in amounts equivalent to that removed in the hay crop. ESE, effective standard errors of the estimated means. Degrees of freedom=22.
Significance of treatment effects: $*P<0.05$; $**P<0.01$; $***P<0.001$.

approximately six species per plot; in the four years following cessation of inputs, species richness had increased by only two species per plot in these treatments.

RESIDUAL EFFECTS OF FERTILIZERS ON HERBAGE YIELD AND NUTRIENT CONCENTRATION

Where fertilizer applications were discontinued in the large-scale experiment, there was no residual effect of previous fertilizer input on hay yield in any of the subsequent years (Mountford et al. 1994). Only the yield of K was found to be enhanced in all former treatment plots (compared with the unfertilized controls) for two years after the cessation of inputs (Table 17.2).

In all treatment plots in the small-scale experiment, except for those which had received no P, herbage dry-matter yields remained enhanced compared with the control two years after fertilizer inputs ceased (Figure 17.4) (Mountford et al. 1994). The least significant difference ($P=0.05$) for each year following cessation of fertilizer input was ±0.98 tonnes ha^{-1} (1989), ±0.90 tonnes ha^{-1} (1990 and 1991), ±1.20 tonnes ha^{-1} (1992) and ±1.42 tonnes ha^{-1} for 1993. The total yields of N in the herbage, although apparently enhanced after four years without fertilizer input, were in fact not significantly elevated (Figure 17.5). Yields of P in the herbage from plots that had previously received 75 kg fertilizer P ha^{-1} yr^{-1} were significantly enhanced compared with the unfertilized control in each of the three years following the cessation of inputs (Figure 17.6). The least significant difference ($P=0.05$) for each year was ±2.19, ±2.13, ±5.55 and ±2.35 kg ha^{-1} for 1990, 1991, 1992 and 1993 respectively. A significant enhancement in total herbage K yields was found in all the former fertilized plots relative to the unfertilized control in the year following cessation of inputs (Figure 17.7). The least significant difference ($P=0.05$) for each year was ±6.69, ±8.49, ±8.19 and ±10.52 kg ha^{-1} for 1990, 1991, 1992 and 1993

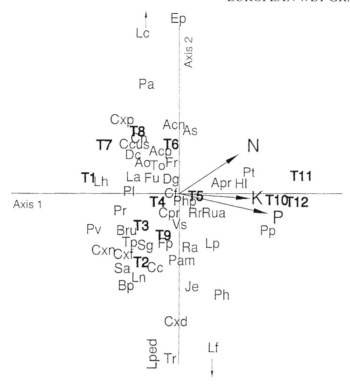

Figure 17.2 Species–treatment biplot (Canonical Correspondence Analysis) on 1990 vegetation data related to the mean amounts of N, P and K applied each year, 1986–90, in the small-scale experiment. Symbols in bold type represent mean ordination positions of each treatment T1–12 (treatments as listed in Table 17.1). The large arrows indicate the direction and strength of the fertilizer gradients in relation to Axes 1 and 2 of the ordination. The remaining symbols are species codes: Acn, *Agrostis canina*; Acp, *A. capillaris*; Ao, *Anthoxanthum odoratum*; Apr, *Alopecurus pratensis*; As, *Agrostis stolonifera*; Bp, *Bellis perennis*; Bru, *Brachythecium rutabulum* (a moss); Cc, *Cynosurus cristatus*; Ccus, *Calliergon cuspidatum* (a moss); Cf, *Cerastium fontanum*; Cn, *Centaurea nigra*; Cpr, *Cardamine pratensis*; Cxd, *Carex disticha*; Cxf, *Carex flacca*; Cxn, *Carex nigra*; Cxp, *Carex panicea*; Dc, *Deschampsia cespitosa*; Dg, *Dactylis glomerata*; Ep, *Eleocharis palustris*; Fp, *Festuca pratensis*; Fr, *F. rubra*; Fu, Filipendula ulmaria; Hl, *Holcus lanatus*; Je, *Juncus effusus*; La, *Leontodon autumnalis*; Lc, *Luzula campestris*; Lf, *Lychnis flos-cuculi*; Lh, *Leontodon hispidus*; Ln, *Lysimachia nummularia*; Lp, *Lolium perenne*; Lped, *Lotus uliginosus*; Pa, *Potentilla anserina*; Pam, *Polygonum amphibium*; Ph, *Poa subcaerulea*; Php, *Phleum pratense*; Ph, Pl, *Plantago lanceolata*; Pp, *Poa pratensis*; Pr, *Potentilla reptans*; Pt, *Poa trivialis*; Pv, *Prunella vulgaris*; Ra, *Ranunculus acris*; Rr, *R. repens*; Rua, *Rumex acetosa*; Sa, *Senecio aquaticus*; Sg, *Stellaria graminea*; To, *Taraxacum* spp.; Tp, *Trifolium pratense*; Tr, *T. repens*; Vs, *Veronica serpyllifolia*

respectively. Significant residual effects of fertilization on K yield persisted only in former high P plots for a further two years in the herbage cut in the autumn. There were no significant residual effects of previous fertilizer application on K yield in the third year.

The availability/uptake of nutrients other than those applied as fertilizers was also affected by the high P fertilizer treatments in the small-scale experiment. In the autumn cut, herbage calcium (Ca), magnesium (Mg) and sodium (Na) yields remained enhanced two years after fertilizer inputs had ceased in those plots which had received 75 kg $P ha^{-1} yr^{-1}$ (Table 17.3). A significant residual effect of previous high P input on Mg

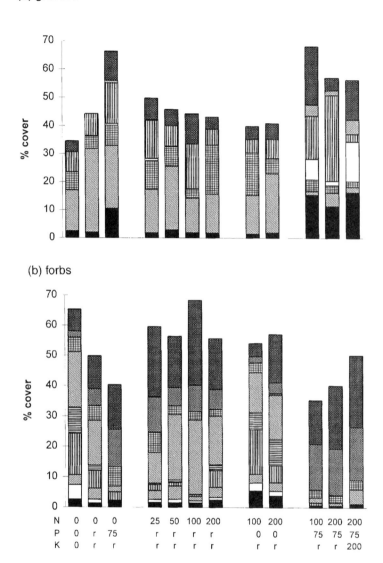

Figure 17.3 Species abundance (% cover) in May 1993 in the small-scale experiment. (a) Grasses: � *Holcus lanatus*; ▨ *Anthoxanthum odoratum*; ▦ *Agrostis capillaris*; ▭ *Alopecurus pratensis*; ||||||||| *Festuca rubra*; ▨ *Poa trivialis*; ▨ other grasses. (b) Forbs: ■ *Taraxacum officinale*; ▭ *Leontodon autumnalis*; ▨ *Trifolium pratense*; ||||||||| *Centaurea nigra*; ≡ *Leontodon hispidus*; ▨ *Plantago lanceolata*; ▦ *Ranunculus acris*; ▨ *Rumex acetosa*; ▨ other forbs. r, replacement: e.g. K r = K applied in amounts equivalent to that removed in the hay crop

Table 17.2 Yield of potassium (kg K ha^{-1}) in herbage cut for hay in the large-scale experiment in which the regrowth was grazed in the three years following cessation of fertilizer inputs

Treatment	1990	1991	1992
N0	37.0	36.7	29.6
N25	54.8	50.3	38.8
N50	63.5	70.2	58.8
N100	60.9	50.5	37.0
N200	52.7	54.3	40.6
ESE	4.52	5.29	7.42
	*	*	

ESE, effective standard errors of the estimated means. Significance of treatment effects: *$P<0.05$.
Degrees of freedom=8

yield persisted for three years in the herbage cut in the autumn. Significant residual effects of previous fertilizer treatments on these other nutrients were present for only one year in herbage cut in July.

DISCUSSION

The application of inorganic P fertilizer was a major factor in causing large changes in the botanical composition of the species-rich wet hay meadows on the experimental site on the Somerset Levels (Kirkham *et al.*, 1996). The decline in floristic diversity associated with high P fertilizer inputs involved the loss of distinctive semi-natural grassland species. The meadows fertilized with high amounts of N and P became dominated by competitive species. The grasses *Holcus lanatus* and *Lolium perenne* and the dicotyledon *Rumex acetosa* became key components of these fertilized communities, changing the community association from a predominantly species-rich *Centaureo–Cynosuretum* (MG5) to *Holco–Juncetum/Lolio–Cynosuretum* (MG10/MG6) types of grassland (Rodwell 1992), the latter being equivalent to CORINE biotopes C37.217 and C38.111 respectively. There is an abundant resource of species-poor MG10 and MG6 grassland types in ESAs and throughout the UK, so with regard to ESA objectives fertilization led to undesirable loss of floristic diversity in this landscape.

Upon cessation of fertilizer treatments in the small-scale experiment there was an initial large drop in total herbage dry-matter yield from about 8.5 tonnes ha^{-1}. This decline in yield was consistent with studies on different soil types elsewhere (Oomes 1990; Berendse *et al.* 1992). However, over the subsequent three years no further decreases in dry-matter yield occurred and yields from these former high N with high P plots remained enhanced compared with the control.

Associated with the enhanced dry-matter yields was a greater availability of macronutrients on plots that had received the high P fertilizer inputs. Of particular interest was the increased yield of N on plots that had previously received high inputs of P with replacement amounts of K but without any fertilizer N input. The P (and possibly K) availability in the peat soil clearly limited N mineralization/availability. This yield/N response to P fertilization would explain why floristic change was so strongly associated with P input rather than N. This would also explain why recovery in floristic diversity was

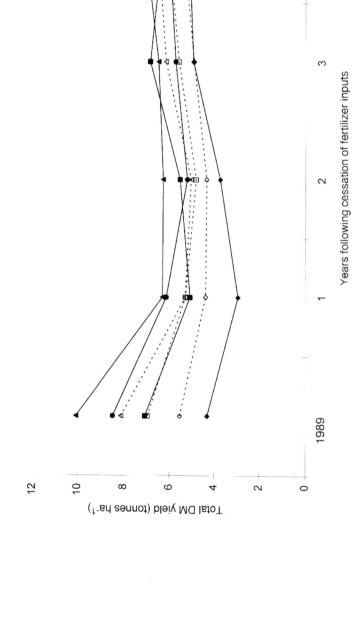

Figure 17.4 Dry-matter (DM) yields (tonnes ha^{-1}) from different former fertilizer treatments in the small-scale experiment in the four years 1990–93, after cessation of fertilizer inputs. ——◆—— N0, P0, K0; ----□---- N0, P and K replaced; ----△---- N200, P and K replaced; ----◇---- N200, P0, K replaced; ———■——— N0, P75, K replaced; ———▲——— N200, P75, K200

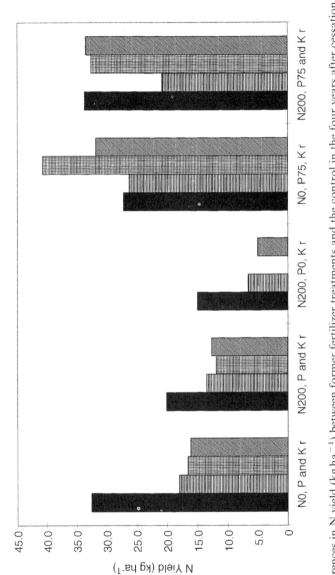

Figure 17.5 Differences in N yield (kg ha^{-1}) between former fertilizer treatments and the control in the four years after cessation of inputs to the small-scale experiment. ■ 1990; ▥ 1991; ▦ 1992; ▨ 1993. r = replacement: e.g. K r = K applied in amounts equivalent to that removed in the hay crop

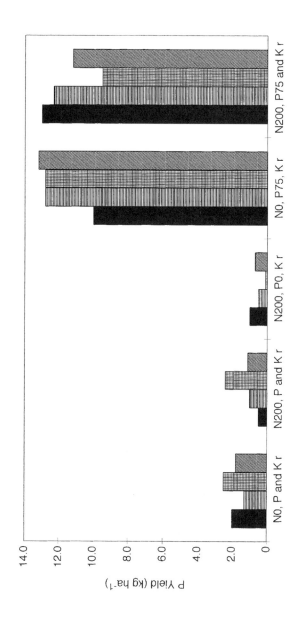

Figure 17.6 Differences in P yield (kg ha^{-1}) between former fertilizer treatments and the control in the four years after cessation of inputs to the small-scale experiment. ■ 1990; ▥ 1991; ▦ 1992; ▨ 1993. r = replacement: e.g. K r = K applied in amounts equivalent to that removed in the hay crop

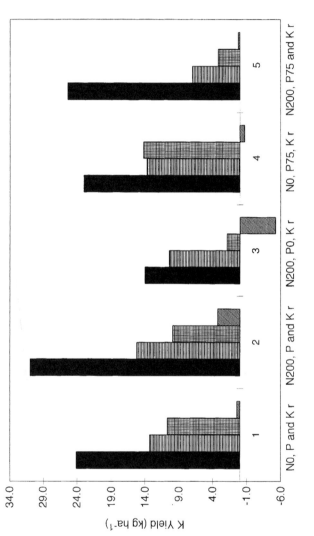

Figure 17.7 Differences in K yield (kg ha^{-1}) between former fertilizer treatments and the control in the four years after cessation of inputs to the small-scale experiment. ■ 1990; ▥ 1991; ▦ 1992; ▨ 1993. r = replacement: e.g. K r = K applied in amounts equivalent to that removed in the hay crop

Table 17.3 Yield (kg ha^{-1}) of calcium (Ca), magnesium (Mg) and sodium (Na) in herbage cut in September (second cut) in the small-scale experiment in the three years (1991–93) following cessation of fertilizer inputs

Treatment	1991			1992			1993		
	Ca	Mg	Na	Ca	Mg	Na	Ca	Mg	Na
N0, P0, K0	2.0	0.4	0.8	1.6	0.4	–	10.3	3.3	5.9
N0, Pr, Kr	9.0	1.7	2.8	3.5	0.9	2.4	12.6	4.7	7.8
N25, Pr, Kr	7.4	1.6	2.5	5.7	1.9	3.3	11.1	4.5	7.3
N50, Pr, Kr	6.0	1.4	2.3	4.8	1.4	3.1	10.9	4.2	7.9
N100, Pr, Kr	6.2	1.7	2.6	4.0	1.3	3.1	9.0	4.2	5.8
N200, Pr, Kr	9.3	2.1	3.3	2.9	0.9	1.9	12.9	4.4	7.0
N100, P0, Kr	6.5	1.2	1.8	2.9	0.7	1.5	15.5	5.1	7.9
N200, P0, Kr	5.6	1.3	2.1	2.3	0.6	1.5	11.3	4.4	7.9
N0, P75, Kr	14.9	3.5	6.2	7.3	2.6	4.3	15.3	7.3	11.3
N100, P75, Kr	11.6	3.0	4.8	6.8	2.8	5.7	14.1	7.5	9.0
N200, P75, Kr	11.4	2.7	4.4	7.8	3.2	5.5	14.5	7.2	11.7
N200, P75, K200	12.4	3.3	4.9	7.9	3.6	6.3	13.4	6.6	11.6
ESE	1.81	0.42	0.71	0.93	0.30	0.52	2.11	0.93	1.19
	**	***	**	***	***	***	*		

r, replacement: e.g. Kr = K applied in amounts equivalent to that removed in the July hay crop. ESE, effective standard errors of the estimated means. Degrees of freedom=22. Significance of treatment effects: *$P<0.05$; **$P<0.01$; ***$P<0.001$.

so slow in the plots that had previously received the high P fertilizer inputs (Mountford *et al.* 1994; Tallowin *et al.* 1994).

No decline in P yield occurred in the former high P plots over the four years since the cessation of inputs. On the former N200 with 75 kg P and K-replaced treatment plot, for example, an average of 17.7 kg P ha^{-1} were harvested in the cut herbage each year. This amount cannot be assumed to represent P derived solely from fertilizer residues. The unfertilized control plots yielded about 6 kg P ha^{-1} yr^{-1} during the course of the eight-year experiment. Therefore, this 'background' amount of available P from the existing total pool of soil P should be subtracted from the 17.7 kg P ha^{-1} to provide an estimate of the 'extra' P derived from the high P fertilizer input. In total, 300 kg P ha^{-1} were applied to the high P treatment plots, 93.1 kg P ha^{-1} were harvested during the period of fertilizer input (this excludes an estimated 23.2 kg P ha^{-1} from the soil pool) and 45.9 kg P ha^{-1} (excluding 24.9 kg P ha^{-1} contributed from the soil pool) were harvested during the four years after cessation of fertilization. Overall, it is estimated that in 1993, four years after the cessation of fertilization, there would have been a maximum of about 160 kg P ha^{-1} that could be termed residual or 'extra' P derived from the fertilizer inputs. If this amount of 'extra' P continues to be harvested from this treatment each year (rather than assuming a more likely exponential decay relationship occurring with time) and assuming that a similar amount of P continues to be derived from the soil P pool, it is estimated that it will take at least a further 13 years under the current cutting management to deplete the 'residual' P in the former high P fertilized plots. This conclusion accords with published findings that the depletion of fertilizer 'residues' by cutting alone to achieve comparability with the nutrient status of species-rich grassland can be a lengthy and uncertain process (Gough and Marrs 1990; Oomes 1990; Berendse *et al.* 1992).

It is apparent, therefore, that once changed by high inputs of inorganic P fertilizer, recovery of species-rich wet hay meadows on the Somerset Levels will take many years.

ACKNOWLEDGEMENTS

The research was commissioned by the Ministry of Agriculture, Fisheries and Food, English Nature and the Department of the Environment.

SUMMARY

Fertilizer experiments were carried out on species-rich wet hay meadows on peat soils on the Somerset Levels, UK, over an eight-year period. Experiments were designed to establish the impact of a range of nitrogen (N), phosphorus (P) and potassium (K) inputs on floristic diversity and then to investigate the rate and extent of floristic recovery when fertilizer applications ceased. High inputs of inorganic P together with N and K fertilization had the most damaging effects on the floristic composition of the meadows. However, high P input with K in the absence of any fertilizer N also caused significant and undesirable floristic changes. Upon cessation of fertilization, recovery of species richness was particularly slow on former high P plots. Significant residual effects of high P input, expressed by enhanced dry-matter and nutrient yields, were present after two to three years of no input, with P yields remaining significantly enhanced four years after inputs ceased. It was predicted that it would take a minimum of 13 years to eliminate the enhanced P availability on those plots that received high P inputs over a four-year period. Floristic recovery on these plots is likely to be both protracted and uncertain.

Keywords: Fertilizer, Floristic diversity restoration, Nitrogen, Phosphorus, Potassium, Residual effects of fertilizer, Species-rich grassland.

REFERENCES

Al-Mufti, M.M., Sydes, C.L., Furness, S.B., Grime, J.P. and Band, S.R. (1977) A quantitative analysis of shoot phenology and dominance in herbaceous vegetation. *Journal of Ecology*, **65**, 759–791.

Bakker, J.P. (1989) *Nature management by grazing and cutting*. Kluwer Academic, Dordrecht.

Baldock, D. (1990) *Agriculture and habitat loss in Europe*. WWF International CAP Discussion Paper No. 3. Gland, Switzerland.

Berendse, F., Oomes, M.J.M., Altena, H.J. and Elberse, W.Th. (1992) Experiments on the restoration of species-rich meadows in The Netherlands. *Biological Conservation*, **62**, 59–65.

Brenchley, W.E. and Warington, K. (1958) *The park grass plots at Rothamsted 1856–1949*. Rothamsted Experimental Station, Harpenden.

Devillers, P., Devillers-Tershuren, J. and Ledant, J.P. (1991) *CORINE biotope manual*, Vol. 2, *Habitats of the European Community. Section 38 Mesophile grasslands*. Commission of the European Communities, Luxembourg.

GENSTAT V Committee (1987) *GENSTAT V reference manual*. Clarendon Press, Oxford.

Glaves, D. (1988) Environmental monitoring of grassland management in the Somerset Levels and Moors Environmentally Sensitive Area, England. In: Joyce, C.B. and Wade, P.M. (eds), *European wet grasslands: biodiversity, management and restoration*, pp. 73–94. John Wiley, Chichester.

Gough, M.W. and Marrs, R.H. (1990) Trends in soil chemistry and floristics associated with the establishment of a low-input meadow system on an arable clay soil in Essex, England. *Biological Conservation*, **52**, 135–146.

Grime, J.P. (1979) *Plant strategies and vegetation processes.* Wiley, Chichester.

Kirkham, F.W. and Wilkins, R.J. (1994a) The productivity and response to inorganic fertilizers of species-rich wetland hay meadows on the Somerset Moors: nitrogen response under hay cutting and aftermath grazing. *Grass and Forage Science*, **49**, 152–162.

Kirkham, F.W. and Wilkins, R.J. (1994b) The productivity and response to inorganic fertilizers of species-rich wetland hay meadows on the Somerset Moors: the effect of nitrogen, phosphorus and potassium on herbage production. *Grass and Forage Science*, **49**, 163–175.

Kirkham, F.W., Mountford, J.O. and Wilkins, R.J. (1996) The effects of nitrogen, potassium and phosphorus addition on the vegetation of a Somerset peat moor under cutting management. *Journal of Applied Ecology*, **33**, 1013–1019.

Marrs, R.H. (1990) *Soil fertility and conservation: review.* Report to the Nature Conservancy Council. CSD 1185. English Nature, Peterborough.

Ministry of Agriculture, Fisheries and Food (1994) *Environmentally Sensitive Areas.* MAFF, London.

Mountford, J.O., Lakhani, K.H. and Kirkham, F.W. (1993) Experimental assessment of the effects of nitrogen addition under hay-cutting and aftermath grazing on the vegetation of meadows on a Somerset peat moor. *Journal of Applied Ecology*, **30**, 321–332.

Mountford, J.O., Tallowin, J.R.B., Kirkham, F.W. and Lakhani, K.H. (1994) Effects of inorganic fertilizers in flower-rich hay meadows on the Somerset Levels. In: Haggar, R.J. and Peel, S. (eds), *Grassland management and nature conservation*, pp. 74–85. Occasional Symposium No. 28. British Grassland Society, Reading.

Oomes, M.J.M. (1990) Changes in dry matter and nutrient yields during the restoration of species-rich grasslands. *Journal of Vegetation Science*, **1**, 333–338.

Rodwell, J. (1992) *British plant communities*, Vol. 3, *Grassland and montane communities.* Cambridge University Press, Cambridge.

Smith, A.J.E. (1978) *The moss flora of Britain and Ireland.* Cambridge University Press, Cambridge.

Tallowin, J.R.B., Kirkham, F.W., Brookman, S.K.E. and Patefield, M. (1990) Response of an old pasture to applied nitrogen under steady-state continuous grazing. *Journal of Agricultural Science, Cambridge*, **115**, 179–194.

Tallowin, J.R.B., Mountford, J.O., Kirkham, F.W., Smith, R.E. and Lakhani, K.H. (1994) The effect of inorganic fertilizer on a species-rich grassland – implications for nature conservation. In: t' Mannetje, L. and Frame, J. (eds), *Grassland and society*, pp. 332–335. Proceedings of the XVth General Meeting of the European Grassland Federation, June 1994. Wageningen Pers, Wageningen.

ter Braak, C.J.F. (1987–92) *CANOCO – a FORTRAN program for Canonical Community Ordination.* Microcomputer Power, Ithaca, New York.

Tilman, D. (1982) *Resource competition and community structure.* Princeton University Press, Princeton.

Vermeer, J.G. and Berendse, F. (1983) The relationship between nutrient availability, shoot biomass and species richness in grassland and wetland communities. *Vegetatio*, **53**, 121–126.

Wheeler, B.D. and Shaw, S.C. (1991) Above-ground crop mass and species-richness of the principal types of herbaceous rich-fen vegetation of lowland England and Wales. *Journal of Ecology*, **79**, 285–301.

18   Restoration of Soil Chemical Conditions of Fen-Meadow Plant Communities by Water Management in The Netherlands

DICK VAN DER HOEK and WIM BRAAKHEKKE
Wageningen Agricultural University, The Netherlands

INTRODUCTION

Until 50 years ago, nutrient-poor, base-rich fens and fen meadows were common in The Netherlands. Fens comprise minerotrophic wetland vegetation on peaty soils. Fen meadows are semi-natural ecosystems derived from fens by extensive agricultural exploitation for hay-making without fertilization but with some degree of drainage (Wheeler 1995). One of the typical plant communities in fen meadows is *Cirsio–Molinietum caeruleae* (nomenclature of plant communities follows Westhoff and den Held 1969). This community corresponds to the British National Vegetation Classification community type M24 (Rodwell 1992) and CORINE biotope *Junco–Molinion* 37.312 (Moss *et al.* 1991). When the agricultural use is intensified by fertilization and deep drainage, a fen meadow develops into a hay meadow with an *Arrhenaterion* or eventually a *Poo–Lolietum* type of vegetation.

Almost everywhere in The Netherlands the relics of oligotrophic fens and fen meadows have become increasingly eutrophicated and acidified. An important factor in this process has been the lowering of groundwater, caused by exploitation of deep groundwater for drinking water and by drainage for agriculture. Lowering of groundwater levels has resulted in an increased availability of nitrogen and phosphorus in the soil. Water infiltrating as precipitation and nitrogen deposition from the air have further contributed to the eutrophication and acidification. Consequently, biomass production in these vegetation types has increased, but as a result many species have become endangered or extinct.

In most fen nature reserves in The Netherlands it is relatively simple to maintain a high groundwater level during the summer, because there is usually a large storage capacity for winter precipitation facilitated by a system of sluices for controlling water level. Experiments with additional fertilizer and studies of nutrient balances in rich-fen quagmires (*Caricion davallianae*, CORINE biotope 54.231; Moss *et al.* 1991) have shown that nitrogen is the main limiting factor for plant growth. Despite the high rate of nitrogen deposition of 40 kg ha^{-1} yr^{-1}, management by hay-making appears to have been suffi-

See Glossary, p. 305, for explanation of technical terms. Scientific names of vascular plants follow Tutin, T. G. *et al.* (1964–80) *Flora Europaea* Volumes 1–5. Cambridge University Press. See p. 319.

European Wet Grasslands: Biodiversity, Management and Restoration. Edited by Chris B. Joyce and P. Max Wade.
© 1998 John Wiley & Sons Ltd.

cient for the conservation of species-rich plant communities in fens (Verhoeven *et al.* 1994).

In fen meadows on peaty and sandy soils, the situation is different. Despite hay-cutting management for many years, grasses have become dominant over sedges and herbs in most Dutch fen meadows. Introduction of surface water is no solution, because it is usually too eutrophicated due to over-manuring of the surrounding agricultural land. Special water management is needed to maintain or restore the characteristic abiotic site conditions. It is known that the availability of nitrogen and phosphorus should be low for these wet, nutrient-poor grassland communities.

In this chapter we describe the application of some water-management measures for restoration of the characteristic site conditions for *Cirsio–Molinietum caeruleae* fen-meadow vegetation. We also present results of a study into the effects of groundwater level on the availability of nitrogen and phosphorus in soil and their uptake by the vegetation.

THE IMPACT OF GROUNDWATER ON THE AVAILABILITY AND UPTAKE OF NUTRIENTS

Kemmers (1986), Grootjans (1985) and others have shown that the depth and quality of groundwater have important effects on the availability of nitrogen and phosphorus in wet, nutrient-poor grassland communities. To gain insight into these key factors we conducted descriptive and experimental research in *Cirsio–Molinietum* communities in the Bennekomse Meent, a nature reserve near Wageningen, in the centre of The Nether-lands (Figure 18.1). The relationships between groundwater level, groundwater quality, availability and uptake of nitrogen and phosphorus, and species composition of the vegetation, were studied.

SITE DESCRIPTION

The Bennekomse Meent is a 14-ha remnant of formerly extensive communal hay fields consisting of a number of nutrient-poor, species-rich fen meadows. It is situated in the centre of a valley, the Gelderse Vallei, a glacial trough enclosed by ice-pushed ridges. In the valley fill, two aquifers can be distinguished, separated by less-permeable layers of peat and clay. The groundwater in the deepest aquifer is mainly supplied by the infiltra-tion area in the east. Movement from the middle aquifer towards the upper one is about 0.6 mm day^{-1} and is generally greater in summer than in winter. The influence of calcareous, vertical seepage water from this artesian aquifer is more or less predominant over precipitation water in the peaty topsoils of the nature reserve.

Nowadays the Bennekomse Meent is protected as a nature reserve surrounded by heavily fertilized agricultural land. The border zones of the reserve are influenced by extra nutrient inputs and drainage (van der Hoek and van der Schaaf 1988). Historical data on the species composition of the area show a shift over the last 50 years from species that indicate nutrient-poor habitats and calcareous conditions, such as *Carex panicea* and *Carex hostiana*, to fast-growing species that indicate less calcareous and more nutrient-rich conditions, in particular *Holcus lanatus*. We have studied three sites in this nature reserve in more detail (Figure 18.2): site A near the east corner of the reserve; site B

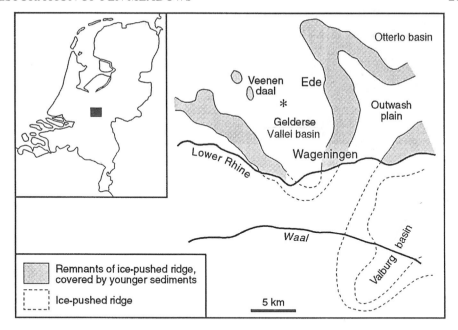

Figure 18.1 Location of the Bennekomse Meent study area, shown by asterisk, and basins and ridges of the Saale glacial tongues

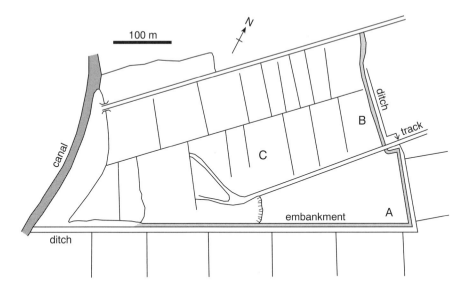

Figure 18.2 Position of study sites (A, B and C) in the Bennekomse Meent

near the north-east border; and site C in the centre of the reserve, which is the least disturbed site, with the deepest peat layer and the strongest influence of upward seepage.

Site A has thin peaty soils (about 20-cm thickness) and the groundwater level is influenced by the drainage of the surrounding agricultural area. It has been enclosed by an embankment to accumulate precipitation water and reduce desiccation of the vegetation. Species like *Anthoxanthum odoratum*, *Danthonia decumbens*, *Festuca rubra* and *F. ovina* have increased here.

Site B has an intermediate position in between site A and C. The peaty topsoil extends down to a depth of 20–50 cm and there is some drainage caused by the border ditch. The influence of calcareous seepage water is stronger than at site A, but less than at site C. *C. panicea*, *C. hostiana*, *Cirsium dissectum* and *Gentiana pneumonanthe* are decreasing at site B, while species like *H. lanatus*, *Plantago lanceolata* and *Filipendula ulmaria* are increasing.

Site C has peaty soils up to 1.5-m thick and calcareous seepage water reaches into the root zone. The most abundant species are *C. panicea*, *C. hostiana*, *Molinia caerulea* and *C. dissectum*. Rare species like *Carex pulicaris* and *G. pneumonanthe* are abundant. During the last 50 years, there has been no notable shift in species composition or dominance at this site.

FERTILIZATION EXPERIMENT

To investigate which element is limiting in the centre and border zone (at sites C and B), a 2×2 factorial fertilization experiment was carried out, in which, for a period of one year, nitrogen (N) was applied at rates of 0 and 200 kg N ha^{-1} and phosphorus (P) at rates of 0 and 40 kg P ha^{-1} (van Mierlo *et al.*, in prep.). The rates were calculated to be sufficiently high to eliminate any limitations and were based on soil conditions and previous experiments. The response of the vegetation was monitored over four years.

At site C, N fertilization resulted in increased biomass production, but no shift in the relative contributions of different groups of species was observed. P fertilization did not result in increased biomass production, but did cause a shift in botanical composition from a sedge-dominated vegetation, with high abundance of *C. panicea*, towards a grass-dominated vegetation with high abundance of *H. lanatus*. This shift started only in the second year after P fertilization. At site B, N fertilization had no effect, while P fertilization resulted in an increased biomass in the first year after fertilization and in a shift in vegetation composition similar to that in site C.

The lack of response of the biomass production to N fertilization at site B shows that N availability was not limiting at this site, probably because of extra N input from the adjacent pasture or from high N mineralization rates in spring when the groundwater level is dropping. In contrast, at site C, the increase in biomass production after N fertilization shows that N availability was limiting growth.

The lack of response of the biomass production at site C to application of P does not disprove that P availability is a limiting factor. Since P is easily adsorbed to the soil it is possible that the applied P rates were insufficient to saturate the capacity of the soil to fix P. Even though biomass production was not increased, P application did change the species composition of the vegetation. Therefore it is concluded that P supply was limiting growth at both sites, which means that there was co-limitation by N and P at site C, which supports the most typical *Cirsio–Molinietum* community. The *Cirsio–Molinietum* community could have been derived from a *Caricion davallianae* community by selective P depletion due to long-continued mowing management, resulting in biomass production and species composition that are probably equally or even more limited by P availability than by N (Verhoeven *et al.* 1994).

It is possible that the species composition at site C changed whilst the biomass production did not increase, because, although most of the added P was fixed, the equilibrium P concentration in the soil solution was increased just enough for the less P-efficient and more P-demanding species to benefit, but not enough to increase total biomass production. This would imply that biomass production in these *Cirsio–Molinietum* communities is mainly determined by the annual P flux that results from decomposition of organic matter, while the species composition is mainly determined by the P concentration in the soil solution, which is in turn determined by the exchange equilibrium with the P adsorbed to the soil. The quality of the water in the root zone (pH and concentrations of Ca^{2+}, Fe^{2+} and Al^{3+} ions) may regulate the P concentration, while water quantity, especially the groundwater level, may regulate the P flux.

AVAILABILITY OF NITROGEN AND PHOSPHORUS AT *CIRSIO–MOLINIETUM* SITES

To test the above hypotheses, we studied several processes that determine the availability of N and P in soil and their uptake by the vegetation at 44 plots evenly distributed over sites A, B and C. Depth of the groundwater table was measured twice a month at each site. Mineralization of N was measured over a period of three years, by means of *in situ* incubation of undisturbed soil cores at 0–15 cm depth (according to Raison *et al.* 1987). In soil samples the following variables were measured: pH (H_2O), pH (KCl), percentage organic matter, P fractions (total P, soluble P, P-FeOOH, P-$CaCO_3$, organic P; according to de Groot and Golterman 1990). During the growing season in 1994, the vegetation was sampled five times; biomass and mineral content were measured.

A positive correlation between depth of groundwater table and nitrification of the plots was found in the *Cirsio–Molinietum* at sites A, B and C. The fluctuations during the

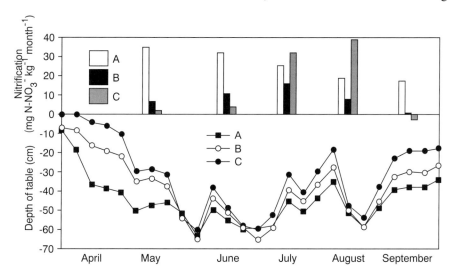

Figure 18.3 Groundwater depth (cm below surface) and net nitrification (mg N-NO_3^- kg dry soil^{-1} month^{-1}) at three hydrologically different sites (study sites A, B and C) of a *Cirsio–Molinietum* ecosystem

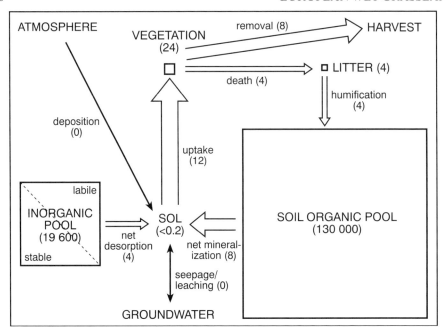

Figure 18.4 Phosphorus (P) flows (kg ha^{-1} yr^{-1}) and P pools (kg ha^{-1}) in a *Cirsio–Molinietum* ecosystem at a maximum depth of 15 cm; (in)organic pools are represented at a scale of 1:10

growing season are illustrated in Figure 18.3. At site C, high groundwater levels (up to the soil surface) retarded nitrification at the beginning of the growing season. However, as a consequence of the high percentage of organic matter (56%) and the low water table in summer, annual nitrification may still have been considerable. At sites A and B, water levels started to fall earlier, leading to stronger nitrification in spring. These results are in accordance with Grootjans (1985), who stated that site conditions of *Cirsio–Molinietum* communities are characterized by temporary inundation, high water tables in spring and a maximum groundwater depth of approximately 60 cm below surface in summer. Koerselman and Verhoeven (1995) have shown that a change in the trophic status of fens cannot always be explained by external inputs. It may also result from a change in the rates of soil nutrient release, associated with changed environmental conditions, such as a lowering of the water table. This 'internal eutrophication' can be very important, particularly in older successional stages with large organically and chemically complexed nutrient pools in the soil.

Where upward seepage has been reduced, storage of precipitation water by means of an embankment (as in site A) is not effective in restoring the required groundwater levels. In spite of the embankment, the groundwater level at site A falls relatively rapidly in early spring due to drainage. Moreover, conservation of rainwater is deleterious to ground-water quality. In the topsoil of the enclosed site A, rainwater had a considerable influence as evidenced by a relatively low pH and a high content of sulphate. At site C, which is not embanked, higher pH values and a dominance of Ca^{2+} and HCO_3^- ions occurred in the root zone, caused by the influence of the upward seepage of calcareous groundwater (van der Hoek and van der Schaaf 1988).

The soil processes affecting the availability of P are complex. Using information provided in Bolt and Bruggenwert (1978), we constructed a conceptual diagram of P flow, which consisted of pools and fluxes, emphasizing the role of P in the soil solution (Figure 18.4). The pool sizes and rates were based on the results of the investigation. They represent an average of the three sites A, B and C. In constructing Figure 18.4 it was inferred that, at these *Cirsio–Molinietum* sites, the decomposition of soil organic matter (specific rate of P mineralization × the amount of organic matter) determines the net flux of phosphate to the soil solution, and that the concentration of phosphate in the soil solution is regulated by sorption processes in combination with an efficient uptake by the vegetation, as hypothesized above.

The total amount of phosphate in the topsoil (0–15 cm) appeared to be more than 5000 times as high as the amount present in maximum standing crop and litter. Figure 18.4 shows that this is mainly organic P. The turnover rate of this pool is very slow. Since there is also a prominent pool of inorganic P, desorption of labile, inorganic P can also be important for the supply of phosphate to plants, especially under changing soil chemical conditions. Desorption of P will proceed much faster than P mineralization, but will eventually lead to a new dynamic equilibrium with a negligible annual net P desorption.

It is concluded from the available data that a change in solubility of P compounds (P-FeOOH and P-CaCO$_3$) due to changes in soil pH has not been a key factor in determining the changes in availability of P in the Bennekomse Meent. This is because at all *Cirsio–Molinietum* sites the pH appeared to be buffered at about 5.5 by cation exchange of the adsorption complex in the soil. The presence of calcareous groundwater in the root zone is a prerequisite to maintain this buffer capacity. Consequently changes in N availability were considered to be the main cause of changes in vegetation composition in the border sites A and B.

UPTAKE OF NITROGEN AND PHOSPHORUS BY *CIRSIO–MOLINIETUM* COMMUNITIES

The uptake of N and P by *Cirsio–Molinietum* communities differed considerably between the investigated sites, especially at the beginning of the growing season. This was expressed in the concentration in the plants (Figure 18.5) as well as in the uptake of N and P m^{-2} yr^{-1} (not shown). In May, the N concentration of plant material at site A exceeded that at site B and the concentration at site B was greater than at site C. Based on data for the whole growing season as well as just May, the P concentration in plant material also differed between the sites. The P concentration of plant material at site C was less than at sites A and B.

N : P ratios in plant material, measured in July for each of the three years, were relatively low at all three sites. The mean N : P ratio at sites A and B fluctuated from 12 to 19, while the values were somewhat higher (13–22) at site C, caused by a low P concentration in spring. Pegtel *et al.* (1996) and Koerselman and Verhoeven (1995) showed that production of biomass of plant communities is limited by P when the N : P ratio in the biomass is higher than 11 and 16 respectively. These data indicate that, at all sites of *Cirsio–Molinietum* investigated, biomass production is limited by P availability. Moreover the results of the fertilizer experiment showed a co-limitation of N and P at site C.

The main conclusion of these investigations is that continuous upward seepage of

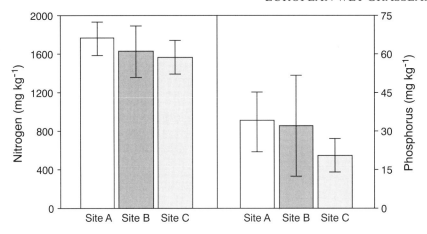

Figure 18.5 Concentration of nitrogen (N) and phosphorus (P) in above-ground biomass of the total vegetation at the *Cirsio–Molinietum* study sites A, B and C in May 1994. The error bars indicate ± 1 SE

calcareous groundwater in the root zone is essential for the maintenance of two conditions that are typical for *Cirsio–Molinietum* communities: a high buffer capacity in the root zone and high groundwater levels in spring and summer, ensuring a low mineralization of N and P. Moreover, high water levels restrict the depth of rooting, leading to greater competition between plant species and a lower nutrient uptake per m² (Aber and Melillo 1991). Deep-rooting species, like many grasses, are hampered, and shallow-rooting, stress-tolerant species, like the sedges, are favoured under these wet and nutrient-poor conditions.

EFFECTIVENESS OF WATER MANAGEMENT FOR RESTORATION

Restoration of site conditions for *Cirsio–Molinietum* communities requires a combination of measures. First, measures have to be taken to decrease losses of water due to pumping of deep groundwater and drainage of phreatic groundwater. However, a reduction of groundwater extraction for drinking water may be difficult to establish, in view of the demands of the drinking-water industry. Secondly, measures are needed inside the nature reserve to diminish the undesirable effects of internal eutrophication and acidification caused by a decrease in upward seepage. These may include turf-stripping and trench-cutting (Marrs 1993). Turf-stripping removes the eutrophicated topsoil, and has the additional favourable effect that wetter conditions are created in the remaining topsoil by lowering of the surface level. In order to minimize the adverse effects of the removal of topsoil, which include destruction of vegetation and removal of the seedbank, this should only be practised after careful consideration. Shallow trenches increase the surface runoff of surplus precipitation in winter and in this way prevent downward seepage and acidification processes. Over the short term, adaptations of the drainage system in a buffer zone around nature reserves appear to be very promising. Several measures are possible to decrease the discharge of groundwater in such a buffer zone, for example maintaining higher water levels in ditches and partial or total infilling of ditches.

We used hydrological models of the study area to predict the impact of various combinations of the above measures on the extent of upward seepage and rainwater infiltration in the nature reserve. Although integrated non-steady-state models of groundwater flows applied to the saturated and the unsaturated zone are preferred, in practice simpler models had to be used, due to the limited availability of data. For predicting hydrological impacts of some planned measures in the surroundings of the Bennekomse Meent, we used MICROFEM, a steady-state model for the saturated zone (Hemker and van Elburg 1988), and ONZAT, a non-steady-state model for the unsaturated zone (van Drecht 1983).

The model calculations predicted that the infilling of the bordering ditch would result in a 15-cm elevation of the groundwater level at site A in summer. The contribution of calcareous seepage water to the soil water in the root zone was predicted to increase from 0 to 25% at a depth of 10 cm and from 20 to 90% at a depth of 30 cm in summer. The increased upward discharge in combination with the capillary rise would bring HCO_3^--rich water into the root zone and restore the buffer capacity against acidification (Figure 18.6). This means that the infilling of the 1-m deep ditch would be effective in restoring the characteristic soil chemical conditions in the margins of the reserve.

Other predictions were that the infilling of ditches in the surroundings of the nature reserve would result in the required increase in upward seepage and in higher water levels over the whole reserve. In the surroundings, the higher water levels would be accompanied by an increase in infiltration and acidification. In summary, this study showed that such hydrological models can be very useful in planning drainage systems for nature restoration projects. Based on these predictions the recommended measures were implemented in 1994–95 and a monitoring study has been initiated to follow the effects.

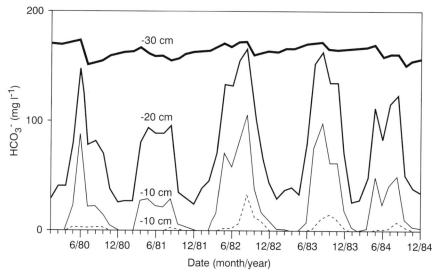

Figure. 18.6 Calculated course of the HCO_3^- concentrations in the soil solution at 10-cm depth without ditch-infilling (- - - - -) and at various depths in the soil with ditch-infilling in the surroundings of the reserve (————). The calculations were carried out with a combination of the hydrological models MICROFEM and ONZAT

Effects of restoration measures in wet, nutrient-poor fens and fen meadows have already been monitored in several other test projects in The Netherlands. They indicate that the discharge of precipitation by shallow trenches reduces acidification of meadow soils and stimulates upward seepage (Kemmers *et al.* 1994; van der Hoek *et al.* 1994). Particularly at plots where trenching was combined with turf-stripping, the characteristic wet, nutrient-poor site conditions were restored and, within four years, some typical plant species, *Parnassia palustris, Carex panicea, Carex oederi* and *Juncus tenageia*, were re-established. If the exploitation of deep groundwater for drinking water and industrial purposes is not reduced in the near future, it will be necessary to extend such measures gradually to whole nature reserves and repeat them at regular intervals to conserve the last few *Cirsio–Molinietum* communities that were once so common in The Netherlands.

ACKNOWLEDGEMENTS

The authors thank J.E.M. van Mierlo, T.J. Hiemstra, P. Roos and H. Schaap for their comments and for the use of the results of their investigations.

SUMMARY

The availability of nitrogen (N) and phosphorus (P) in most of the nutrient-poor, species-rich fen-meadow ecosystems in The Netherlands has increased due to lowering of the groundwater level. Consequently biomass production has increased and many species have become locally extinct. The effects of the availability of N and P on *Cirsio–Molinietum caeruleae* plant communities were studied in fertilizer trials in the centre and on the border of a nature reserve. It is concluded that there was co-limitation by N and P in the relatively wet centre of the reserve, while there was only P limitation on the border where groundwater level had been lowered by drainage. The species composition of these communities appeared to be determined by the inorganic P concentration in the soil solution, which is known to depend on the chemical composition of the soil water, while the biomass production is determined by the annual P flux resulting from turnover of the large pool of organic P, which is known to depend on the groundwater level. A continuous upward seepage of calcareous groundwater into the root zone is considered to be essential for maintaining the chemical conditions in the root zone and for reducing fluctuations in the groundwater level in the growing season. Hydrological models were used to predict the impact of restoration measures on the quantity and quality of water in the root zone. It is concluded that the characteristic site conditions for a *Cirsio–Molinietum* fen meadow can be restored by reduction of deep drainage in the surroundings of the reserve, for example by partial infilling of ditches, in combination with measures inside the reserve such as the cutting of shallow trenches to remove surplus rainwater and turf-stripping to lower the surface level.

Keywords: *Cirsio–Molinietum*, Fen meadows, Groundwater level, Groundwater quality, Hydrological models, Nitrogen mineralization, Phosphate fixation, Upward seepage, Water management.

REFERENCES

Aber, J.D. and Melillo, J.M. (1991) Biological modification of nutrient availability. In: Aber, J.D. and Melillo, J.M. (eds), *Terrestrial ecosystems*, pp. 139–152. Saunders College, Philadelphia.
Bolt, G.H. and Bruggenwert, M.G.M. (eds) (1978) *Soil chemistry,* Vol. A, *Basic elements*. Develop-

ments in soil science No. 5A. Elsevier Scientific, Amsterdam.

de Groot, C.J. and Golterman, H.L. (1990) Sequential fractionation of sediment phosphate. *Hydrobiologica*, **192**, 143–148.

Grootjans, A.P. (1985) *Changes of groundwater regime in wet meadows.* Unpublished Ph.D. Thesis. State University of Groningen, The Netherlands.

Hemker, C.J. and van Elburg, H. (1988) *MICROFEM users manual (version 2.0). Microcomputer multilayer steady state finite element groundwater modelling.* Free University, Amsterdam.

Kemmers, R.H. (1986) Perspectives in modelling of processes in the root zone of spontaneous vegetation at wet and damp sites in relation to regional water management. In: *Water management in relation to nature, forestry and landscape management*, pp. 91–116. Proceedings of the TNO Committee on Hydrological Research, Vol. 34. The Hague, The Netherlands.

Kemmers, R.H., van Delft, S.P.J. and Jansen, P.C. (1994) *Effecten van hydrologische maatregelen tegen verzuring en vermesting op vegetatie, bodem en grondwater in Groot-Zandbrink.* Report No. 319. DLO-Staring Centrum, Wageningen.

Koerselman, W. and Verhoeven, J.T.A. (1995) Eutrophication of fens ecosystems: external and internal nutrient sources and restoration strategies. In: Wheeler, B.D., Shaw, S.C., Fojt, W.J. and Robertson, R.A. (eds), *Restoration of temperate wetlands*, pp. 91–112. Wiley, Chichester.

Marrs, R.H. (1993) Soil fertility and nature conservation in Europe: theoretical considerations and practical management solutions. *Advances in Ecological Research*, **24**, 241–300.

Moss, D., Wyatt, B., Cornaert, M.-H. and Roekaerts, M. (1991) *Corine biotopes. The design, compilation and use of an inventory of sites of major importance for nature conservation in the European Community.* EUR 13231. Commission of the European Communities, Luxemburg.

Pegtel, D.M., Bakker, J.P., Verweij, G.L. and Fresco, L.F.M. (1996) N, K and P deficiency in chronosequential cut summer-dry grasslands on gley podzol after the cessation of fertilizer application. *Plant and Soil*, **178**, 121–131.

Raison, R.J., Cannell, M.J. and Khanna, P.K. (1987) Methodology for studying fluxes of soil mineral-N *in situ. Soil Biology and Biochemistry*, **19**, 521–530.

Rodwell, J.S. (ed.) (1992) *British plant communities*, Vol. 3, *Grasslands and montane communities.* Cambridge University Press, Cambridge.

van der Hoek, D. and van der Schaaf, S. (1988) The influence of water level management and groundwater quality on vegetation development in a small nature reserve in the southern Gelderse Vallei (The Netherlands). *Agricultural Water Management*, **14**, 423–437.

van der Hoek, D., van Mierlo, J.E.M. and van Walsum, J.D. (1994) *Effekten van maatregelen tege verzuring in een schraalgrasland van het Korenburgerveen.* Vakgroep Terrestrische Oecologie en Natuurbeheer, Landbouwuniversiteit Wageningen.

van Drecht, G. (1983) *Simulatie van het verticale, niet-stationaire transport van water en een daarin opgeloste stof in de grond.* RIVM, Bilthoven.

van Mierlo, J.E.M., van Groenendael, J.M. and van der Hoek, D. (in prep.) *A nutrient-driven shift in plant species dominance in a fen-meadow ecosystem.*

Verhoeven, J.T.A., Wassen, M.J., Meuleman, A.F.M. and Koerselman, W. (1994) Op zoek naar de bottleneck. (Nutrient limitation in fens and dune slacks: in search of the bottleneck; English summary.) *Landschap*, **11**, 25–38.

Westhoff, V. and den Held, A.J. (1969) *Plantengemeenschappen van Nederland.* Thieme, Zutphen.

Wheeler, B.D. (1995) Introduction: restoration and wetlands. In: Wheeler, B.D., Shaw, S.C., Fojt, W.J. and Robertson, R.A. (eds), *Restoration of temperate wetlands*, pp. 1–19. Wiley, Chichester.

19 Restoration of a Target Wet Grassland Community on Ex-Arable Land

SARAH MANCHESTER, JO TREWEEK, OWEN MOUNTFORD, RICHARD PYWELL and TIM SPARKS

Institute of Terrestrial Ecology, Monks Wood, UK

INTRODUCTION

The ongoing decline of semi-natural habitats in Britain is well documented (Ratcliffe 1977; Fuller 1987; Wells 1989), even for habitats and communities that were once a common feature of the farmed countryside. Many extensively managed wet meadows and pastures, for example, persisted on heavy clay soils with impeded drainage or on floodplain sites subject to periodic inundation, long after agricultural intensification had degraded semi-natural communities and habitats on other, more tractable soil types. Eventually, however, improved drainage and mechanical cultivation techniques made even wet sites amenable to intensification, and semi-natural wet grassland communities began to disappear rapidly from the farmed landscape. The wildlife species (plants, birds and invertebrates) characteristic of these grasslands have become increasingly rare and threatened, to the point where protection of remaining areas has become important and mechanisms for restoration of degraded areas may be necessary to prevent irrevocable loss.

Opportunities to reinstate some of the semi-natural habitats that evolved under less intensive systems of agriculture have multiplied as agricultural overproduction in the European Union has reduced the pressure to increase output through continued intensification. Production support mechanisms are now accompanied by schemes intended to remove some arable land from agricultural use through set-aside. More importantly, the Ministry of Agriculture, Fisheries and Food has made funds available to support methods of land management more likely to maintain or enhance the wildlife and landscape values of designated Environmentally Sensitive Areas (ESAs) in the UK (Ministry of Agriculture, Fisheries and Food 1989). The ESA scheme is designed to help conserve areas of high value in terms of landscape and/or wildlife that are vulnerable to changes in farming practices. Farmers voluntarily enter land into the scheme and receive payments for the maintenance or introduction of environmentally beneficial farming practices.

Lowland wet grasslands, including grazing marshes and flood meadows, are an important feature of a number of the ESAs of England and Wales (Glaves 1998). Although

See Glossary, p. 305, for explanation of technical terms. Scientific names of vascular plants follow Tutin, T. G. *et al.* (1964–80) *Flora Europaea* Volumes 1–5. Cambridge University Press. See p. 319.

European Wet Grasslands: Biodiversity, Management and Restoration. Edited by Chris B. Joyce and P. Max Wade.
© 1998 John Wiley & Sons Ltd.

the ESAs tend to include a relatively high proportion of land under more traditional management, there are many areas eligible for support under the ESA scheme that have been drained, cultivated or otherwise improved and used for intensive arable production for a number of years. The changes in soil structure, fertility and seedbank brought about by intensive arable use over prolonged periods make it necessary to adopt approaches to habitat restoration that are quite different from those required on sites which have been less intensively used and have remained 'under grass'.

ESTABLISHING TARGETS FOR RESTORATION

By definition, the term 'restoration' implies the existence of a defined endpoint. The first stage in the process of habitat restoration is therefore to determine exactly what is to be restored to a specific site. Sometimes historical records can be used to identify the composition of plant communities that previously occupied a site, but it is rare for agricultural records to provide sufficient ecological detail for successful duplication. Where fields are released from arable cultivation in an area still dominated by arable agriculture, it may be very difficult to infer the composition of any semi-natural communities that may have occurred. Where possible, however, existing semi-natural communities occurring in similar conditions, and with known management histories, should be used to provide a 'template'. Regional or geographic variation in the distributions of native species and communities should always be taken into account when attempting habitat restoration or creation. The National Vegetation Classification (NVC) (Rodwell 1992) has made it possible to define restoration 'targets' in terms of vegetation community composition and therefore to adopt uniform approaches to restoration on different sites.

RESTORATION TECHNIQUES

If restoration is to be widely implemented on sites released from arable agriculture, techniques are required that are both straightforward and cost-effective. Appropriate techniques will depend on the physical conditions of a site, for example its location relative to potential sources of colonizing species, and on the time and financial resources available.

One possible approach to habitat restoration is to rely on 'natural' or spontaneous regeneration from existing sources of colonizing species such as the soil seedbank and the 'seed rain'. Soil seedbanks generally contain seeds of species present in the above-ground flora, seed of previous vegetation, and also seed that has arrived at the site from other sources. Replacement of semi-natural vegetation with agricultural crops results in alterations to the composition of the soil seedbank. Seeds of earlier communities, if still present in a viable condition in the soil, may be buried deeper by cultivation or may germinate only for the plants to be removed by agricultural practices before themselves shedding seed, whilst seeds of annual arable weeds will tend to increase in numbers (Graham and Hutchings 1988). On sites that have remained under arable cropping for a number of years, where soil seedbanks are more likely to be degraded, seed dispersal from other sites with existing semi-natural vegetation may be important. However, fragmentation, and hence isolation, of remaining wetlands means that natural colonization is a slow and uncertain process. It is also possible that propagules of wetland plants may be carried on to wetland areas by floodwater from much more distant locations upstream (though this is very difficult to test). There are two main methods for the investigation of seedbank composition: extraction and identification of seed from the soil or recording of seedling emergence. Direct counting of extracted seed provides information on total seed numbers in the soil, but further experimentation is required to establish viability (Leck *et al.* 1989). Seedling emergence techniques, although requiring more space and a greater length of time to complete, will give information on seed viability and seasonality of germination as well as the species composition of the soil seedbank. Once the soil seedbank has been characterized for a potential restoration site, it is possible to determine whether or not deliberate reintroduction of species is likely to be necessary. Previous work has suggested that, on ex-arable sites, the soil seedbank alone is unlikely to be an adequate source of 'desirable' recolonizers.

Another possible technique for restoration of communities involves the use of seed remaining in hay bales harvested from species-rich grassland. In the UK, Wells *et al.* (1986) investigated the use of the technique for creation of species-rich vegetation and concluded that, although the method appeared to have potential, it tended to give unpredictable results. The time of hay-cutting in relation to flowering times will obviously determine whether or not the species present in the standing crop will actually contribute to the seed component of the bale. The length of time the hay lies on the ground will also affect the seed composition of bales: too long and much seed may be shed; too short and seed may not ripen. The main advantage of this technique is its relatively low cost, and the fact that the seed reintroduced to a site is derived from standing vegetation of known composition and known origin.

In order to accelerate the restoration process, it is common for species to be deliberately reintroduced, most commonly using mixtures of seed. There are a number of potential problems associated with this approach. For example, on ex-arable sites that are characterized by high nutrient availability, ruderal and agricultural species tend to be at a competitive advantage over the slower-growing herbaceous perennials that often characterize semi-natural vegetation (Hodgson 1989). However, seeding is considerably cheaper than the introduction of species as transplants or turves (Byrne 1990) and can be carried out using standard agricultural techniques.

Ideally, only seed of local provenance should be used for habitat restoration. This will ensure that species are not introduced into areas outside their natural distribution range, and that localized genetic diversity is preserved. In fragmented agricultural landscapes, however, local seed of the appropriate species may not be available in sufficient quantities. If local seed is unavailable, native seed is nevertheless preferable to foreign seed, but the origin of seed bought commercially is not always clear. There is currently much concern over the use of foreign strains of species for conservation purposes. Akeroyd (1994) voices these potential concerns as confusion of native ranges and landscape patterns, competitive exclusion of native flora by possibly more vigorous foreign strains and erosion of native genetic diversity by interbreeding of native and foreign strains. In many cases where both native and foreign strains are available, procurement of native seed may well incur as much as a tenfold increase in the price of seed. There are also considerable difficulties in obtaining seed of some species commercially. Fluctuations in seed harvests often mean that some species are in very limited supply, or are totally unavailable in some years.

The work described here explored the need for deliberate reintroduction of plant

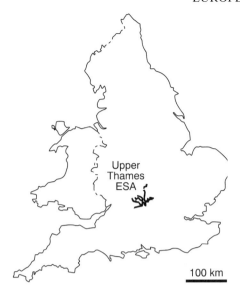

Figure 19.1 The location of the Upper Thames Tributaries Environmentally Sensitive Area in England

species to an ex-arable site in order to restore wet grassland plant communities modelled on the vegetation of an adjacent wet meadow Site of Special Scientific Interest (SSSI). SSSIs are areas of land containing plant or animal communities, geological features or landforms that are judged to be 'special' and are notified to ensure that nature conservation considerations are taken into account when development and land-use changes are proposed (Nature Conservancy Council 1989). A suitable site was selected in 1992 and the field experiment was established in 1993, to assess a variety of potential sources of seed for restoration, ranging from the soil seedbank to sown seed mixtures. The use of a 'nurse' crop to promote establishment was also investigated, as the traditional method for establishment of grass leys on ex-arable sites in this area was to undersow spring barley crops or to sow with a nurse crop of *Lolium multiflorum* (R. Lambourne, pers. comm.). The present study focuses on restoration techniques based on the introduction of potential colonizers as seed, as seeding is most commonly recommended for habitat restoration funded under the Countryside Stewardship Scheme (Countryside Commission 1993) and the ESA scheme. Preliminary results of the study are presented.

THE STUDY AREA

The study site occupies approximately 4 ha of the floodplain of the River Ray in the Upper Thames Tributaries ESA, close to the border between Oxfordshire and Buckinghamshire (Grid Reference SP 65100 20290; 51°52′ 36′ latitude, 1°3° 15′ longitude) in southern England (Figure 19.1). The site is subject to seasonal flooding and is largely underlain by Oxford Clay with soils of the Denchworth and Fladbury series. These soils are impermeable and subject to frequent inundation, making them unusually difficult to drain and cultivate. For this reason, wet meadows and pastures persisted in this area and farmers have been swift to take advantage of the set-aside, Countryside Stewardship and

ESA schemes to release these marginal arable sites from intensive use. Grasslands subject to varying degrees of intensification occur in the immediate vicinity of the experimental site and an SSSI wet meadow borders it on one side. The study site itself was in arable use for at least 15 years before being set-aside in 1993. It was selected for this study because of its proximity to the main water course (the Ray) and a source of potential colonizing species (the adjacent SSSI).

MATERIALS AND METHODS

ESTABLISHING TARGETS FOR RESTORATION

Vegetation communities

The SSSI wet meadow adjacent to the study site was used to provide a template for restoration. This meadow is known to have been unimproved and uncultivated for at least 60 years (R. Lambourne, pers. comm.). The NVC (Rodwell 1992) was used to characterize the target vegetation of the SSSI, in order to permit subsequent extrapolation to other sites. A series of relocatable transects with 1-m^2 quadrats located at 10-m intervals were recorded in the summer of 1993. Species were recorded on a presence/absence basis and their percentage cover in the quadrats was estimated by eye. The results were analysed using TABLEFIT (Hill 1991) to classify the vegetation. TABLEFIT has been written to identify the vegetation types of the NVC, assigning a goodness-of-fit value between samples of vegetation and the expected species composition of NVC vegetation types. The use of such a computer package to classify vegetation provides consistency and removes the subjectivity associated with the classification of vegetation by individuals.

It became clear that the SSSI wet meadow vegetation used to define restoration targets in the present study would not give an obvious 'fit' to any one NVC type. In fact it approximated to a mixture of Mesotrophic Grassland (MG) 4 (*Alopecurus pratensis–Sanguisorba officinalis* grassland), MG5a (*Lathyrus pratensis* subcommunity of *Cynosurus cristatus–Centaurea nigra* grassland) and MG8 (*Cynosurus cristatus–Caltha palustris* grassland). These communities are all characterized by a species-rich, somewhat variable sward of grasses and herbaceous dicotyledons, with no single, constantly dominant species (Rodwell 1992). MG4 and MG8 are both communities characteristic of sites subject to seasonal inundation. All three communities tend to evolve (in lowland Britain) under more traditional agricultural management regimes and are damaged by agricultural improvement. MG4 and MG8 are also damaged by drainage. Species typical of these types of communities include *Alopecurus pratensis, Cynosurus cristatus, Festuca rubra, Holcus lanatus, Lolium perenne, Agrostis capillaris, Anthoxanthum odoratum, Poa trivialis, Cerastium fontanum, Filipendula ulmaria, Lathyrus pratensis, Leontodon autumnalis, Rumex acetosa, Sanguisorba officinalis, Plantago lanceolata, Ranunculus acris, Trifolium pratense* and *T. repens*.

Selection of Potential Constituent Species

The total species list for the SSSI (Appendix A) was reclassified according to F (moisture) and N (fertility) indicator values as assigned by Ellenberg (1988) (Appendix B). F values are based on the correlation between species distributions in relation to site wetness, representing a continuum from dry rocky slopes with shallow soils, to marshy ground, and then shallow and deep, open water. An F value of < 5 indicates a preference for drier sites, and of > 5, for wetter sites. Similarly N values represent the distribution of species in relation to available soil nitrogen. An N value of < 5 indicates a preference for sites with below-average available nitrogen, while a value > 5 indicates a preference for sites with above-average nitrogen availability. Selection of species for inclusion in the seed mixtures to be used in the experiment was based on the use of these two indicator values, those species characteristic of wetter and less fertile conditions being selected in preference to those characteristic of more well-drained and nitrogen-rich conditions. Once species considered suitable in terms of soil moisture and nitrogen preferences had been selected, the list was further restricted by exclusion of species generally regarded as ruderal or 'weedy', for example *Cirsium arvense*. The remaining species can thus be considered as the 'Potential Constituent Species' for any seed mixtures derived.

Experimental treatments: seed mixtures

The list of Potential Constituent Species was used to derive three seed mixtures for inclusion in the experiment (Appendix C). The first mixture (SM1) was based on the recommendations of the Countryside Stewardship Scheme for restoration of waterside landscapes. Current prescriptions require the selection of at least four grass species from a standard list and are based on the assumption that 'desirable' species will ultimately invade an established grass matrix. The four grasses chosen from the management prescription list were: *Alopecurus pratensis, Anthoxanthum odoratum, Cynosurus cristatus* and *Festuca rubra*, all of which were recorded in the SSSI. Unfortunately, *A. odoratum* was commercially unavailable at the time and was replaced by *Phleum pratense* subsp. *bertolonii*. The second mixture (SM2) represented a compromise between the inclusion of a high proportion of potential species and financial cost. In terms of species composition this mixture was intermediate between the simple four-species mixture described above and the more comprehensive third mixture (SM3). The SM3 mixture was modelled as closely as possible on the target community and included all 'potential' species that were commercially available. A relatively large proportion of species were commercially unavailable, including *Cardamine pratensis, Carex nigra, C. disticha, C. panicea, Juncus acutiflorus, Cirsium dissectum, Lotus uliginosus, Lysimachia nummularia, Oenanthe fistulosa, Oenanthe silaifolia, Ranunculus flammula, Serratula tinctoria* and *Thalictrum flavum*. Seeds of *O. fistulosa* and *T. flavum* were therefore hand-picked due to their local availability, as were those of *Sanguisorba officinalis* and *Filipendula ulmaria*. The mean percentage cover values of the constituent species of the seed mixtures (as recorded in the SSSI) were used to calculate their relative proportions for the experimental mixtures, which were based on a ratio of grasses to forbs of 80 : 20 (by weight).

SITE PREPARATION

The above-ground vegetation was cut and removed from the field in the summer of 1993. In early September 1993, the field was ploughed and harrowed to prepare the seedbed and plots were marked out.

Table 19.1   Experimental treatments

Treatment (seed source)	Nurse	No nurse
Natural regeneration (NR)	x	x
Hay bales (HB)	x	x
Basic seed mix (SM1)	x	x
Intermediate seed mix (SM2)	x	x
Comprehensive seed mix (SM3)	x	x

EXPERIMENTAL DESIGN AND ESTABLISHMENT

The experiment consisted of a randomized block design with three replicate blocks, each consisting of ten 38×18 m plots, to which the 2×5 factorial combination of treatments was allocated at random. Each of the five main seed sources was used with and without a nurse crop of *Lolium multiflorum,* giving 10 experimental treatments in total (Table 19.1).

The 'natural regeneration' treatment was used as a control to determine whether or not the deliberate reintroduction of 'target' species to this site was likely to be necessary. This treatment also provided a field-based estimate of soil seedbank composition for comparison with the laboratory/greenhouse tests.

The 'hay bales' treatment used hay baled in 1993 from the SSSI.

Seeds of species were bought individually packaged, and the seed mixtures were made up separately for each plot. Seed of each species was retained for germination tests. Seed was broadcast by hand and raked into the topsoil.

SEEDBANK SAMPLING

Once the experimental design had been decided upon and the plots marked out, the soil seedbank was sampled. It is generally accepted that the majority of seed present in soils beneath grassland occurs within the surface 5 cm (Thompson and Grime 1979; Roberts 1981; Wells, pers. comm. 1993). Thus, samples were taken to a depth of 5 cm using a soil corer with a 4-cm diameter, giving a volume of 63 cm^3 per core. Five cores were taken from each plot, at measured, relocatable points arranged on two diagonal, crossed transects. The five cores per plot were pooled to give a total volume of 315 cm^3 per sample.

Soil samples were air-dried, thoroughly mixed and divided into two equal parts. One half was cold-stored over winter, while the other half was sprinkled on to sterile compost in 20×15 cm paper-lined plant trays. A thin layer of compost was then sprinkled over the samples, which were watered and placed in a glasshouse at a temperature of approximately 20°C. Germinating seedlings were identified and removed fortnightly, and the species, number of seedlings and date of emergence recorded. After removal of seedlings from trays, the soil was stirred gently to encourage further germination. The remaining half-samples were removed from cold storage in the spring of 1994, and the experiment was repeated.

SOWING OF EXPERIMENTAL TREATMENTS

The seed mixtures were broadcast by hand and raked into the topsoil in mid-September 1993. Hay bales were shaken vigorously and spread on the ground by hand.

Table 19.2 Summary of species recorded at the experimental site

Sample	TS	GS	FS	NS	PCS
June 1993	56	16	40	40.7	9.3
Dec 1993	33	13	20	60.6	24.2
June 1994	79	22	54	54.4	22.0
Seedbank	40	13	27	42.5	15.0

TS, total no. of species; GS, total no of grass species; FS, total no of forb species; NS, species also recorded in the Site of Special Scientific Interest (SSSI), as a percentage of the SSSI species (Appendix A); PCS, number of Potential Constituent Species as a percentage of the total Potential Constituent Species (Appendix B).

MONITORING OF SEEDLING ESTABLISHMENT AND SURVIVAL

The field was surveyed in December 1993 and June 1994, to monitor seedling establishment and survival; three and five permanent 1-m^2 quadrats were recorded respectively per plot. Quadrats were located at least 3 m from the edge of the plots to avoid sampling vegetation contaminated by adjacent plots. In the December survey, percentage cover of species was estimated for the quadrat as a whole from the presence/absence of species in each of 25 constituent grid cells. In June, two transects were laid out diagonally in each plot and five quadrats recorded.

ANALYSIS OF RESULTS

Results were analysed using analysis of variance (ANOVA) to determine whether or not there were significant differences between treatments in terms of the numbers of species established by December 1993 or surviving until June 1994. If ANOVA indicated significant differences, they were further investigated using the Least Significant Difference (LSD) test. Statistical analysis for the percentage cover of different species was carried out following an angular transformation of the data. Unless otherwise stated, results presented from the treatment plots are pooled, i.e. with and without a nurse crop.

RESULTS

TOTAL NUMBER OF SPECIES RECORDED: WHOLE SITE

The total number of species recorded on the experimental site as a whole is shown in Table 19.2. Results are given for the baseline vegetation survey in June 1993, for seedling establishment and survival in December 1993 and June 1994 and for the soil seedbank.

Baseline vegetation

Prior to restoration the experimental site supported a total of 56 species (Appendix D), the vegetation being dominated by *Triticum aestivum* (the last crop sown in 1992), and *Lolium multiflorum*. Of the 40 herbaceous species recorded, 17 were annual arable weeds of agriculture. Twenty-two of the species recorded were also recorded in the SSSI during 1993 (Appendix A), and five of the species were Potential Constituent Species (Appendix C).

Seedling establishment and survival

Only 33 species were recorded in December 1993, including 13 grass species; *Lolium* spp. were dominant. Twenty herbaceous species were recorded, three of which had been sown in September 1993 (*Trifolium pratense, Lotus corniculatus* and *Vicia cracca*), whilst four were neither sown, nor recorded in the baseline survey (*Heracleum sphondylium, Galium aparine, Myosotis* sp. and *Matricaria perforata*).

By June 1994, 79 species were present in the vegetation, 21 of which were sown species. Three of the sown grasses had failed to establish (*Agrostis capillaris, Trisetum flavescens* and *Briza media*). Two sown forb species were unrepresented in the recorded vegetation (*Thalictrum flavum* and *Oenanthe fistulosa*). Of the 79 species recorded overall, 43 were also present in the SSSI, and 22 were Potential Constituent Species (out of the possible total of 36; Appendix B). There were three unsown Potential Constituent Species, namely *Cardamine pratensis, Ranunculus flammula* and *Anthoxanthum odoratum.*

Investigation of the seedbank

The seedbank studies are still ongoing, so the results presented here should not be regarded as final. By December 1994, 40 species had germinated for the site as a whole. Thirteen species were grasses, with *Lolium* sp., *Poa trivialis, Alopecurus myosuroides* and *Triticum aestivum* being the most abundant species recorded. Most species had been recorded in the standing vegetation, except *Elymus repens,* which only appeared in the seedbank.

TOTAL NUMBERS OF SPECIES PER TREATMENT

Results for individual treatments are summarized in Table 19.3.

Seed mixtures

In the basic seed mixture (SM1) three of the four species sown were recorded in June 1994, *Phleum pratense* subsp. *bertolonii* having been recorded in December 1993, but not in 1994. In the intermediate seed mixture (SM2) 10 of the 11 sown species were recorded, only *Agrostis capillaris* being missing. For the third (comprehensive) seed mixture (SM3) 17 of the 23 species sown had established by June 1994. The exceptions were *Agrostis capillaris, Briza media, Trisetum flavescens, Thalictrum flavum, Oenanthe fistulosa* and *Hordeum secalinum.*

MEAN NUMBERS OF SPECIES PER TREATMENT

Analysis of variance for the total number of species per plot showed that the comprehensive seed mixture (SM3) resulted in a significantly higher number of species than for any other treatment by June 1994 ($P<0.001$) (Figure 19.2). Significantly more Potential Constituent Species were recorded in plots of the SM3 treatment than of any other treatment ($P<0.001$). The intermediate seed mixture (SM2) also resulted in significantly higher numbers of Potential Constituent Species than did the natural regeneration, hay bales or basic seed mixture (SM1) treatments.

Table 19.3 Summary of June 1994 survey results

Treatment	TS	NS	PCS(%)	SS	SP	GS	NG	GPCS(%)	FS	NF	FPTS(%)
Natural regeneration	36	22	8(22.2)	1	1	14	10	4(44.4)	21	12	4(14.8)
Hay bales	40	23	9(22.5)	1	1	15	10	4(44.4)	24	13	5(18.5)
Seed mix 1	44	25	11(30.6)	4	3	15	10	4(44.4)	27	15	7(25.9)
Seed mix 2	47	321	14(38.9)	11	10	18	13	6(66.6)	27	18	8(29.6)
Seed mix 3	55	37	20(55.6)	23	17	18	14	6(66.6)	34	23	14(51.9)

Data are pooled over the with nurse/without nurse treatments. TS, total no. of species; NS, no. of species; PCS, no. of Potential Constituent Species (as a % of total Potential Constituent Species); SS, no. of species also recorded in the Site of Special Scientific Interest (SSSI); PCS, no. of Potential Constituent Species (as a % of total Potential Constituent Species); SS, no. of species sown; SP, no. of sown species present; GS, total no. of grass species; NG, no. of grass species also in SSSI; GPCS, no. of grass species that are Potential Constituent Species (as a % of total grass Potential Constituent Species); FS, total no. of forb species; NF, no. of forb species also in SSSI; FPTS, no. of forb species that are Potential Constituent Species (as a % of total forb Potential Constituent Species).

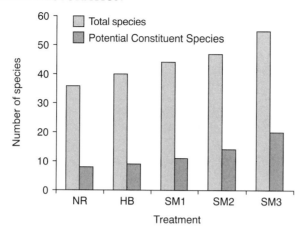

Figure 19.2 Total number of species per treatment, pooled over the with nurse/without nurse treatments (with the number of those that are Potential Constituent Species).Treatments: NR, natural regeneration; HB, hay bales; SM1, basic seed mix; SM2, intermediate seed mix; SM3, comprehensive seed mix

ESTABLISHMENT OF SOWN SPECIES

The relative establishment success differed for the three seed mixtures. The percentage establishment of sown species was highest for the intermediate seed mixture and lowest for the most complex seed mixture (approximately 91% and 74% respectively). Establishment success was influenced by the presence of a nurse crop, being significantly higher in all cases when seed mixtures were sown without a nurse ($P<0.05$).

EFFECTS OF THE NURSE CROP

Percentage covers of *Alopecurus myosuroides*, *Ranunculus acris* and *Lolium multiflorum* were all influenced by the presence of the nurse crop. *R. acris* had a significantly higher percentage cover ($P<0.05$) in treatments without a nurse crop, whilst *L. multiflorum* and *A. myosuroides* were more abundant ($P<0.05$) with a nurse crop. Other species were not significantly affected by the nurse crop. The dominance of *L. multiflorum* can be seen from Figure 19.3. The total percentage cover of *L. multiflorum* was greater in plots where it had also been sown as a nurse crop ($P<0.05$), where its percentage cover exceeded the combined ground cover of all other species present.

DISCUSSION

Ex-arable sites are potentially difficult to restore to semi-natural grassland communities, often having relatively high residual soil fertility (Gough and Marrs 1990) and an impoverished soil seedbank (Leck *et al.* 1989). Chances of success are generally considered to be greatest on sites that have been under cultivation for a relatively short time (Kropac 1966; Leck *et al.* 1989), or which are close to existing semi-natural vegetation that can act as a source of colonizing species. Restoration of lowland wet grassland is

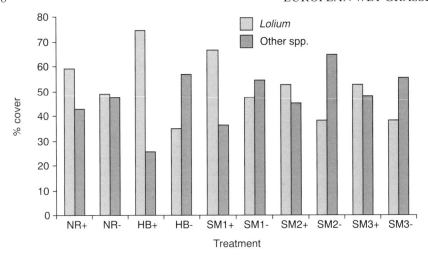

Figure 19.3 Mean percentage cover of *Lolium multiflorum* compared with the mean cover of all other species. Treatments: NR, natural regeneration; HB, hay bales; SM1, basic seed mix; SM2, intermediate seed mix; SM3, comprehensive seed mix. +, with a nurse crop; −, without a nurse crop

further complicated where there have been major alterations to hydrological regimes through under-drainage. In some cases, therefore, reinstatement of an appropriate hydrological regime is necessary for restoration to be successful. The site used in this experiment has remained subject to regular seasonal inundation despite drainage improvements in the area and was therefore considered to have appropriate physical conditions for wet grassland restoration. On such a site, cessation of fertilizer use alone is likely to result in significant changes in species composition with time. For example, the dominance of species like *Lolium multiflorum* and *Triticum aestivum* is likely to be reduced under frequent flooding and without fertilizer input.

A major aim of this experiment was to determine whether or not the deliberate introduction of desirable wet grassland species as seed would improve the likelihood of restoring a target vegetation type. Although only one year's data are currently available from the experiment, some tentative conclusions may be drawn.

Soil seedbank studies suggested that a very small number of 'desirable' species might be available to recolonize the site spontaneously (notably *Cardamine pratensis* and *Ranunculus flammula*). The reinstatement of species to a site through natural regeneration depends on the soil seedbank and 'seed rain' from other sites. The probability of viable seeds remaining from an earlier community generally decreases with increasing time under arable cultivation (Kropac 1966; Leck *et al.* 1989), but the chances of successful restoration through natural regeneration on this site are likely to have been increased by its immediate proximity to a rich source of potential colonizers (the SSSI wet meadow) (Baines 1989). In this respect, the study site may be somewhat atypical of ex-arable sites. In general, intensive agriculture has resulted in a highly fragmented landscape, with islands of semi-natural vegetation remote from one another.

The use of hay bales as a source of seed was relatively unsuccessful in this experiment, there being no significant difference in the number of species establishing from this method and from natural regeneration. Further research into this technique would be beneficial because of its relatively low cost, and the ability to reintroduce species of known, local provenance in appropriate proportions (taking account of flowering times). The limited success of the technique in this experiment may be attributable to factors such as the hay-cut taking place before or after peak flowering periods for most species, or excessive seed-shed during the harvest.

In this experiment, numbers of 'target' or 'desirable' species established after one year were greater on plots where species had been deliberately sown as seed mixtures. Sowing a relatively comprehensive proportion of the species represented in the 'model' community produced a more diverse sward and a higher cover of 'desirable' species.

The importance of selecting species that are adapted to prevailing physical conditions and which are locally represented cannot be overemphasized. It is important to note that the lack of reliable commercial suppliers for many of the species that form vital components of semi-natural lowland wet grassland communities (notably *Anthoxanthum odoratum*) is likely to hamper many restoration attempts. Hand collection of local seed is time-consuming and may have significant effects on the community harvested (though research is required to quantify such potential impacts), but may be necessary in a number of cases in order to ensure that vital component species can be reintroduced.

Results from the first year of the experiment suggest that the chances of restoring semi-natural grassland on ex-arable sites are unlikely to be improved by sowing species with a nurse crop. The species that showed the most positive response to sowing with *Lolium multiflorum* was *Lolium multiflorum* (also recorded in the baseline survey and the seedbank study).

Following only one year of botanical recording, it would be premature to conclude which of the five main sources of seed will be most effective in the long term. Semi-natural communities can take many years to develop, and considerable fluctuations in species composition can take place in the first phase of restoration. Longer-term results are required before it will be possible to rank the various treatments in terms of both establishment success and cost and to determine whether or not the use of more comprehensive seed mixtures can be economically justified.

ACKNOWLEDGEMENTS

The work described was funded by the Ministry of Agriculture, Fisheries and Food.

SUMMARY

Lowland wet grassland has declined within Britain, mainly as a result of agricultural intensification. Opportunities are now available for the restoration/re-creation of plant communities through initiatives such as the Environmentally Sensitive Area (ESA) scheme. Current recommendations for the reinstatement of these habitats assume natural processes of dispersal and colonization will function to ensure the arrival and establishment of characteristic species. However, where natural processes do not function efficiently, appropriate plant species will need reintroducing to ensure their presence at a site. The targeting of restoration, and possible techniques, are discussed with reference to an experiment established in 1993, within the Upper Thames Tributaries ESA, which explored the need for deliberate reintroduction of plant species as seed to an ex-arable site. An adjacent Site of Special Scientific Interest was used to provide a regionally appropriate 'template' for the restoration, from which experimental treatments were derived. Preliminary results, one year after sowing, are presented and discussed.

Keywords: Environmentally Sensitive Area, Habitat restoration, Lowland wet grassland, Restoration targets.

REFERENCES

Akeroyd, J. (1994) *Seeds of destruction? Non-native wildflower seed and British floral biodiversity.* Plantlife, London.

Baines, J.C. (1989) Choices in habitat re-creation. In: Buckley, G.P. (ed.), *Biological habitat reconstruction*, pp. 5–8. Belhaven Press, London.

Byrne, S. (1990) *Habitat transplantation in England. A review of the extent and nature of the practice and the techniques employed.* English Field Unit, Nature Conservancy Council, Peterborough.

Countryside Commission (1993) *Handbook for Countryside Stewardship.* Countryside Commission, Cheltenham.

Ellenberg, H. (1988) *Vegetation ecology of central Europe.* Cambridge University Press, Cambridge.

Fuller, R.M. (1987) The changing extent and conservation interest of lowland grasslands in England and Wales: a review of grassland surveys 1930–1984. *Biological Conservation*, **40**, 281–300.

Glaves, D. (1998) Environmental monitoring of grassland management in the Somerset Levels and Moors Environmentally Sensitive Area, England. In: Joyce, C.B. and Wade, P.M. (eds), *European wet grasslands: biodiversity, management and restoration*, pp. 73–94. John Wiley, Chichester.

Gough, M.W. and Marrs, R.H. (1990) A comparison of soil fertility between semi-natural and agricultural plant communities: implications for the creation of species-rich grassland on abandoned agricultural land. *Biological Conservation*, **51**, 83–96.

Graham, D.J. and Hutchings, M.J. (1988) Estimation of the seed bank of a chalk grassland ley established on former arable land. *Journal of Applied Ecology*, **25**, 241–252.

Hill, M.O. (1991) TABLEFIT *version 0.0, for identification of vegetation types.* Institute of Terrestrial Ecology, Abbots Ripton.

Hodgson, J.G. (1989) Selecting and managing plant materials used in habitat construction. In: Buckley, G.P. (ed.), *Biological habitat reconstruction*, pp. 45–67. Belhaven Press, London.

Kropac, Z. (1966) Estimation of weed seeds in arable soil. *Pedobiologia*, **6**, 105–128.

Leck, M.A., Parker, V.T. and Simpson, R.L. (1989) *Ecology of soil seed banks.* Academic Press, London.

Ministry of Agriculture, Fisheries and Food (1989) *Environmentally Sensitive Areas.* HMSO, London.

Nature Conservancy Council (1989) *Guidelines for selection of biological SSSIs. Detailed guidelines for habitats and species-groups.* Nature Conservancy Council, Peterborough.

Ratcliffe, D.A. (ed.) (1977) *A nature conservation review. The selection of biological sites of national importance to nature conservation in Britain.* Cambridge University Press, Cambridge.

Roberts, H.A. (1981) Seed banks in soils. *Advances in Applied Biology*, **6**, 1–55.

Rodwell, J.S. (1992) *British plant communities*, Vol. 3, *Grasslands and montane communities.* Cambridge University Press, Cambridge.

Thompson, K. and Grime, J.P. (1979) Seasonal variation in the seed banks of herbaceous species in ten contrasting habitats. *Journal of Ecology*, **67**, 893–921.

Wells, T.C.E. (1989) The re-creation of grassland habitats. *The Entomologist*, **108**, 97–108.

Wells, T.C.E., Frost, A. and Bell, S. (1986) *Wildflower grasslands from crop-grown seed and hay bales.* Focus on Nature Conservation No. 15. Nature Conservancy Council, Peterborough.

APPENDIX A

SPECIES RECORDED IN SITE OF SPECIAL SCIENTIFIC INTEREST WET MEADOW (JUNE 1993)

Herbaceous species

Achillea ptarmica
Cardamine pratensis
Centaurea nigra
Cerastium fontanum
Cirsium arvense
Cirsium dissectum
Filipendula ulmaria
Galium aparine
Galium palustre
Geranium dissectum
Heracleum sphondylium
Lathyrus pratensis
Leontodon autumnalis
Leucanthemum vulgare
Lotus corniculatus
Lotus uliginosus
Lychnis flos-cuculi
Lysimachia nummularia
Myosotis discolor
Myosotis laxa subsp. caespitosa
Oenanthe fistulosa
Oenanthe silaifolia
Ophioglossum vulgatum
Plantago lanceolata
Polygonum amphibium
Potentilla reptans
Prunella vulgaris
Ranunculus acris
Ranunculus flammula
Ranunculus repens
Rhinanthus minor
Rumex acetosa
Rumex conglomeratus
Rumex crispus
Rumex X pratensis
Sanguisorba officinalis
Serratula tinctoria
Silaum silaus
Stellaria graminea
Succisa pratensis
Taraxacum agg.

Thalictrum flavum
Trifolium pratense
Trifolium repens
Vicia cracca
Vicia sativa

Grasses

Agrostis canina
Agrostis capillaris
Agrostis stolonifera
Alopecurus geniculatus
Alopecurus pratensis
Anthoxanthum odoratum
Arrhenatherum elatius
Briza media
Bromus commutatus
Bromus hordeaceus.
Bromus racemosus
Cynosurus cristatus
Dactylis glomerata
Deschampsia cespitosa
Festuca pratensis
Festuca rubra
Holcus lanatus
Hordeum secalinum
Lolium perenne
Phleum pratense
Poa annua
Poa trivialis
Trisetum flavescens

Sedges and rushes

Carex disticha
Carex flacca
Carex hirta
Carex nigra
Carex panicea
Juncus acutiflorus
Juncus conglomeratus
Juncus effusus

APPENDIX B

ELLENBERG *F* (MOISTURE) AND *N* (FERTILITY) VALUES FOR
'POTENTIAL CONSTITUENT SPECIES'

Species name	*F*	*N*
Achillea ptarmica	8	2
Agrostis capillaris	x	3
Alopecurus pratensis	6	7
Anthoxanthum odoratum	x	x
Briza media	x	2
Cardamine pratensis	7	x
Carex disticha	9	5
Carex nigra	8	2
Carex panicea	7	3
Centaurea nigra	5	?
Cirsium dissectum	8	2
Cynosurus cristatus	5	4
Festuca pratensis	6	6
Festuca rubra	x	x
Filipendula ulmaria	8	4
Holcus lanatus	6	4
Hordeum secalinum	6	5
Lathyrus pratensis	6	6
Leucanthemum vulgare	4	3
Lotus corniculatus	4	3
Lotus uliginosus	8	4
Lychnis flos-cuculi	6	x
Lysimachia nummularia	6	x
Oenanthe fistulosa	9	5
Oenanthe silaifolia	8	5
Ranunculus acris	x	x
Ranunculus flammula	9	2
Rhinanthus minor	x	2
Rumex acetosa	x	5
Sanguisorba officinalis	7	3
Serratula tinctoria	x	5
Silaum silaus	7	2
Thalictrum flavum	8	2
Trifolium pratense	x	x
Trisetum flavescens	x	5
Vicia cracca	5	x

APPENDIX C

SEED MIXTURES

Commercially unavailable species are in parentheses; species for which seed was hand-picked are marked by an asterisk.

Seed mixture 1 (basic)

Alopecurus pratensis, (Anthoxanthum odoratum), Cynosurus cristatus, Festuca rubra, Phleum pratense subsp. *bertolonii*

Seed mixture 2 (intermediate)

Grasses

Agrostis capillaris
Alopecurus pratensis
(Anthoxanthum odoratum)
Cynosurus cristatus
Festuca pratensis
Festuca rubra
Holcus lanatus

Herbaceous species

(Cardamine pratensis)
Filipendula ulmaria
Leucanthemum vulgare
Lotus corniculatus
Ranunculus acris
Trifolium pratense

Seed mixture 3 (comprehensive)

Grasses

Agrostis capillaris
Alopecurus pratensis
(Anthoxanthum odoratum)
Briza media
Cynosurus cristatus
Festuca rubra
Holcus lanatus
Hordeum secalinum
Trisetum flavescens

Sedges and rushes

(Carex disticha or *C. panicea)*
(Carex nigra)
(Juncus acutiflorus)

Herbaceous species

Achillea ptarmica
(Cardamine pratensis)
Centaurea nigra
(Cirsium dissectum)
*Filipendula ulmaria**
Lathyrus pratensis
Leucanthemum vulgare
(Lotus uliginosus)
Lychnis flos-cuculi
(Lysimachia nummularia)
*(Oenanthe fistulosa)**
(Oenanthe silaifolia)
Ranunculus acris
(Ranunculus flammula)
Rhinanthus minor
Rumex acetosa
*Sanguisorba officinalis**
(Serratula tinctoria)
Silaum silaus
*(Thalictrum flavum)**
Trifolium pratense
Vicia cracca

APPENDIX D

SPECIES RECORDED PRIOR TO THE EXPERIMENT

Herbaceous species

Anthriscus sylvestris
Atriplex prostrata
Brassica rapa
Calystegia sepium
Cardamine pratensis
Centaurea nigra
Chamomilla recutita
Chamomilla suaveolens
Chenopodium album
Cirsium arvense
Cirsium vulgare
Conium maculatum
Coronopus squamatus
Crepis biennis
Dipsacus fullonum
Epilobium ciliatum
Epilobium hirsutum
Epilobium montanum
Galeopsis tetrahit
Geranium dissectum
Lactuca serriola
Leontodon autumnalis
Picris echioides
Plantago major
Polygonum amphibium
Polygonum hydropiper
Polygonum persicaria
Ranunculus flammula
Ranunculus repens
Rorippa islandica
Rumex conglomeratus

Rumex crispus
Senecio vulgaris
Sonchus asper
Stellaria media
Taraxacum agg.
Trifolium dubium
Trifolium repens
Veronica catenata
Vicia faba

Grasses

Agrostis stolonifera
Alopecurus geniculatus
Alopecurus myosuroides
Alopecurus pratensis
Avena fatua
Bromus commutatus
Bromus hordeaceus
Bromus sterilis
Dactylis glomerata
Festuca pratensis
Lolium multiflorum
Lolium perenne
Poa annua
Poa trivialis
Triticum aestivum

Rushes

Juncus effusus

20 Five Years of Restoration of Alluvial Meadows: A Case Study from Central Europe

JANA STRAŠKRABOVÁ[1] and KAREL PRACH[2]

[1]*Agricultural University, Prague, Czech Republic*
[2]*University of South Bohemia, České Budějovice; and Academy of Sciences, Třeboň, Czech Republic*

INTRODUCTION

There are no large natural alluvial grasslands in central Europe (Ellenberg 1988). The present grasslands originated as pastures or meadows and need management to ensure their continued existence.

Central European wet meadows have been subjected to great alterations in the past several decades, reflecting not only changes in their direct management but also changes in the land use of the surrounding landscape. This is especially the case for alluvial meadows in the Czech Republic, where the economics of the communist era did not respect the natural and social conditions connected with river systems. If not ploughed, meadows were damaged through both overexploitation and neglect, resulting in loss of biodiversity and disturbed ecological functions (Rychnovská 1985; Banásová et al. 1994; Prach et al. 1996).

The meadows in the floodplain of the Lužnice River in the Třeboň Biosphere Reserve are in the southern part of the Czech Republic near the border with Austria (Figure 20.1). They are among 10 of the most valuable complexes of alluvial meadows recently recognized in the Czech Republic (Straškrabová et al. 1996). The total extent of the meadows is approximately 10 km², extending a further 3 km² into the adjoining part of Austria. The meadows along the whole of the Lužnice River were described together with other vegetation types by Prach et al. (1990).

In the best preserved part of the Lužnice River floodplain, a long-term ecological project was launched in 1986 within the Unesco Man and Biosphere (MAB) project 'Role of Wetlands in the Temperate Forest Biome'. An important task of the project was the description of rate and directions of vegetation changes. Vegetation pattern was related to three main river-induced environmental gradients: (i) moisture gradient; (ii) nutrient gradient; and (iii) gradient of disturbance intensity (Day et al. 1988). Relationships between vegetation and site moisture were described in detail by Prach (1992); the nutrient gradient was evaluated by Prach and Rauch (1992). The gradient of disturbance

See Glossary, p. 305, for explanation of technical terms. Scientific names of vascular plants follow Tutin, T. G. et al. (1964–80) *Flora Europaea* Volumes 1–5. Cambridge University Press. See p. 319.

European Wet Grasslands: Biodiversity, Management and Restoration. Edited by Chris B. Joyce and P. Max Wade.

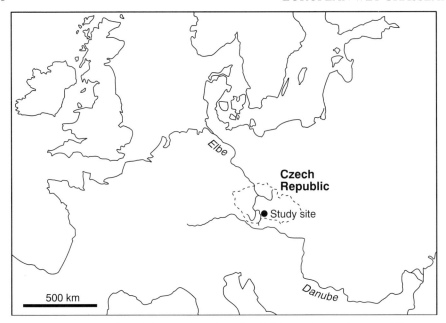

Figure 20.1   Location of the Lužnice River floodplain study area in central Europe

intensity, i.e. frequency and regularity of mowing, has not yet been quantified, but it is possible to compare the vegetation pattern in the floodplain segments unmanaged for 40 years with the pattern in those mown regularly (Figure 20.2). The scheme in Figure 20.2 can serve as a conceptual base for an understanding of vegetation pattern in the floodplain under study (see also Prach *et al*. 1996). It was constructed partly on the basis of quantitative data (see the references cited above) and partly on field experience. A basically similar pattern was observed in other river systems over the Czech Republic (Straškrabová *et al*. 1996).

   One of the questions that emerged early in the Unesco MAB project related to the length of time taken for degraded meadows left unmanaged to return to a stage acceptable to both ecologists and agriculturalists. An experiment was established in 1989 to explore this question, and the results from the following five years are presented here.

SITE DESCRIPTION AND METHODS USED

The Lužnice River originates in Austria at 990 m above sea level. After flowing for about 40 km in the solid bedrock of an upland landscape, the river enters a flat sedimentary area of the Třeboň basin in the Czech Republic. Here it meanders for nearly 100 km, between an altitude of 500 and 403 m above sea level, before flowing for about 60 km in a canyon-like valley and finally emptying into the Vltava River at 350 m above sea level. The floodplain section that forms the subject of this chapter is located at 135.5 stream km measured from the mouth, at 455 m altitude. The average discharge near the study site is $5 \text{m}^3 \text{s}^{-1}$.

   The part of the floodplain between the river and the nearest terrace, where the

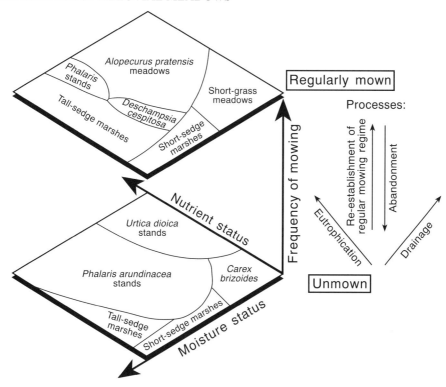

Figure 20.2 Vegetation pattern in regularly mown, and long-abandoned parts of the Lužnice River floodplain, related to the main environmental gradients. Adapted from Prach (1992)

experiment was established, had been left unmown for approximately 20 years. *Phalaris arundinacea* dominated over most of the area, together with *Urtica dioica* in the elevated parts.

In 1989 a transect with a length of 150 m was laid between the river bank and the top of the terrace. The whole transect was surveyed by a level-meter to measure its elevation. Phytosociological records were made in each 1 m² along the transect in the beginning of June, comprising a visual estimate of the percentage cover of each species present in the plots. This procedure was repeated in the following five years (until 1994). Vegetation was cut along the transect in a strip 150 m long and 5 m wide. This was done three times a year in the early part of the experiment (1989, 1990, 1991), and twice a year thereafter because of the insufficient increase in biomass later in the 1992 and 1993 seasons.

From the vegetation data, changes in average cover of species were estimated, as well as the number of species along the transect and species density in the 1-m² plots. The data were processed by the ordination technique Canonical Correspondence Analysis (CCA) (CANOCO/CANODRAW program; ter Braak and Šmilauer 1993). The cover data were transformed into the ordinal scale 1–8, corresponding to the degrees of the Braun-Blanquet scale (see van der Maarel 1979). The symbol *r* was not distinguished from +. The set of samples was subdivided into two equal subsets according to elevation, corresponding to the wet and dry parts of the transect. Due to the high number of samples, centroids (weighted averages) were used in the graphic output (Figure 20.3), each representing the position of the wet and dry plots in six successive years.

Above-ground biomass samples were taken from randomly selected plots of 0.5×0.5 m, five each in both cut and uncut sites. Sampling took place just before cutting of the whole transect. Above-ground biomass was sorted to particular species and dead biomass, dried at 90°C, and weighed. Differences between cut and uncut sites were evaluated using t-tests. Biomass analyses were performed only in the elevated (dry) part of the transect because only elevated parts of the floodplain have been recently used for hay production.

RESULTS AND DISCUSSION

VEGETATION CHANGES ALONG THE EXPERIMENTAL TRANSECT

Restoration of the cutting regime induced rapid changes in vegetation cover, as shown in Table 20.1. The dominant species typical of abandoned meadows, namely *Phalaris arundinacea* and *Urtica dioica* decreased dramatically during the observation period. In contrast, species typical of the regularly managed meadows in the area (*Alopecurus pratensis, Deschampsia cespitosa, Poa* sp. div., *Ranunculus repens*) started to increase. Instead of a monotonous cover of *P. arundinacea* over the major portion of the transect, the vegetation began to differentiate with regard to moisture conditions reflected by the elevation. *Carex* communities prevailed in the lowest elevation (the wettest part of the moisture gradient), being followed by *D. cespitosa*-dominated communities and *A. pratensis* communities. In the highest elevation, species appeared that are typical of the driest alluvial meadows in the Lužnice River floodplain, such as *Avenula pubescens, Holcus lanatus* and *Festuca rubra* (Prach 1992; Prach *et al*. 1996). For a complete list of species noted before and additional ones observed during the experiment see the Appendix.

Similarly rapid changes were observed with respect to species diversity (Table 20.2). The number of species per m² (species density) almost doubled after one year. The observed maximum in the second year of the experiment can be explained by the fact that, in the period between the decline of the dominants typical of abandoned meadows

Table 20.1   Changes in average cover (%) of principal species along the transect across the Lužnice River floodplain after a regular mowing regime was re-established (in 1989, after a period of approximately 20 years without mowing). After Prach *et al*. (1996)

Species	Average cover (%)					
	1989	1990	1991	1992	1993	1994
Alopecurus pratensis	14.4	20.3	16.3	26.5	26.8	30.4
Deschampsia cespitosa	0.0	0.0	0.4	0.6	1.6	1.7
Phalaris arundinacea	28.0	35.1	12.3	4.4	1.0	0.9
Poa spec. div.	0.0	0.7	1.5	2.5	2.9	4.7
Ranunculus repens	0.0	5.8	10.8	29.2	42.4	43.5
Urtica dioica	18.4	7.8	2.0	0.0	0.0	0.0

Table 20.2 Average number of species per m² and the total number of species along the whole transect. After Prach *et al.* (1996)

	1989	1990	1991	1992	1993	1994
No. of species m⁻²	4.0	7.3	8.9	6.9	8.1	8.2
Total no. of species	28	48	61	71	79	70

and expansion of grasses typical of managed meadows, there was an opportunity available for the establishment of various other species including some ruderals. The average species density in regularly managed meadows nearby reaches 8.6 species m⁻². Thus, it can be concluded that the meadows were restored with respect to species diversity in only two years. The increase in the number of species along the whole transect was similarly dramatic, reaching almost three times the number recorded before the experiment started.

Results of ordination (CCA) demonstrate that the quickest changes in floristic composition occurred during the first three years following the re-establishment of mowing in both wet and dry parts of the transect (Figure 20.3). Differences between the last two years seem to be small, especially in the case of plots representing the dry part of the transect. Thus, the results of the ordination support the conclusion that five years was a sufficient period for the restoration of the meadows. If the species composition of the experimental transect is compared with that of the regularly mown meadows nearby (Prach 1992; Prach *et al.* 1996), it is almost the same, especially in the dry parts.

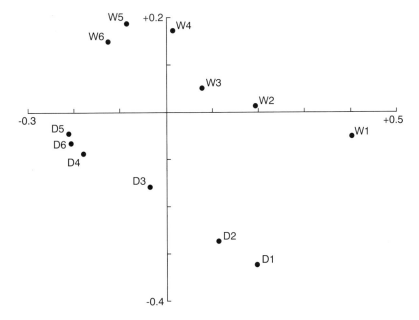

Figure 20.3 Results of ordination Canonical Correspondence Analysis (CCA). The positions of centroids are indicated, representing samples in the dry (D) and wet (W) parts of the experimental transect in the successive years 1–6 (from 1989, just before the experiment started, to 1994). The longer the distance between two successive years, the greater the change in species composition. From Prach *et al.* (1996), reproduced with permission from SPB Academic Publishing, Amsterdam

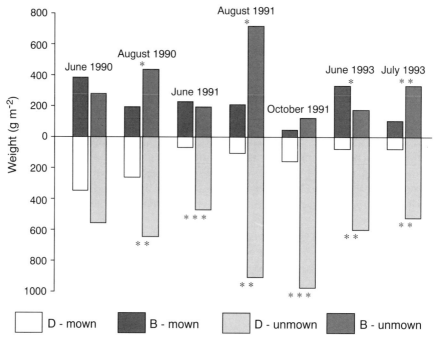

Figure 20.4   Changes in living (B) and dead (D) biomass of mown and unmown stands in three of the years during which the experiment was conducted. Significant differences (t-test) between mown and unmown variants are indicated: *$P < 0.05$; **$P < 0.01$; ***$P < 0.001$

CHANGES IN PRODUCTIVITY: A POTENTIAL FOR AGRICULTURAL USE

The potential harvest of alluvial meadows is remarkably high and regular cutting generally supports the productivity of the grasslands (Rychnovská 1985). This is recognizable from the comparison of above-ground living biomass of the mown and unmown variants sampled at the time of the first annual harvest at the beginning of June (Figure 20.4). After four years of cutting, the biomass of the mown variant was significantly greater than the biomass of the unmown one. In the second cut of the year, the biomass of the unmown variant was naturally greater than that of the mown one. The higher biomass of the mown variant in the first harvest was partly caused by the growth earlier in the season of *Alopecurus pratensis* in comparison with *Phalaris arundinacea* (Rychnovská 1985), the two dominant species of the respective stands.

As expected, cutting gradually decreased the amount of dead biomass (Figure 20.4). The low amount of dead biomass apparently enabled new species to establish, mainly heliophytes of shorter stature and with a lower competitive ability. Some of these species would probably not be able to germinate in soil covered with several centimetres of litter and a closed cover of tall grasses such as *Phalaris arundinacea* (see Grime *et al.* 1988).

Approximate estimates of productivity can be obtained by summing the biomass figures obtained at the time of each harvest in the mown variant, and using the maximum biomass of the unmown variant, respectively. These are summarized in Table 20.3 for

Table 20.3 Approximate annual production of above-ground biomass estimated for the mown and unmown variant in the elevated part of the floodplain section

Year	Above-ground biomass (g dry mass m^{-2})	
	Mown	Unmown
1990	593	436
1991	495	701
1993	444	344

each year and variant. It is evident that except for the wet year of 1991 the productivity of the mown variant was higher than that of the unmown one. Although the measurement of biomass was repeated only two to three times a season, so the estimates of production are very approximate, it was found that the productivity of the restored meadow in the elevated part of the transect was comparable with that of regularly mown meadows nearby (Šmilauer et al. 1996).

The change from *Phalaris arundinacea* stands to *Alopecurus pratensis* meadow is desirable from the point of view of potential agricultural exploitation. *A. pratensis* is more palatable to animals than *P. arundinacea* and is the most productive species in the area (Tetter et al. 1988).

CONCLUSIONS

It was demonstrated that restoration of alluvial meadows in the Lužnice River floodplain was relatively easy and rapid even when the meadows had been left without cutting for two decades. Similar trends, albeit perhaps slower, can be expected for those meadows abandoned for the longest time in the floodplain under study (over 40 years). Unfortunately, at present, there are no economic incentives that would persuade local agriculturalists to re-establish regular mowing over the whole floodplain. If they were to do so, they could have, relatively quickly, the most productive meadows in the area, which are also of value from an ecological perspective. The conclusions described here seem to be applicable to the majority of alluvial meadows of similar species composition in central Europe.

ACKNOWLEDGEMENTS

We thank Marek Bastl for computation of CCA, Petr Šmilauer for consultancy, Chris Joyce for English revision and two anonymous reviewers for their helpful comments. The research was supported by the grant of the Czech Academy of Sciences no. 60520 and partly by the grant of the Grant Agency of the Czech Republic no. 204/94/0395.

SUMMARY

An experiment was conducted to assess the feasibility of restoring a complex of alluvial meadows in the Lužnice River floodplain, in the Třeboň Biosphere Reserve, Czech Republic. The meadows had been abandoned for approximately 20 years. A regular cutting regime was re-established along a transect from the river to the nearest terrace. Rapid changes in species composition were observed during the next five years, from monotonous, species-poor stands to diverse meadows comparable with those subjected to uninterrupted management. The total number of species along the transect increased nearly threefold. Species density (the number of species m^{-2}) more than doubled. Biomass and production were comparable with those of regularly cut meadows by the third year of the experiment. Thus, a potential agricultural use could easily be restored to abandoned parts of the floodplain.

Keywords: Alluvial meadows, Biomass, Production, Restoration, Vegetation.

REFERENCES

Banásová, V., Otahelová, H., Jarolímek, I., Zaliberová, M. and Husák, Š. (1994) Morava River floodplain vegetation in relation to limiting ecological factors. *Ekológia (Bratislava)*, **13**, 247–262.

Day, R.T., Keddy, P.A. and McNeill, J. (1988) Fertility and disturbance gradients: a summary model for riverine marsh vegetation. *Ecology*, **69**, 1044–1054.

Ellenberg, H. (1988) *Vegetation ecology of central Europe*. Cambridge University Press, Cambridge.

Grime, J.P., Hodgson, J.G. and Hunt, R. (1988) *Comparative plant ecology. A functional approach to common British species*. Unwin Hyman, London.

Prach, K. (1992) Vegetation, microtopography and water table in the Lužnice River floodplain, South Bohemia, Czechoslovakia. *Preslia, Praha*, **64**, 357–367.

Prach, K. and Rauch, O. (1992) On filter effects of ecotones. *Ekológia, Bratislava*, **11**, 293–298.

Prach, K., Kučera, S. and Klimešová, J. (1990) Vegetation and land use in the Lužnice River floodplain and valley in Austria and Czechoslovakia. In: Whigham, D.F., Good, R.E. and Květ, J. (Ed.), *Wetland ecology and management: case studies*, pp. 117–125. Kluwer, Dordrecht.

Prach, K., Jeník, J. and Large, A.R.G. (eds) (1996) *Floodplain ecology and management. The Lužnice River in the Třeboň Biosphere Reserve, central Europe*. SPB Academic, Amsterdam.

Rychnovská, M. (ed.) (1985) *Ecology of grasslands*. Academia, Praha [in Czech].

Šmilauer, P., Prach, K. and Rauch, O. (1996) Biomass and nutrient allocation in the main vegetation types. In: Prach, K., Jeník, J. and Large, A.R.G. (eds), *Floodplain ecology and management. The Lužnice River in the Třeboň Biosphere Reserve, central Europe*, pp. 181–190. SPB Academic, Amsterdam.

Straškrabová, J., Prach, K., Joyce, C. and Wade, M. (eds) (1996) Floodplain meadows in the Czech Republic – ecological functioning, contemporary state and possibilities for restoration. *Příroda, Praha*, **4**, 1–176.

ter Braak, C.J.F. and Šmilauer, P. (1993) *CANOCO 3.1/CANODRAW 3.0*. Šmilauer, České Budějovice.

Tetter, M., Květ, J., Suchý, K. and Dvořáková, H. (1988) Productivity of grassland communities in the Upper Lužnice River floodplain. *Sborník VŠZ České Budějovice*, **5**, 119–129 [in Czech].

van der Maarel, E. (1979) Transformation of cover-abundance values in phytosociology and its effects on community similarity. *Vegetatio*, **38**, 85–96.

APPENDIX

Species present along the transect before the experiment started: *Alisma plantago-aquatica, Alopecurus pratensis, Angelica sylvestris, Barbarea stricta, Cardamine pratensis, Carex acuta, Carex vesicaria, Carex vulpina, Cerastium fontanum* subsp. *vulgare, Deschampsia cespitosa, Galeopsis tetrahit, Galium aparine, Galium palustre, Glechoma hederacea. Glyceria fluitans, Iris pseudacorus, Lysimachia vulgaris, Myosotis scorpioides, Oenanthe aquatica, Phalaris arundinacea, Poa palustris, Poa trivialis, Ranunculus repens, Rorippa amphibia, Rumex crispus, Rumex obtusifolius, Symphytum officinale, Urtica dioica.*

Species that appeared during the experiment: *Achillea millefolium, Achillea ptarmica, Aegopodium podagraria, Agrostis capillaris, Anemone nemorosa, Arrhenatherum elatius, Artemisia vulgaris, Avenula pubescens, Betula pendula* juv., *Capsella bursa-pastoris, Cardaminopsis halleri, Carex ovalis, Centaurea jacea, Cirsium arvense, Cirsium palustre, Cuscuta europaea, Dactylis glomerata, Dianthus deltoides, Elymus repens, Epilobium roseum, Festuca rubra, Filipendula ulmaria, Galium album, Galium uliginosum, Heracleum sphondylium, Hieracium pilosella, Holcus lanatus, Hypericum humifusum, Hypericum perforatum, Knautia arvensis, Leontodon autumnalis, Lotus corniculatus, Lotus uliginosus, Lychnis flos-cuculi, Lycopus europaeus, Lysimachia nummularia, Lythrum salicaria, Matricaria perforata, Mentha arvensis, Myosoton aquaticum, Pimpinella major, Plantago lanceolata, Poa pratensis, Prunella vulgaris, Ranunculus auricomus, Ranunculus flammula, Rumex acetosa, Salix cinerea* juv., *Sanguisorba officinalis, Scutellaria galericulata, Stellaria graminea, Stellaria media, Taraxacum officinale, Trifolium hybridum, Trifolium pratense, Trifolium repens, Verbascum* sp. juv., *Veronica arvensis, Veronica chamaedrys, Veronica scutellata, Veronica serpyllifolia.*

Glossary

Italicized terms are defined in the Glossary.

abiotic   non-biological; the physical and chemical factors affecting an organism.

Above Ordnance Datum (AOD)   height above the mean sea level used as a datum for calculating absolute height of land on British maps.

ADAS   see **Agricultural Development Advisory Service**

adsorption   the accumulation of ions at the surfaces of mineral and *humic* particles in soils.

adventive   referring to a species unintentionally introduced into a region by human activities that is at least partly naturalized; it reproduces for a while, then becomes extinct.

aeolian   a term pertaining to the wind; hence wind-blown, wind-borne or wind-deposited. Aeolian processes are most common in arid environments.

aerenchyma   air-filled spaces in the roots and stems of *hydrophytic* plants.

aftergrazed   refers to field that livestock graze after the main crop (e.g. hay) has been harvested.

aftermath   grass regrowth after the cutting of a crop of hay or silage that can subsequently be used for grazing or a second grass crop.

aggradation   a rise in ground level caused by the accumulation of sediment often brought about by the deposition of fluvial deposits.

Agricultural Development Advisory Service (ADAS)   a leading consultancy to land-based industry, including agriculture, in Britain.

alkalinophyte   a plant adapted to growing in alkaline groundwater and soils.

allochthonous   applied to material that originated elsewhere.

alluvium   the sedimentary deposits resulting from the action of rivers, thus including those laid down in river channels, *floodplains*, estuaries, lakes and deltas, hence **alluvial**.

amphiphyte   a plant growing in the border zone of wetland and water, with amphibious characteristics.

anthropogenous   refers to influences caused by human activity, e.g. cultivation, hence **anthropogenic**.

anthropophyte   a plant that grows in proximity to humans, such as *weeds* in cultivated land or in paths, or all non-native species (*adventive* and cultivated).

AOD   see *Above Ordnance Datum*

archaeophyte   an alien plant species introduced to a region through human activity before 1500 AD (see *neophyte*).

arkose   a coarse-grained sandstone.

association   a category in the classification of vegetation: an assemblage of plants living in close interdependence. They exhibit similar habitat and growth requirements, although there are usually one or more dominant species.

autecology   the study of the ecology of an individual species, hence **autecological.**
avifauna   the bird fauna of a region.

benthos   (1) the bottom of a water body or the ocean; (2) organisms that live on or in
the bottom of water bodies or the ocean, from the water's edge down to the greatest
depths, hence **benthic**.
biocide   any agent that kills living things. Sometimes used as a synonym for pesticide.
biodiversity   the variety of living organisms and the ecological complexes of which they
are a part. This includes diversity within species, between species and of ecosystems. It
is a concept that applies at all levels, from landscapes and ecosystems down to
individual species and their gene pools.
Biological Oxygen Demand (BOD)   a measure of the amount of pollution by organic
substances in water. It is expressed as the number of milligrams of oxygen required by
the micro-organisms to oxidize the organic substances in a litre of water.
biomass   any quantitative estimate of the total weight of living biological organisms
within a specified unit (area, community, population or individual).
biotope   an area in which all the faunal and floral elements are adapted to the
environment in which they occur.
BOD   see *Biological Oxygen Demand*
boreal   applied to a climatic zone with short, warm summers and cold, snowy winters.
bund   an artificial embankment, dam, dyke or causeway.

CA   see *Correspondence Analysis*
calcicole   plant that grows best on calcareous (chalk and limestone) soils, hence
calcicolous.
CANOCO   a computer program that enables various types of multivariate ordination
analysis to be undertaken. Such techniques allow the composition of species
communities to be related to environmental variables often through an ordination
diagram consisting of species and/or site scores plotted against two or more axes.
Ordination techniques include *Correspondence Analysis, Canonical Correspondence
Analysis, Detrended Correspondence Analysis* and *Principal Components Analysis.*
Canonical Correspondence Analysis (CCA)   an ordination technique (see *CANOCO*) in
which the axes are constrained to the linear combinations of environmental variables.
CCA expresses the main relations between species and each of the observed
environmental variables.
carabid   ground beetle (Carabidae), most species of which are carnivorous.
carr   a wooded *fen* in waterlogged terrain where the pH is not too acid and where the
soils are not too deficient in mineral elements. Characteristic trees include *Alnus* (alder)
and *Salix* (willow).
catchment or **catchment area**   in British usage the term refers to the total area from
which a single river collects water. In the USA the term watershed is used in this
context.
CCA   see *Canonical Correspondence Analysis*
character (or characteristic) species   used in *phytosociology* for a plant species that has
a phytosociological optimum within a given *syntaxon*. Such species provide the most
reliable floristic expression of the ecology of the group.
charadriiforms   a large group of mostly aquatic or marsh-loving birds including the

waders.

click beetle   beetle of the family Elateridae; larvae are wire-worms.

climax   the terminal community in ecological *succession* that maintains itself, more or less unchanged, for a long period of time under the prevailing climatic and *edaphic* conditions; the terminal stage of a *sere*.

cochylid micromoth   *microlepidoptera* (moths) belonging to the family Cochylidae (= Phaloniidae).

coenogeography   the geography of natural communities, hence **coenogeographical**.

collectivization   the process by which collective farms were established during the communist era in countries such as Estonia and the Czech Republic (see *decollectivization*).

Common Agricultural Policy   the framework on which the agricultural policy of the European Economic Community is built, facilitating trade in food products by establishing common prices throughout the Community.

Continentality Index   a measure of the continentality of a climate calculated by taking mean July air temperature and subtracting mean January air temperature.

coppice   the cutting of trees or shrubs near to the ground at regular time intervals, causing several shoots to arise from the stump. **Coppicing** to provide wood was formerly a common practice in woodland management.

CORINE   CO-ordination of INformation on the Environment, a data management system for environmental information developed by the European Commission. The system covers various relevant media, including information on nature (see *CORINE biotopes*).

CORINE biotopes   a standardized site-based database for information on nature conservation areas of European importance. Sites include designated and non-designated areas and vary in size and nature. The site description includes a reference to the *Palaearctic* classification (see *CORINE habitat classification*).

CORINE habitat classification   a system developed to describe *phytosociological* aspects of *CORINE biotopes*. The classification has a hierarchical structure and has been developed for the whole of the *Palaearctic*. The *TABLEFIT* program can be used to identify *CORINE biotopes*.

Correspondence Analysis (CA)   an ordination technique (see *CANOCO*) where the axes are unconstrained, representing theoretical variables that explain species and/or sample dispersion.

corvid   a member of the bird family Corvidae (crows and ravens).

Countryside Stewardship Scheme (CSS)   a voluntary, discretionary scheme operating in England aimed at protecting and enhancing threatened landscapes and habitats including chalk and limestone grassland, lowland heaths and waterside landscapes. It is administered by the *Ministry of Agriculture, Fisheries and Food* and co-funded by the European Union. Landowners joining the scheme receive annual and capital payments for adhering to specified management prescriptions.

CSS   see *Countryside Stewardship Scheme*

Culm grassland   local name given to unimproved pastures that support a distinctive flora on the underlying Carboniferous sandstones, slates and shales (Culm measures) of mid and north Devon and north-east Cornwall, UK, which give rise to acidic, poorly draining soils most conspicuously in the many valley bottoms of this region.

DCA   see *Detrended Correspondence Analysis*

decollectivization   the process by which land acquired in the establishment of collective farms in countries such as Estonia and the Czech Republic was returned to private ownership, often to former owners (see *collectivization*).

dehesas   the Spanish term for Iberian pastoral woodlands composed of grasslands and Mediterranean scrub, interspersed with trees in a savannah-like landscape. Dehesa management is typically of low intensity.

desorption   the removal of molecules, ions, etc. from the surface of a solid; the reverse of *adsorption*.

Detrended Correspondence Analysis (DCA)   an ordination technique (see *CANOCO*) where the axes are unconstrained, representing theoretical variables that explain species and/or sample dispersion. Detrending is a mathematical procedure that removes a perceived fault of *Correspondence Analysis*, namely particular types of distortion in the species/sample points.

detritivorous   feeding on freshly dead or partially decomposed organic matter.

diaspore   any spore, seed, fruit or other portion of a plant when being dispersed and able to produce a new plant (synonymous with *propagule*).

dicotyledon   those flowering plants whose seeds have two seed-leaves (cotyledons), whose leaves are usually broad, often stalked and nearly always net-veined and whose flower parts are usually in multiples of four or five.

dipwell   a lined auger hole for the measurement of *water-table* levels.

ditch   a long narrow channel to hold or conduct water as in drainage or irrigation; a boundary or waterway bordering fields; sometimes called a drain.

Domin   a scale of percentage values used to record the cover of vegetation, i.e. an estimate of the area of coverage of the foliage of the plant species in a vertical projection on to the ground, in which + = a single individual; 1 = 1–2 individuals; 2 = <1%; 3 = 1–4%; 4 = 5–10%; 5 = 11–25%; 6 = 26–33%; 7 = 34–50%; 8 = 51–75%; 9 = 76–90%; 10 = 91–100%.

drove   a road along which horses or cattle are driven.

ECNC   see *European Centre for Nature Conservation*

ecotone   the transitional zone between two habitats, e.g. between grassland and woodland, which has characteristics of both as well as properties of its own.

ecotope   a particular kind of habitat within a region.

edaphic   relating to soil.

EEA   see *European Environment Agency*

EIONET   see *European Information and Observation Network*

EN   see *English Nature*

endemic   belonging or native to a given, usually small, geographic region; not introduced or naturalized.

endemism   the occurrence of *endemics* in an area.

English Nature (EN)   the government-funded body the purpose of which is to promote the conservation of England's wildlife and natural features. Formerly part of the Nature Conservancy Council.

Environment Agency   a powerful and independent body that combines the regulation and management of land, air and water resources in England and Wales by bringing together the expertise of the *National Rivers Authority*, Her Majesty's Inspectorate of Pollution, the Waste Regulation Authorities and certain parts of the Department of the Environment.

Environmentally Sensitive Area (ESA)   defined area of high landscape, nature conservation or archaeological value in the UK administered by the *Farming and Rural Conservation Agency* for which farmers may voluntarily enter into agreements for 10 years to manage land according to specific prescriptions in return for annual and capital payments. Payments are tiered according to the nature and level of the management required for enhancing landscape or conservation value.

epilittoral   the shore zone immediately above high water level influenced by wave splash and spray.

equitability index   an index used as an assessment of diversity. It is a measure of the evenness of the abundances of species in a community or sample. Low evenness, or equitability, usually equates to low diversity since this means that one or few species are dominant or abundant. In contrast, maximum diversity would be found in a situation where all species are equally abundant.

ESA   see *Environmentally Sensitive Area*

estuarine   referring to a broad brackish portion of a low river course where the tide meets the downstream current, and mixing of salt and freshwater occurs.

ETC/NC   see *European Topic Centre on Nature Conservation*

European Centre for Nature Conservation (ECNC)   a non-profit based organization under Dutch law, operating with a network of institutes throughout Europe. The ECNC secretariat is located in Tliburg, The Netherlands.

European Environment Agency (EEA)   a service organization of the European Union with a pan-European scope and located in Copenhagen.

European Information and Observation Network (EIONET)   a key information system currently being developed at the *EEA*.

European Topic Centre on Nature Conservation   a part of the *EEA* and responsible for managing information on nature for the European Commission and the *EEA*, located in Paris.

eutrophic   applied to habitats, especially wetlands (e.g. *fens*) and freshwater bodies, that have an excess of nutrients, hence **eutrophication**.

evapotranspiration   the combined evaporation from the soil surface and transpiration of plants.

Farming and Rural Conservation Agency   an executive agency of the *Ministry of Agriculture, Fisheries and Food* and the Welsh Office, which provides technical support in managing agri-environment schemes in England and Wales and has responsibility for rural economy, land and planning, environmental protection, wildlife management and milk hygiene.

fen, fenland   waterlogged peat soil that may be alkaline, neutral or slightly acid.

Fens, the   a large coastal basin in eastern England that has been progressively occupied and drained for agriculture over several thousand years.

floodplain   that part of a river valley adjacent to the channel over which a river flows in times of flood. It is a zone of low relief and gentle gradients and composed of *alluvium*, which generally buries the rock floor of the valley to variable depths; often characterized by high *biodiversity* and production.

fluvisols   *alluvial* soils.

forb   an herbaceous plant that is not a grass, nor grasslike such as *Carex* (*sedge*). Examples are *Sanguisorba officinalis* and *Galium palustre*.

geobotany   the study of the global distribution of plants, hence **geobotanical**; synonymous with *phytosociology*.

geolittoral   the shore zone bounded by mean and high water level.

geophyte   perennial *herb* with buds below soil surface.

gleysols   a major group of soil types characteristically affected by periodic or permanent saturation by water in the absence of effective drainage, hence **gleyic**.

glycophyte   a plant growing in soil that is low in salt content, in contrast to *halophyte*.

grip   a shallow channel within a field formed to drain surface water into a *ditch*.

groundwater   water occupies pores, cavities, cracks and other spaces in rock formations. It includes water precipitated from the atmosphere that has percolated through the soil, water that has risen from deep sources, and fossil water retained in sedimentary rocks since their formation.

guild   a group of closely related species in competition for the same resource and using the same ecological strategy.

gutter   a channel for draining away water, typically small and narrow as compared to a *ditch*.

halite   mineral, rock salt.

halocalcicole   a plant adapted to growing in *saline groundwater* and soils with a high calcium salt content.

halophilous   having an affinity for *saline* conditions.

halophyte   a plant adapted to growing in *saline* soils and *groundwater*.

harvestman   an order of often very long-legged carnivorous arachnids (Opiliones).

headage   the stocking rate of livestock, usually expressed as numbers of animals per unit area.

heliophyllous   refers to organisms that grow best in full sunlight, hence **heliophytes**.

helophyte   *herb* whose buds lie in mud.

hemicryptophyte   a biennial or perennial plant whose surviving buds are on or in the surface soil, protected by the soil itself and by the dead portions of the plant.

herb   a plant with one or more stems that, in the temperate zone, dies back to the ground each year; comprising grasses and *forbs* as distinct from shrubs and trees.

Holarctic   the landmass of the entire northern region of the Northern Hemisphere including the *Palaearctic* region (most of Eurasia north of the Tropic of Cancer) and the *Nearctic* region (North America north of the Mexican Plateau).

humic   derived from, or containing, *humus*.

humus   decomposing organic matter in the soil.

hydrochoric   dispersed by water; dependent on water for dissemination, hence **hydrochory**.

hydrolittoral   the shore zone below mean water level, free from sea water during low tide.

hydrophyte   a plant that grows wholly or partly immersed in water, hence **hydrophytic**.
hygrophyte   a plant that grows in moist or wet places, hence **hygrophytic**.
hygrophilous   growing or living in moist or wet places.
hydrosere   a collective term that includes all the stages in a *succession* beginning in open water.
hygromesophyte   a plant that grows in moist or wet to average but not excessive water-supply conditions (see *hygrophyte* and *mesophyte*).
hygrotype   a plant species adapted to conditions of a regular and plentiful supply of water (see *mesophyte* and *xerophyte*).
hypogean, hypogeal   living or growing underground.
hypogenic   relating to geological processes that originate at depth within the Earth.

ichthyological   of, or pertaining, to fish; **ichthyology**   the branch of natural history that deals with fishes.
intertidal   the zone between mean high and mean low water levels.
inundation   the process of flooding or swamping with water.

JNCC   see *Joint Nature Conservation Committee*
Joint Nature Conservation Committee   the body constituted by the UK Environmental Protection Act 1990 to be responsible for research and advice on nature conservation at both UK and international levels.

lacustrine   pertaining to lakes.
leatherjacket   larva of a family of Diptera (flies), the Tipulidae (craneflies).
lemnid   species of the plant genus *Lemna* (duckweeds) or other species with a similar lifeform.
levee   a naturally formed elevated bank bordering the channel of a river that stands above the level of the *floodplain*.
ley   grass sown as a crop.
limnology   the scientific study of physical, chemical and biological conditions in lakes, hence **limnological**.
lithology   the study of rock characteristics, particularly their grain size, particle size and their chemical and physical character, hence **lithological**.
litter   dead plant material such as leaves and stems lying loose on the surface of the soil and destined to be transformed into *humus*.
littoral   inhabiting the bottom of a lake, sea or such water body, near the shore, or the shore zone bounded by high and low water level.
loam   an easily worked permeable soil comprising an almost equal mix of sand and silt but with less than 30% clay.
long-horn beetle   beetle of the family Cerambycidae.
lysimeter   equipment for measuring the rate and amount of water storage and percolation through a soil. If rainfall is recorded, the rate of *evapotranspiration* can be estimated.

MAB   see *Man and Biosphere Programme*

MAFF   see *Ministry of Agriculture, Fisheries and Food*

Man and Biosphere Programme (MAB)   an international *Unesco* programme that carries out, through its national committees, projects focusing on problems of resource management involving both natural and social scientists.

mesophile   a plant growing in intermediate or moderate environmental conditions, e.g. neutral soil, or constant moderately moist conditions (see *mesophilous*).

mesophilous   (1) applied to plants or plant communities associated with neutral soil conditions; (2) pertaining to constant moderately moist conditions.

mesophyte   a plant that grows under average but not excessive water-supply conditions, demanding less than a *hygrophyte* but more than a *xerophyte*. Typically such plants have wide and soft leaf blades.

mesotrophic   applied to freshwater bodies that contain moderate amounts of plant nutrients and are therefore moderately productive. Also applied to grasslands and wetlands that are moderately rich in nutrients.

microclimate   the climate of a very local area.

microlepidoptera   an unscientific division of the Lepidoptera (butterflies and moths), consisting of a selection of the smaller moths.

mineralization   the production of inorganic ions in the soil by the decomposition of organic compounds, e.g. nitrates from proteins.

minerotrophic   referring to an organism nourished by minerals.

Ministry of Agriculture, Fisheries and Food (MAFF)   administers the UK Government's agriculture, horticulture and fisheries policies including agri-environment schemes in England and has responsibilities for food, trade and animal health throughout the UK.

monocotyledon   those flowering plants whose seedlings have only one seed-leaf (cotyledon), and whose flower parts are nearly always in multiples of three. The principal families are the Gramineae (grasses), Cyperaceae (*sedges*), Juncaceae (rushes), Liliaceae (lilies) and Orchidaceae (orchids).

montane   mountainous; mountain-dwelling.

moor, moorland   open ground with acid-peaty soil. Although usually referring to high ground typically dominated by *Calluna* and *Erica* (heathers), the terms are also used for low-lying land dominated by grass and generally used for grazing, e.g. Somerset Moors, and Tadham and Tealham Moors.

National Nature Reserve (NNR)   land of national/international importance for nature conservation declared as a nature reserve by the Great Britain statutory nature conservation agencies (*English Nature*, Countryside Council for Wales, Scottish Natural Heritage) under Act of Parliament. In Great Britain all NNRs are also notified as *Sites of Special Scientific Interest*.

National Rivers Authority (NRA)   former environmental protection agency with responsibility for safeguarding and improving the natural water environment in England and Wales. Became part of the *Environment Agency* in 1996.

National Vegetation Classification (NVC)   summarizes variation within the British flora by describing over 240 communities (each of which is subdivided into up to six subcommunities). The communities were identified from a classification of the frequency and abundance of plant species in over 35 000 *quadrats* distributed throughout Britain. It is possible to classify species data into NVC groups using the *TABLEFIT* program.

Nearctic   see *Holarctic*

neoindigenophyte   an alien plant introduced unintentionally by human activity that has become naturalized having invaded semi-natural and natural plant communities (see *neophyte*).

neophyte   an alien plant species introduced to a region through human activity after 1500 AD (see *archaeophyte* and *neoindigenophyte*).

nidifugous   of those birds which hatch in a relatively advanced and mobile state and are capable of leaving the nest immediately and of searching for food, often assisted by a parent or parents.

nitrophilous   refers to plants that grow well in soil that is rich in nitrogen, hence **nitrophyte**.

noctuid moth   a large family of stoutly-built mostly rather drab-coloured moths (Caradrinoidea: Caradrinidae).

NRA   see *National Rivers Authority*

NVC   see *National Vegetation Classification*

oligotrophic   applied to habitats, especially mires (e.g. *fens*) and freshwater bodies that are poor in nutrients.

Ordnance Survey   the government body in Great Britain responsible for the survey, production and publication of topographical maps.

ORNIS   see *Ornithological Information System*

Ornithological Information System (ORNIS)   the database developed by the European Commission in 1993 that provides information about birds in Europe including population sizes and trends; temporal and spatial aspects of reproduction, migration and wintering; hunting; and also habitats.

outwash   material of glaciofluvial origin deposited by meltwater streams beyond the margins of ice-sheets and glaciers.

Palaearctic   see *Holarctic*

palaeoecosystem   an ecosystem of a former geological period.

Palaeozoic   the first of the eras of geological time after the Precambrian and prior to the Mesozoic. It lasted from about 600 million years BP to 240 million years BP.

paludal   pertaining to marshes; marshy; *water table* above ground level for most of the year, hence **paludified**.

Pannonia   a region of central Europe between the Danube and Sava Rivers including parts of south-west Slovakia, eastern Austria, Hungary and the northern region of former Yugoslavia.

pedogenesis   the natural process of soil formation.

penetrometer   an instrument designed to measure the vertical resistance of soil or snow to penetration by a rod or cone to a particular depth at a specified rate.

phaeozems   meadow soils.

phenology   the study of the timing of recurring natural phenomena, e.g. annual flowering and dormancy, hence **phenological**.

phreatic   pertaining to *groundwater*, wells and springs, or to the soil or rocks containing such water.

phreatophyte   a plant that obtains water from *groundwater*, often at great depths.

phytocoenology   the study of plant communities, more widely termed *phytosociology*.

phytocoenosis   the assemblage of plants inhabiting a particular area; the entire plant community.

phytogeography   the science that deals with the geographic distribution of plants and the causes of their distribution and dispersal.

phytophagous   refers to an organism that feeds on plants.

phytosociology   the study of plant communities including their origin, composition, structure, characteristics, distribution, dynamics and classification; or the branch of botany comprising ecology, biogeography and genetics of plant *associations*, hence **phytosociological.**

phytotoxic   poisonous to green plants.

plagioclimax   any plant community whose composition is more or less stable and in equilibrium under existing conditions but which, as a result of human intervention, has not achieved the natural *climax*, e.g. grassland under continuous pasture.

poaching   the penetration of the soil surface by the hooves of grazing animals (or trafficking vehicles) causing damage to the sward. It is usually the consequence of grazing when the soil surface is too wet.

pollard   a tree having the whole crown cut off, hence **pollarded**, leaving it to grow new branches from the top of the stem.

Principal Components Analysis (PCA)   an ordination technique (see *CANOCO*) where axes are constructed representing theoretical variables that explain species and/or sample dispersion by fitting straight lines to the data by means of regression.

propagule   any spore, seed, fruit or other portion of a plant when being dispersed and able to produce a new plant (synonymous with *diaspore*).

psammophyte   a plant that is adapted to live in unconsolidated sand, hence **psammophilous**.

psammosere   the *succession* of vegetation that develops in a sand environment, typically associated with sand dunes.

psychromesophyte   a plant that exhibits characteristics intermediate between a *mesophyte* and a *psychrophyte*.

psychrophyte   a plant growing on moist and cold soils (0–10°C) having a *xeromorphic* habit and developing mainly vegetatively.

pyralid moth   a large family of *microlepidoptera* (Pyraloidea: Pyralidae).

quadrat   a sampling area, usually square, used for surveying or monitoring vegetation or as the basis for sampling other elements of the biota.

Raised Water-Level Areas (RWLAs)   coherent blocks of land in *Environmentally Sensitive Areas* in which raised water levels are maintained for nature conservation objectives through the payment of supplements.

Ramsar site   site designated as a Wetland of International Importance under the

Ramsar Convention. This requires signatory countries to protect internationally important wetlands, particularly those used by migratory waterfowl. The convention was signed in Ramsar, a town in Iran.

rank   coarsely overgrown, typically of grasses.

Red Data Book (RDB)   catalogues published by the International Union for the Conservation of Nature and Natural Resources (IUCN) or by national nature conservation authorities listing species that are rare or in danger of extinction globally or nationally.

reedbed   a large often monospecific stand of tall emergent grass or grass-like species, usually *Phragmites australis*.

refugium (pl. **refugia**)   an area that has escaped great changes undergone by the region as a whole, and so often provides conditions in which relic colonies of plants and/or animals can survive.

regosol   a major group of soil types, typically thin, immature and developing on newly deposited unconsolidated deposits of wind-blown sand.

remediation   the action of repairing or rehabilitating.

RES   see *Reserves Enhancement Scheme*

Reserves Enhancement Scheme (RES)   a grant scheme administered by *English Nature* to assist non-governmental conservation organizations, such as Wildlife Trusts, with the management of land managed as nature reserves that have been designated as *Sites of Special Scientific Interest*.

rhizosphere   zone of soil immediately surrounding a root and modified by it, which is characterized by enhanced microbial activity.

rhyne (rhine)   the name for a large open *ditch* or drain in Somerset, England.

riverine   of, on, or dwelling in or near a river.

Royal Society for the Protection of Birds (RSPB)   a non-governmental organization that has worked since 1889 to protect birds in Britain. It maintains a number of bird reserves, presses for new laws and policies to protect birds, advises landowners and planners, researches the needs of birds, and educates and fosters a concern for wildlife among young people.

ruderal   plants able to tolerate frequent disturbance often due to their rapid growth and development.

RWLA   see *Raised Water-Level Areas*

ryegrass   *Lolium perenne*, a productive and palatable grass common in natural grasslands and frequently sown in *leys*. Other species of cultivated rye-grass occur, e.g. *Lolium multiflorum*.

SAC   see *Special Area of Conservation*

salinity   the degree to which water contains dissolved salts, hence **saline**. It is most commonly expressed in parts per 1000 or in grams per 1000 grams; standard sea water has a salinity of 34.33‰.

saltmarsh   an area of sand and mud covered by the sea at high tide and exposed at low tide with a distinctive flora.

Schedule 8 species   plant species listed on Schedule 8 of the British *Wildlife and Countryside Act* 1981. Under the Act, it is illegal to pick, uproot or destroy species listed on this Schedule.

sedge a large genus of grass-like plants, *Carex*, growing in grassy places and by freshwater.

seedbank those seeds laid down in the soil and remaining dormant until suitable conditions for germination occur.

seepage (1) the slow emission of *groundwater* at the surface where the rock structures permit, but with insufficient volume for it to constitute a spring; (2) the gradual soaking away of surface water into the soil.

sensu lato in a broad sense.

sere a series of temporary communities that develop a *successional* sequence in a given area and lead to a *climax* community (see *succession*)

shrubland a habitat dominated by shrubs, i.e. low-growing woody perennials, which, unlike small trees, branch from the base. Shrubs are typically less than 6 m in height.

Site of Special Scientific Interest (SSSI) site designated by the Great Britain statutory nature conservation agencies under the 1981 *Wildlife and Countryside Act* as supporting habitats and species of special nature conservation interest. Designation provides a measure of protection against damaging activities and allows for positive management agreements with landowners and occupiers to benefit the wildlife value.

slub mild form of dredging to remove sludge and accumulated sediment typically from a *ditch*. The spoil (or **slubbing**) is usually deposited to the side of the channel.

slug a land mollusc with the shell absent or rudimentary.

soldier fly flies, mostly brightly coloured, of the family Stratiomyidae.

sp. abbreviation for species (singular).

Special Area of Conservation (SAC) a site of European importance for nature conservation designated by European Union Member States under European Community Directive 92/43 on the conservation of natural habitats and of wild flora and fauna.

Special Protection Area (SPA) a European site designated under Article 4 of European Community Directive 79/409 on the conservation of wild birds. Sites support important populations of vulnerable wintering and/or breeding bird species.

Species Recovery Programme initiated in England in 1991 by *English Nature*, the programme consists of a range of projects that aim to restore, maintain or enhance populations of plants and animals that are in severe decline or under threat of extinction.

spp. abbreviation for species (plural).

springtail small wingless insects of the order Collembola found in abundance in the soil.

SSSI see *Site of Special Scientific Interest*

sub-atlantic pertaining to a biogeographic zone (or climate) characterised by mild and wet conditions.

sub-boreal pertaining to a biogeographic zone (or climate) approaching frigid or boreal conditions.

succession the process of community change; the sequence of populations or communities that replace one another in a given area culminating in a *climax* community (see *sere*).

sward a stretch or expanse of *turf* or grass.

synanthrope a plant or animal often found associated with humans, human dwellings or other artefacts, hence **synanthropic** and **synanthropization**.

syntaxon   a plant community of any rank, equivalent to taxon in traditional taxonomy.

synusia (pl. **synusiae**)   an aggregation of plants belonging to the same lifeform having similar environmental requirements and occurring in a similar habitat.

TABLEFIT   a computer program that classifies species data into vegetation groups according to the British *National Vegetation Classification*, and identifies habitat types according to the European Union *CORINE* system.

taxon (pl. **taxa**)   the organisms comprising a particular taxonomic entity, e.g. a particular class, family or genus.

Tertiary   the first period of the Cenozoic era, preceding the Quaternary period. It lasted from about 6.5 million years BP to 2 million years BP.

theodolite   a surveying instrument for measuring horizontal and vertical angles.

therophyte   an annual plant; a plant that completes its life cycle in a single season and survives the unfavourable season as seeds.

transect   a long narrow sample area or a line used for sampling or monitoring.

trophic   refers to nutrition or nutrient status.

turf   the layer of low, dense grassland comprising the above-ground portions and the upper roots and rhizomes with attached soil particles (pl. **turves**; see *sward*).

tussock   a bunched clump of grasses or sedges, hence **tussocky**.

TWINSPAN   a computer program for the hierarchical classification of species and samples.

Unesco   see *United Nations Educational, Scientific and Cultural Organization*

United Nations Educational, Scientific and Cultural Organization (Unesco)   since 1946 Unesco has promoted collaboration among nations through education, science and culture. Its three main objectives are the extension of education, the improvement of education and life-long education in a world community.

wader   a group of birds often frequenting shallow water and mostly long-legged and long-billed, including the *charadriiforms* (Charadriiformes). Comprises Charadriidae (plovers), Scolopacidae (snipes and sandpipers), Haematopidae (oystercatchers) and Recurvirostridae (avocet and stilts).

wainscot moth   a type of noctuid moth (Lepidoptera) in two subfamilies comprising 10 genera, mainly frequenting *fens* and *reedbeds*.

washland   a river *floodplain* surrounded by artificial embankments into which the river is diverted in times of flood in order to alleviate further flooding downstream. In the USA the term is synonymous with *water meadow*.

water beetle   beetle of one of several families, mainly Dytiscidae, Hydrophilidae, Haliplidae, Gyrinidae, Hydraenidae and Elmidae.

water meadow   grassland along the *floodplain* of a river where the growth of herbage is stimulated by periodic flooding, often by artificial means by use of channels and sluices. It is termed a *washland* in the USA.

water table   the upper surface of the zone of saturation in a soil or rock formation.

weed   a general term for any troublesome or otherwise undesirable plant, usually introduced, that grows without intentional cultivation.

weevil   beetle of the family Curculionidae (Rhynchophora).

wildfowl   the birds of the Anatidae (ducks, geese and swans); game birds.

wildfowling   the pursuit of *wildfowl.*

Wildlife and Countryside Act 1981   Act of Parliament covering Great Britain amending the law with respect to the protection of wildlife habitats and selected species, and making provision for other countryside matters including public rights of way and National Parks.

Wildlife Enhancement Scheme (WES)   a scheme run by *English Nature* that provides financial incentives to landowners for positive management of habitats on *Sites of Special Scientific Interest*. At present the scheme covers selected habitat types in specific geographical areas.

Wildlife Trust   charitable non-governmental organizations covering specific administrative English regions that foster and promote nature conservation by the acquisition and management of threatened habitats, through the provision of education and advice and the running of wildlife campaigns. There are currently 37 Trusts in England.

withy bed   land in which a shrubby species of willow (*Salix viminalis*) and hybrids are grown. The species is noted for its long narrow leaves and very flexible twigs, which are much valued for basket making.

xerohalophyte   a plant adapted to growing in *saline groundwater* and soils that has also developed means of combating drought.

xeromorph   an organism structurally and functionally modified so as to withstand drought, hence **xeromorphic**.

xerophyte   a plant species living in a dry habitat typically showing *xeromorphic* or succulent adaptation, hence **xerophytic** and **xerophilous** (see *hygrophyte* and *mesophyte*).

xerotherm   a plant surviving in conditions of drought and heat, hence *xerothermic*.

zonation   arrangement or distribution in zones, e.g. organisms or physical factors, hence **zonal** and **zonally**.

The following texts were useful in explaining some of the above terms:

Allaby, M. (1985). *Macmillan dictionary of the environment.* Macmillan, London.

Fitter, R. (1967). *The Penguin dictionary of British natural history*. Penguin, London.

Hanson, H.C. (1962). *Dictionary of ecology*. Philosophical Library, New York.

Holmes, S. (1979) *Henderson's dictionary of biological terms*, 9th edn. Longman, London.

International Commission on Irrigation and Drainage (1967) *Multilingual technical dictionary on irrigation and drainage. English–French*. ICID, New Delhi, India.

Jirásek, V. (1967) Zur Vereinheitlichung der Terminologie in der Phytogeography. *Folia Geobotanica et Phytotaxonomica*, **1**, 69–113.

Lewis, W.H. (1977) *Ecology field glossary. A naturalist's vocabulary*. Greenwood Press, Westport, Connecticut.

Lincoln, R.J., Boxhall, G.A. and Clark, P.P. (1982) *A dictionary of ecology, evolution and systematics*. Cambridge University Press, Cambridge.

Ritzema, H.P. (ed.) (1994) *Drainage principles and applications*. ILRI Publication No. 16, 2nd edn. International Institute for Land Reclamation and Improvement, Wageningen, The Netherlands.

Whitlow, J. (1988) *The Penguin dictionary of physical geography*. Penguin, London.

Vascular Plant Nomenclature

Scientific names follow Tutin, T.G., Heywood, V.H., Burges, N.A., Moore, D.D., Valentine, D.H., Walters, S.M. and Webb, D.A. (1964–80) *Flora Europaea*, Vols 1–5. Cambridge University Press, Cambridge.

Further information on *Flora Europaea*, including revisions and synonyms, can be found on the World Wide Web at http://www.rbge.org.uk/forms/fe.html.

Scientific names	English names
Acer negundo	Ashleaf Maple
Achillea millefolium	Yarrow
Achillea ptarmica	Sneezewort
Aegopodium podagraria	Ground-elder
Agrostis canina	Velvet Bent
Agrostis capillaris	Common Bent
Agrostis castellana	Highland Bent
Agrostis gigantea	Black Bent
Agrostis stolonifera	Creeping Bent
Ailanthus altissima	Tree-of-Heaven
Alisma lanceolatum	Narrow-leaved Water-plantain
Alisma plantago-aquatica	Water-plantain
Alliaria petiolata	Garlic Mustard
Allium angulosum	Mouse Garlic
Alnus spp.	Alders
Alopecurus arundinaceus	Reed Foxtail
Alopecurus bulbosus	Bulbous Foxtail
Alopecurus geniculatus	Marsh Foxtail
Alopecurus myosuroides	Black-grass
Alopecurus pratensis	Meadow Foxtail
Althaea officinalis	Marsh-mallow
Anemone nemorosa	Wood Anemone
Angelica palustris	Marsh Angelica
Angelica sylvestris	Wild Angelica
Anthoxanthum odoratum	Sweet Vernal-grass
Anthriscus sylvestris	Cow Parsley
Apium repens	Creeping Marshwort
Arrhenatherum elatius	False Oat-grass
Artemisia caerulescens	A Wormwood
Artemisia vulgaris	Mugwort

Aster novi-belgii	Michaelmas Daisy
Aster tripolium	Sea Aster
Atriplex calotheca	An Orache
Atriplex hastata	Halberd-leaved Orache
Atriplex littoralis	Grass-leaved Orache
Atriplex prostrata	Spear-leaved Orache
Avena fatua	Wild Oat
Avenula pubescens	Downy Oat-grass
Baldellia ranunculoides	Lesser Water-plantain
Barbarea stricta	Small-flowered Winter-cress
Bellis perennis	Daisy
Betula spp.	Birches
Betula pendula	Silver Birch
Bidens frondosa	Beggarticks
Bidens tripartita	Trifid Bur-marigold
Blysmus compressus	Flat-sedge
Blysmus rufus	Saltmarsh Flat-sedge
Brassica rapa	Wild Turnip
Briza media	Quaking-grass
Bromus commutatus	Meadow Brome
Bromus hordeaceus	Soft Brome
Bromus racemosus	Smooth Brome
Bromus sterilis	Barren Brome
Bromus tectorum	Drooping Brome
Bupleureum tenuissimum	Slender Hare's-ear
Butomus umbellatus	Flowering Rush
Calamagrostis stricta	Narrow Small-reed
Calla palustris	Bog Arum
Caltha palustris	Marsh-marigold
Calystegia sepium	Hedge Bindweed
Camphorosma monspeliaca	A Goosefoot
Capsella bursa-pastoris	Shepherd's-purse
Cardamine flexuosa	Wavy Bitter-cress
Cardamine pratensis	Cuckooflower
Cardaminopsis halleri	A Rock-cress
Carex spp.	Sedges
Carex acuta	Slender Tufted-sedge
Carex acutiformis	Lesser Pond-sedge
Carex appropinquata	Fibrous Tussock-sedge
Carex brizoides	Quaking-grass Sedge
Carex cespitosa	Lesser Tufted-sedge
Carex davalliana	Davall's Sedge
Carex diandra	Lesser Tussock-sedge
Carex distans	Distant Sedge
Carex disticha	Brown Sedge

Carex divisa	Divided Sedge
Carex elata	Tufted Sedge
Carex elongata	Elongated Sedge
Carex extensa	Long-bracted Sedge
Carex flacca	Glaucous Sedge
Carex hirta	Hairy Sedge
Carex hostiana	Tawny Sedge
Carex lainzii	A Sedge
Carex lasiocarpa	Slender Sedge
Carex ligerica	French Sedge
Carex melanostachya	Nodding Pond-sedge
Carex nigra	Common Sedge
Carex oederi	A Yellow Sedge
Carex ovalis	Oval Sedge
Carex panicea	Carnation Sedge
Carex paniculata	Greater Tussock-sedge
Carex pediformis subsp. *rhizodes*	A Sedge
Carex praecox	Early Sedge
Carex pulicaris	Flea Sedge
Carex riparia	Greater Pond-sedge
Carex rostrata	Bottle Sedge
Carex tomentosa	Downy-fruited Sedge
Carex vesicaria	Bladder Sedge
Carex vulpina	True Fox-sedge
Centaurea jacea	Brown Knapweed
Centaurea nigra	Common Knapweed
Cerastium dubium	A Mouse-ear
Cerastium fontanum	Common Mouse-ear
Chamomilla recutita	Scented Mayweed
Chamomilla suaveolens	Pineappleweed
Chenopodium album	Fat-hen
Chenopodium botryodes	Saltmarsh Goosefoot
Chenopodium polyspermum	Many-seeded Goosefoot
Chenopodium rubrum	Red Goosefoot
Cirsium spp.	Thistles
Cirsium arvense	Creeping Thistle
Cirsium dissectum	Meadow Thistle
Cirsium palustre	Marsh Thistle
Cirsium vulgare	Spear Thistle
Cistus ladanifer	Gum Cistus
Clematis integrifolia	A Clematis
Cnidium dubium	Cnidium
Colchicum autumnale	Meadow Saffron
Conium maculatum	Hemlock
Convolvulus arvensis	Field Bindweed
Coronopus squamatus	Swine-cress
Crataegus monogyna	Hawthorn

Crepis biennis	Rough Hawk's-beard
Crepis mollis	Northern Hawk's-beard
Crepis paludosa	Marsh Hawk's-beard
Cuscuta europaea	Greater Dodder
Cynosurus cristatus	Crested Dog's-tail
Cyperus fuscus	Brown Galingale
Cyperus longus	Galingale
Cytisus purgans	A Broom
Dactylis glomerata	Cock's-foot
Danthonia decumbens	Heath-grass
Deschampsia cespitosa	Tufted Hair-grass
Dianthus deltoides	Maiden Pink
Dianthus pontederae	A Pink
Dipsacus fullonum	Teasel
Echinocystis lobata	Wild Cucumber
Elatine hydropiper	Eight-stamened Waterwort
Eleocharis spp.	Spike-rushes
Eleocharis palustris	Common Spike-rush
Eleocharis uniglumis	Slender Spike-rush
Elymus spp.	Couches
Elymus curvifolius	A Couch
Elymus repens	Common Couch
Epilobium ciliatum	American Willowherb
Epilobium hirsutum	Great Willowherb
Epilobium montanum	Broad-leaved Willowherb
Epilobium roseum	Pale Willowherb
Equisetum spp.	Horsetails
Equisetum arvense	Field Horsetail
Equisetum fluviatile	Water Horsetail
Erica tetralix	Cross-leaved Heath
Erica vagans	Cornish Heath
Erigeron annus	Tall Fleabane
Eriophorum spp.	Cottongrasses
Erysimum diffusum	A Wallflower
Eupatorium cannabinum	Hemp-agrimony
Euphorbia palustris	Marsh Spurge
Euphrasia spp.	Eyebrights
Festuca ampla	A Fescue
Festuca arundinacea	Tall Fescue
Festuca indigesta	A Sheep's-fescue
Festuca nigrescens	Chewing's Fescue
Festuca ovina	Sheep's-fescue
Festuca pratensis	Meadow Fescue
Festuca rubra	Red Fescue

Filaginella uliginosa	Marsh Cudweed
Filipendula ulmaria	Meadowsweet
Filipendula vulgaris	Dropwort
Fraxinus angustifolia	Narrow-leaved Ash
Fritillaria meleagris	Fritillary
Galeopsis tetrahit	Common Hemp-nettle
Galium album	Upright Hedge-bedstraw
Galium aparine	Cleavers
Galium boreale	Northern Bedstraw
Galium palustre	Common Marsh-bedstraw
Galium uliginosum	Fen Bedstraw
Galium verum	Lady's Bedstraw
Gentiana pneumonanthe	Marsh Gentian
Geranium dissectum	Cut-leaved Crane's-bill
Geranium palustre	Marsh Crane's-bill
Gladiolus imbricatus	A Gladiolus
Glaux maritima	Sea-milkwort
Glechoma hederacea	Ground-ivy
Glyceria fluitans	Floating Sweet-grass
Glyceria maxima	Reed Sweet-grass
Gratiola officinalis	Gratiole
Halimione pedunculata	Pedunculate Sea-purslane
Helianthus decapetalus	Thin-leaved Sunflower
Heracleum sphondylium	Hogweed
Hieracium pilosella	Mouse-ear Hawkweed
Hohenackeria polyodon	An Umbellifer
Holcus lanatus	Yorkshire-fog
Hordeum marinum	Sea Barley
Hordeum secalinum	Meadow Barley
Hydrocharis morsus-ranae	Frogbit
Hypericum humifusum	Trailing St John's-wort
Hypericum perforatum	Perforate St John's-wort
Hypericum undulatum	Wavy St John's-wort
Impatiens glandulifera	Indian Balsam
Impatiens parviflora	Small Balsam
Inula britannica	Meadow Fleabane
Inula salicina	Irish Fleabane
Iris pseudacorus	Yellow Iris
Iris sibirica	Siberian Iris
Isoetes setacea	Spring Quillwort
Juncus spp.	Rushes
Juncus acutiflorus	Sharp-flowered Rush
Juncus acutus	Sharp Rush

Juncus bufonius	Toad Rush
Juncus conglomeratus	Compact Rush
Juncus effusus	Soft Rush
Juncus gerardii	Saltmarsh Rush
Juncus inflexus	Hard Rush
Juncus subnodulosus	Blunt-flowered Rush
Juncus subulatus	Fine-leaved Rush
Juncus tenageia	Sand Rush
Knautia arvensis	Field Scabious
Lactuca serriola	Prickly Lettuce
Lathyrus palustris	Marsh Pea
Lathyrus pannonicus subsp. *pannonicus*	A Pea
Lathyrus pratensis	Meadow Vetchling
Leersia oryzoides	Cut-grass
Lemna minor	Common Duckweed
Leontodon autumnalis	Autumn Hawkbit
Leontodon hispidus	Rough Hawkbit
Lepidium latifolium	Dittander
Lepidotis inundata	Marsh Clubmoss
Leucanthemum vulgare	Oxeye Daisy
Leucojum aestivum	Summer Snowflake
Ligularia sibirica	A Leopard-plant
Limonium costae	A Sea-lavender
Lobelia urens	Heath Lobelia
Lolium spp.	Rye-grasses
Lolium multiflorum	Italian Rye-grass
Lolium perenne	Perennial Rye-grass
Lotus corniculatus	Common Bird's-foot-trefoil
Lotus uliginosus	Greater Bird's-foot-trefoil
Luzula campestris	Field Wood-rush
Lychnis flos-cuculi	Ragged Robin
Lycopus europaeus	Gipsywort
Lysimachia nummularia	Creeping-Jenny
Lysimachia vulgaris	Yellow Loosestrife
Lythrum hyssopifolia	Grass-poly
Lythrum salicaria	Purple-loosestrife
Matricaria perforata	Scentless Mayweed
Medicago sativa	Lucerne
Mentha aquatica	Water Mint
Mentha arvensis	Corn Mint
Mentha pulegium	Pennyroyal
Microcnemum coralloides	A Goosefoot
Molinia caerulea	Purple Moor-grass
Myosotis spp.	Forget-me-nots

Myosotis discolor	Changing Forget-me-not
Myosotis laxa subsp. *caespitosa*	Tufted Forget-me-not
Myosotis scorpioides	Water Forget-me-not
Myosoton aquaticum	Water Chickweed
Myosurus minimus	Mousetail
Myriophyllum spp.	Water-milfoils
Myriophyllum verticillatum	Whorled Water-milfoil
Nardus stricta	Mat-grass
Nuphar pumila	Least Water-lily
Nymphoides peltata	Fringed Water-lily
Oenanthe aquatica	Fine-leaved Water-dropwort
Oenanthe fistulosa	Tubular Water-dropwort
Oenanthe pimpinelloides	Corky-fruited Water-dropwort
Oenanthe silaifolia	Narrow-leaved Water-dropwort
Ophioglossum vulgatum	Adder's-tongue
Ornithogallum orthophyllum	A Star-of-Bethlehem
Parnassia palustris	Grass-of-Parnassus
Peucedanum palustre	Milk-parsley
Phalaris arundinacea	Reed Canary-grass
Phleum pratense	Timothy
Phleum pratense subsp. *bertolonii*	Smaller Cat's-tail
Pholiurus pannonicus	A Hard-grass
Phragmites australis	Common Reed
Picris echioides	Bristly Oxtongue
Pimpinella major	Greater Burnet-saxifrage
Pinguicula vulgaris	Common Butterwort
Pinus sylvestris	Scots Pine
Plantago altissima	A Plantain
Plantago lanceolata	Ribwort Plantain
Plantago major	Greater Plantain
Plantago maritima	Sea Plantain
Poa spp.	Meadow-grasses
Poa angustifolia	Narrow-leaved Meadow-grass
Poa annua	Annual Meadow-grass
Poa palustris	Swamp Meadow-grass
Poa pratensis	Smooth Meadow-grass
Poa subcaerulea	Spreading Meadow-grass
Poa trivialis	Rough Meadow-grass
Polygonum spp.	Knotweeds
Polygonum amphibium	Amphibious Bistort
Polygonum aviculare	Knotgrass
Polygonum bistorta	Common Bistort
Polygonum hydropiper	Water-pepper
Polygonum minus	Small Water-pepper

Polygonum mite	Tasteless Water-pepper
Polygonum persicaria	Redshank
Polypogon monspeliensis	Annual Beard-grass
Populus spp.	Poplars
Populus nigra	Black Poplar
Populus tremula	Aspen
Potamogeton spp.	Pondweeds
Potamogeton coloratus	Fen Pondweed
Potamogeton trichoides	Hairlike Pondweed
Potentilla anserina	Silverweed
Potentilla erecta	Tormentil
Potentilla palustris	Marsh Cinquefoil
Potentilla reptans	Creeping Cinquefoil
Prunella vulgaris	Selfheal
Puccinellia spp.	Saltmarsh-grasses
Puccinellia distans	Reflexed Saltmarsh-grass
Puccinellia fasciculata	Borrer's Saltmarsh-grass
Puccinellia festuciformis	A Saltmarsh-grass
Puccinellia rupestris	Stiff Saltmarsh-grass
Pulicaria vulgaris	Small Fleabane
Quercus pyrenaica	Pyrenean Oak
Quercus rotundifolia	An Oak
Ranunculus spp.	Buttercups
Ranunculus acris	Meadow Buttercup
Ranunculus aquatilis	Common Water-crowfoot
Ranunculus auricomus	Goldilocks Buttercup
Ranunculus bulbosus	Bulbous Buttercup
Ranunculus flammula	Lesser Spearwort
Ranunculus repens	Creeping Buttercup
Rhinanthus minor	Yellow-rattle
Rorippa amphibia	Great Yellow-cress
Rorippa islandica	Northern Yellow-cress
Rorippa sylvestris	Creeping Yellow-cress
Rosa spp.	Roses
Rosmarinus officinalis	Rosemary
Rubus spp.	Brambles
Rumex spp.	Docks
Rumex acetosa	Common Sorrel
Rumex angiocarpus	A Sheep's Sorrel
Rumex conglomeratus	Clustered Dock
Rumex crispus	Curled Dock
Rumex maritimus	Golden Dock
Rumex obtusifolius	Broad-leaved Dock
Rumex palustris	Marsh Dock
Rumex papillaris	A Dock

Rumex × *pratensis*	Curled × Broad-leaved Dock hybrid
Sagina maritima	Sea Pearlwort
Salicornia europaea	Glasswort
Salix spp.	Willows
Salix alba	White Willow
Salix cinerea	Grey Willow
Salix triandra	Almond Willow
Salix viminalis	Osier
Sanguisorba officinalis	Great Burnet
Saussurea alpina subsp. *esthonica*	An Alpine Saw-wort
Schoenus ferrugineus	Brown Bog-rush
Scirpus holoschoenus	Round-headed Club-rush
Scirpus lacustris	Common Club-rush
Scirpus maritimus	Sea Club-rush
Scorzonera humilis	Viper's-grass
Scutellaria galericulata	Skullcap
Selaginella spp.	Clubmosses
Selaginella selaginoides	Lesser Clubmoss
Selinum carvifolia	Cambridge Milk-parsley
Senecio aquaticus	Marsh Ragwort
Senecio vulgaris	Groundsel
Serratula tinctoria	Saw-wort
Sesleria caerulea	Blue Moor-grass
Silaum silaus	Pepper-saxifrage
Silene tatarica	A Campion
Sium latifolium	Greater Water-parsnip
Solanum dulcamara	Bittersweet
Solidago canadensis	Canadian Goldenrod
Solidago gigantea	Early Goldenrod
Sonchus asper	Prickly Sow-thistle
Sparganium spp.	Bur-reeds
Sparganium erectum	Branched Bur-reed
Spergularia salina	A Sea-spurrey
Sphenopus divaricatus	A grass
Spirodela polyrhiza	Greater Duckweed
Stellaria spp.	Stitchworts
Stellaria graminea	Lesser Stitchworts
Stellaria media	Common Chickweed
Suaeda vera	Shrubby Sea-blite
Succisa pratensis	Devil's-bit Scabious
Symphytum officinale	Common Comfrey
Tamarix spp.	Tamarisks
Taraxacum spp.	Dandelions
Taraxacum officinale	Common Dandelion

Thalictrum spp.	Meadow-rues
Thalictrum flavum	Common Meadow-rue
Thalictrum lucidum	Shining Meadow-rue
Thelypteris palustris	Marsh Fern
Thymus spp.	Thymes
Tragopogon pratensis	Goat's-beard
Trapa natans	Water Chestnut
Trifolium dubium	Lesser Trefoil
Trifolium fragiferum	Strawberry Clover
Trifolium hybridum	Alsike Clover
Trifolium pratense	Red Clover
Trifolium repens	White Clover
Trifolium squamosum	Sea Clover
Triglochin maritimum	Sea Arrowgrass
Trisetum flavescens	Yellow Oat-grass
Trisetum paniceum	An Oat-grass
Triticum aestivum	Bread Wheat
Typha spp.	Bulrushes
Typha angustifolia	Lesser Bulrush
Typha latifolia	Great Bulrush
Ulmus spp.	Elms
Urtica dioica	Common Nettle
Valeriana officinalis	Common Valerian
Verbascum spp.	Mulleins
Veronica spp.	Speedwells
Veronica arvensis	Wall Speedwell
Veronica catenata	Pink Water-speedwell
Veronica chamaedrys	Germander Speedwell
Veronica longifolia	Long-leaved Speedwell
Veronica scutellata	Marsh Speedwell
Veronica serpyllifolia	Thyme-leaved Speedwell
Vicia cracca	Tufted Vetch
Vicia faba	Broad Bean
Vicia sativa	Common Vetch
Viola spp.	Violets
Viola pumila	Meadow Violet
Wolffia arrhiza	Rootless Duckweed
Xanthium strumarium	Rough Cocklebur

Index

Note: Page references in *italics* refer to figures; those in **bold** refer to Tables